# Texts and Monographs in Physics

*Series Editors:* R. Balian  W. Beiglböck  H. Grosse  E. H. Lieb
N. Reshetikhin  H. Spohn  W. Thirring

**Springer**
*Berlin
Heidelberg
New York
Barcelona
Budapest
Hong Kong
London
Milan
Paris
Santa Clara
Singapore
Tokyo*

# Texts and Monographs in Physics

*Series Editors:* R. Balian  W. Beiglböck  H. Grosse  E. H. Lieb
N. Reshetikhin  H. Spohn  W. Thirring

**From Microphysics to Macrophysics
I + II**  Methods and Applications of
Statistical Physics  By R. Balian

**Variational Methods in Mathematical
Physics**  A Unified Approach
By P. Blanchard and E. Brüning

**Quantum Mechanics:
Foundations and Applications**
3rd enlarged edition  By A. Böhm

**The Early Universe**
Facts and Fiction  3rd corrected
and enlarged edition  By G. Börner

**Operates Algebras and Quantum
Statistical Mechanics I + II**  2nd edition
By O. Bratteli and D. W. Robison

**Geometry of the Standard Model
of Elementary Particles**
By A. Derdzinski

**Random Walks, Critical Phenomena,
and Triviality in Quantum Field
Theory**  By R. Fernández, J. Fröhlich
and A. D. Sokal

**Quantum Relativity**
A Synthesis of the Ideas of Einstein
and Heisenberg
By D. R. Finkelstein

**Quantum Mechanics I + II**
By A. Galindo and P. Pascual

**The Elements of Mechanics**
By G. Gallavotti

**Local Quantum Physics**
Fields, Particles, Algebras
2nd revised and enlarged edition
By R. Haag

**Elementary Particle Physics**
Concepts and Phenomena
By O. Nachtmann

**Inverse Schrödinger Scattering
in Three Dimensions**
By R. G. Newton

**Scattering Theory of Waves
and Particles**  2nd edition
By R. G. Newton

**Quantum Entropy and Its Use**
By M. Ohya and D. Petz

**Generalized Coherent States
and Their Applications**
By A. Perelomov

**Essential Relativity**  Special, General,
and Cosmological  Revised 2nd edition
By W. Rindler

**Path Integral Approach
to Quantum Physics**  An Introduction
2nd printing  By G. Roepstorff

**Advanced Quantum Theory
and Its Applications Through Feynman
Diagrams**  2nd edition
By M. D. Scadron

**Finite Quantum Electrodynamics**
The Causal Approach  2nd edition
By G. Scharf

**From Electrostatics to Optics**
A Concise Electrodynamics Course
By G. Scharf

**Large Scale Dynamics of Interacting
Particles**  By H. Spohn

**General Relativity and Relativistic
Astrophysics**  By N. Straumann

**The Mechanics and Thermodynamics
of Continuous Media**  By M. Šilhavý

**The Dirac Equation**  By B. Thaller

**The Theory of Quark and Gluon
Interactions**  2nd completely revised
and enlarged edition  By F. J. Ynduráin

**Relativistic Quantum Mechanics and
Introduction to Field Theory**
By F. J. Ynduráin

**Supersymmetric Methods in Quantum
and Statistical Physics**  By G. Junker

Francisco J. Ynduráin

# Relativistic Quantum Mechanics and Introduction to Field Theory

Springer

Professor Francisco J. Ynduráin
Universidad Autónoma de Madrid
Departamento de Física Teórica, C-XI
Canto Blanco, E-28049 Madrid, Spain

*Editors*

Roger Balian
CEA
Service de Physique Théorique de Saclay
F-91191 Gif-sur-Yvette, France

Wolf Beiglböck
Institut für Angewandte Mathematik
Universität Heidelberg
Im Neuenheimer Feld 294
D-69120 Heidelberg, Germany

Harald Grosse
Institut für Theoretische Physik
Universität Wien
Boltzmanngasse 5
A-1090 Wien, Austria

Elliott H. Lieb
Jadwin Hall
Princeton University, P. O. Box 708
Princeton, NJ 08544-0708, USA

Nicolai Reshetikhin
Department of Mathematics
University of California
Berkeley, CA 94720-3840, USA

Herbert Spohn
Theoretische Physik
Ludwig-Maximilians-Universität München
Theresienstraße 37
D-80333 München, Germany

Walter Thirring
Institut für Theoretische Physik
Universität Wien
Boltzmanngasse 5
A-1090 Wien, Austria

This book is a completely revised translation of the Spanish original edition:
F. J. Ynduráin, Mecánica Cuántica Relativista, Alianza editorial s/a, Madrid 1990

Library of Congress Cataloging-in-Publication Data applied for.

Die Deutsche Bibliothek – CIP-Einheitsaufnahme
**Ynduráin, Francisco J.:** Relativistic quantum mechanics and introduction to field theory
Fancisco J. Ynduráin. – Berlin; Heidelberg; New York; Barcelona; Budapest; Hong Kong;
London; Milan; Paris; Santa Clara; Singapore; Tokyo: Springer, 1996
(Texts and monographs in physics)
ISSN 0172-5998
ISBN-13: 978-3-642-64674-4          e-ISBN-13: 978-3-642-61057-8
DOI: 10.1007/978-3-642-61057-8

This work is subject to copyright. All rights are reserved, whether the whole or part of the material
is concerned, specifically the rights of translation, reprinting, reuse of illustrations, recitation,
broadcasting, reproduction on microfilm or in any other way, and storage in data banks. Duplication
of this publication or parts thereof is permitted only under the provisions of the German Copyright
Law of September 9, 1965, in its current version, and permission for use must always be obtained
from Springer-Verlag. Violations are liable for prosecution under the German Copyright Law.

© Springer-Verlag Berlin Heidelberg 1996
Softcover reprint of the hardcover 1st edition 1996

The use of general descriptive names, registered names, trademarks, etc. in this publication does not
imply, even in the absence of a specific statement, that such names are exempt from the relevant
protective laws and regulations and therefore free for general use.

Typesetting: Camera-ready copy from the author using a Springer T$_E$X macro package
SPIN: 10511710          55/3144-543210 - Printed on acid-free paper

# Preface

A fully relativistic treatment of the quantum mechanics of particles requires the introduction of quantum field theory, that is to say, the quantum mechanics of systems with an infinite number of degrees of freedom. This is because the relativistic equivalence of mass and energy plus the quantum possibility of fluctuations imply the existence of (real or virtual) creation and annihilation of particles in unlimited numbers.

In spite of this, there exist processes, and energy ranges, where a treatment in terms of ordinary quantum mechanical tools is appropriate, and the approximation of neglecting the full field-theoretic description is justified. Thus, one may use concepts such as potentials, and wave equations, classical fields and classical currents, etc. The present text is devoted precisely to the systematic discussion of these topics, to which we have added a general description of one- and two-particle relativistic states, in particular for scattering processes.

A field-theoretic approach may not be entirely avoided, and in fact an introduction to quantum field theory is presented in this text. However, field theory is not the object *per se* of this book; apart from a few examples, field theory is mainly employed to establish the connection with equivalent potentials, to study the classical limit of the emission of radiation or to discuss the propagation of a fermion in classical electromagnetic fields.

The bulk of applications of our tools is for electromagnetic interections, as is only natural. Nevertheless, some applications to nuclear physics (Yukawa interactions), weak interactions (parity violation in atoms) and strong interactions (the Bogoliubov bag model of quark bound states) are also presented.

This book is the English version of the author's text *Mecánica Cuántica Relativista* (Alianza Editorial, Madrid 1990). It is not merely a translation. Enough new material has been added that the original 260 pages have become well over three hundred, and a few chapters are entirely new.

To end this introduction, a few words on notation, and units, are in order. Carets over operators, $\hat{A}_\mu, \hat{H}$ etc. and tildes under matrix objects, $\gamma_\mu, \underset{\sim}{\sigma}$ etc. are used whenever there is danger of confusion. When the meaning of an object is clear from the context we dispense with these aids. Three-dimensional vector objects are denoted by boldface characters, e.g., $\mathbf{r}$. The convention

$r = |\mathbf{r}|$ will be used when no confusion (particularly with Minkowski vectors) may arise.

A peculiarity in which our text diverges from most presentations (a notable exception is the treatise of Bogoliubov and Shirkov, 1959, which agrees with us on this) is our almost entire avoidance of contravariant components for Minkowski tensors, so that we also seldom employ Einstein's convention of summation over repeated indices. This makes some expressions a bit clumsy, a price that we consider worth paying in an introductory textbook which, moreover, continuously oscillates between relativistic calculations and non-relativistic (or semirelativistic) limits and concepts: the use of only one kind of component makes it easier to avoid a large number of ambiguities, notably when identifying nonrelativistic component notations with the corresponding relativistic ones. This ease of connection with nonrelativistic quantum mechanical conventions is also the reason why we have kept $\hbar$ and $c$ explicit in many formulas. As for units, we use, unless explicitly stated otherwise, the SI system with Gauss units (so that the potential between charges $e_1, e_2$ is $e_1 e_2/r$). In a number of instances, however, the Heaviside system (in which the Coulomb potential is $e_1 e_2/4\pi r$) and/or natural units, with $\hbar = c = 1$, is employed. Explicit warning is given in such cases. Other matters of notation, which are not very standardized in our subject, may be found throughout the text.

The typesetting of this text has been a rather tough enterprise for a LaTeX beginner like the author. That the job has been completed was due to the invaluable help of Gabriel Martínez, Stéphan Titard, Elena Ynduráin and, above all, Lourdes Rey and Kassa Adel (whose program "Kdraw" was used for the drawings), help which is here gratefully recorded.

Madrid, April 1996                                          *F. J. Ynduráin*

# Table of Contents

# 1. Relativistic Transformations.
## The Lorentz Group

## 1.1 Rotations, and Space and Time Reversal for Particles with Spin

A rotation[1] may be specified by a vector, $\boldsymbol{\theta}$, in such a way that (Fig. 1.1.1) the rotation axis lies along $\boldsymbol{\theta}$, the rotation angle being $\theta = |\boldsymbol{\theta}|$, and the direction of the rotation determined by the corkscrew rule. If we denote the rotation by $R(\boldsymbol{\theta})$, it acts upon a vector $\mathbf{r}$ according to

$$\mathbf{r} \to \mathbf{r}' = R(\boldsymbol{\theta})\mathbf{r} = (\cos\theta)\mathbf{r} + (1 - \cos\theta)\frac{\boldsymbol{\theta}\mathbf{r}}{\theta^2}\boldsymbol{\theta} + \frac{\sin\theta}{\theta}\boldsymbol{\theta} \times \mathbf{r}; \qquad (1.1.1a)$$

for $\boldsymbol{\theta}$ infinitesimal,

$$R(\boldsymbol{\theta})\mathbf{r} = \mathbf{r} + \boldsymbol{\theta} \times \mathbf{r} + O(\theta^2). \qquad (1.1.1b)$$

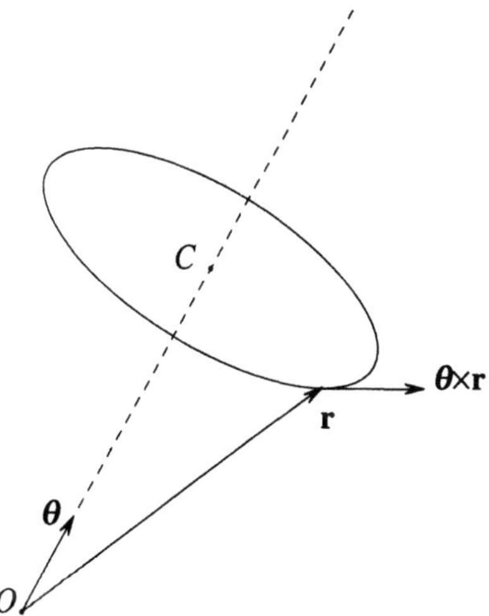


**Fig. 1.1.1.** The rotation $R(\boldsymbol{\theta})$.

---

[1] This and the following section are presented here mainly to establish notation, but with no real proofs or much detail. These can be found in textbooks on nonrelativistic quantum mechanics (Gottfried, 1966; Galindo and Pascual, 1978; Ynduráin, 1988; Wigner, 1959).

Instead of $\boldsymbol{\theta}$ we can characterize rotations by the Euler angles $\alpha$, $\beta$, $\gamma$. We write $R_{\alpha\beta\gamma}$ for such a rotation. Let $R_j(\varphi)$ be a rotation around the $Oj$ axis by the vector $\varphi$, in the positive (corkscrew) direction. Then,

$$R_{\alpha\beta\gamma} = R_z(\alpha)R_y(\beta)R_z(\gamma). \tag{1.1.2}$$

We will seldom use the Euler angles. The advantage of the parametrization by $\boldsymbol{\theta}$ is that the $\boldsymbol{\theta}$ are what are called *normal parameters*. In general, given a group of transformations with elements $g(\alpha_1, \ldots, \alpha_n)$, specified by the parameters $\alpha_1, \ldots, \alpha_n$, we say that these parameters are normal if, whenever the vectors $\boldsymbol{\alpha}$, $\boldsymbol{\beta}$ are parallel, we have

$$g(\alpha_1, \ldots, \alpha_n)g(\beta_1, \ldots, \beta_n) = g(\alpha_1 + \beta_1, \ldots, \alpha_n + \beta_n).$$

This is clearly the case for the $\boldsymbol{\theta}$ of $R(\boldsymbol{\theta})$.

The usefulness of normal parameters is that they allow us to reduce *finite* transformations to *infinitesimal* ones. Indeed, if we have

$$g(\epsilon) = 1 + \epsilon \mathbf{T} + O(\epsilon^2),$$

then we can write a finite transformation as

$$g(\boldsymbol{\alpha}) = [g(\boldsymbol{\alpha}/n)]^n = [1 + \boldsymbol{\alpha}\mathbf{T}/n + O(1/n^2)]^n \underset{n\to\infty}{=} e^{\boldsymbol{\alpha}\mathbf{T}}.$$

Consider now a wave function $\Psi_\lambda(\mathbf{r}, t)$, in general with several components if the particle's spin is nonzero. Then we represent the rotation $R(\boldsymbol{\theta})$ by the operator $U(R(\boldsymbol{\theta})) \equiv U(\boldsymbol{\theta})$, with

$$U(\boldsymbol{\theta})\Psi_\lambda(\mathbf{r}, t) \equiv \sum_{\lambda'} D^{(s)}_{\lambda\lambda'}(R(\boldsymbol{\theta}))\Psi_{\lambda'}(R^{-1}\mathbf{r}, t); \tag{1.1.3a}$$

$s$ is the spin of the particle, and the explicit form of the $D$ may be found in Appendix A.1. If we denote by $\underset{\sim}{\Psi}$ the column matrix with elements $\Psi_\lambda$, and by $\underset{\sim}{D}$ to the matrix with elements $D^{(s)}_{\lambda\lambda'}$, (1.1.3a) can be rewritten as

$$U(\boldsymbol{\theta})\underset{\sim}{\Psi}(\mathbf{r}, t) = \underset{\sim}{D}^{(s)}(R(\boldsymbol{\theta}))\underset{\sim}{\Psi}(R^{-1}\mathbf{r}, t). \tag{1.1.3b}$$

For state vectors $|\mathbf{r}, \lambda\rangle$, $|\mathbf{p}, \lambda\rangle$ ($\mathbf{p}$ is the momentum), (1.1.3) implies that

$$U(\boldsymbol{\theta})|\mathbf{r}, \lambda\rangle = \sum_{\lambda'} D^{(s)}_{\lambda'\lambda}(R(\boldsymbol{\theta}))|R\mathbf{r}, \lambda'\rangle,$$

$$U(\boldsymbol{\theta})|\mathbf{p}, \lambda\rangle = \sum_{\lambda'} D^{(s)}_{\lambda'\lambda}(R(\boldsymbol{\theta}))|R\mathbf{p}, \lambda'\rangle. \tag{1.1.4}$$

For infinitesimal rotations we define the operators $\mathbf{J}$ by[2]

$$U(\epsilon)\Psi(\mathbf{r}) = \left(1 - \frac{i}{\hbar}\epsilon\mathbf{J} + O(\epsilon^2)\right)\Psi(\mathbf{r}). \tag{1.1.5}$$

Then $\mathbf{J}$ may be split as

---

[2] We henceforth suppress the tilde under matrix objects; their matrix character should be clear from the context.

$$\mathbf{J} = \mathbf{L} + \mathbf{S}, \tag{1.1.6}$$

and, in the coordinate (or position) representation in which we are working,

$$\mathbf{L} = \mathbf{r} \times \mathbf{P} = -i\hbar\mathbf{r} \times \boldsymbol{\nabla}, \quad D(\boldsymbol{\epsilon}) = 1 - \frac{i}{\hbar}\boldsymbol{\epsilon}\mathbf{S} + O(\epsilon^2). \tag{1.1.7}$$

$\mathbf{P}$ is the *momentum operator*, $\mathbf{P} = -i\hbar\boldsymbol{\nabla}$ in position space; $\mathbf{L}$ represents the *orbital angular momentum*, and $\mathbf{S}$ the *spin*.

For finite rotations, and because the $\boldsymbol{\theta}$ are normal,

$$\begin{aligned}
U(\boldsymbol{\theta})\Psi &= [U(\boldsymbol{\theta}/n)]^n\Psi = \left(1 - \frac{i}{\hbar}\boldsymbol{\theta}\mathbf{J}/n + O(1/n^2)\right)^n \Psi \\
&\underset{n\to\infty}{=} \left(\exp\frac{-i}{\hbar}\boldsymbol{\theta}\mathbf{J}\right)\Psi :
\end{aligned}$$

we have thus obtained

$$U(\boldsymbol{\theta}) = e^{-\frac{i}{\hbar}\boldsymbol{\theta}\mathbf{J}}, \tag{1.1.8a}$$

and likewise we would find that

$$D(R(\boldsymbol{\theta})) = e^{-\frac{i}{\hbar}\boldsymbol{\theta}\mathbf{S}}. \tag{1.1.8b}$$

We can standardize states and matrices. Following the conventions of Condon and Shortley (1967), we fix the phases of the states $|j, M\rangle$ corresponding to the total angular momentum, $j$, and the third component thereof, $M$, by writing

$$J_+|j, M\rangle = -\hbar\sqrt{\frac{j(j+1) - M(M+1)}{2}}|j, M+1\rangle,$$

$$J_-|j, M\rangle = \hbar\sqrt{\frac{j(j+1) - M(M-1)}{2}}|j, M-1\rangle, \tag{1.1.9}$$

$$J_\pm \equiv \frac{\mp 1}{\sqrt{2}}(J_1 \pm iJ_2).$$

This fixes all phases in terms of that of a single state, say $|j, j\rangle$.

*Space inversion, $I_s$,* is defined by $I_s\mathbf{r} \equiv -\mathbf{r}$. On a wave function, we represent this by the *parity* operator:

$$\mathcal{P}\Psi(\mathbf{r}, t) = \Gamma_P\Psi(-\mathbf{r}, t), \tag{1.1.10}$$

where $\Gamma_P$ is a numerical matrix with the properties

$$\Gamma_P^+\Gamma_P = 1, \quad \Gamma_P^2 = 1. \tag{1.1.11}$$

A state with well-defined parity, $\eta_P$, has a wave function which is an eigenfunction of $\mathcal{P}$:

$$\mathcal{P}\Psi = \eta_P\Psi. \tag{1.1.12}$$

*Time reversal* is represented by an *antiunitary* operator $\mathcal{T}$ with

$$\mathcal{T}\Psi(\mathbf{r}, t) = \Gamma_T\Psi^*(\mathbf{r}, -t); \quad \Gamma_T^*\Gamma_T = 1, \tag{1.1.13}$$

and it corresponds to the transformation $I_t : \mathbf{r} \to \mathbf{r}, t \to -t$.

Acting on states,

$$\mathcal{P}|\mathbf{p},\lambda\rangle = \eta_p|-\mathbf{p},\lambda\rangle;\quad \mathcal{T}|\mathbf{p},\lambda\rangle = \eta_T(-i)^{2\lambda}|-\mathbf{p},-\lambda\rangle, \tag{1.1.14}$$

$\eta_p$ and $\eta_T$ being phases (independent of $\mathbf{p}, \lambda$) called respectively the *intrinsic parity* and *time-reversal parity*, of the system. The seemingly peculiar phases for $\mathcal{T}$ follow from the choice of phases for the states, (1.1.9).

## 1.2 Galilean Transformations

According to the principle of equivalence of inertial reference systems, physical laws do not change when one goes from an inertial system to another moving with a constant velocity, $\mathbf{v}$, with respect to it. In *nonrelativistic* mechanics the corresponding transformations are called *Galilean accelerations* or *boosts*, under which (Fig. 1.2.1)

$$\mathbf{r} \longrightarrow \mathbf{r} + \mathbf{v}t, \tag{1.2.1a}$$

and time does not change,

$$t \longrightarrow t. \tag{1.2.1b}$$

Clearly, the $\mathbf{v}$ are normal parameters.

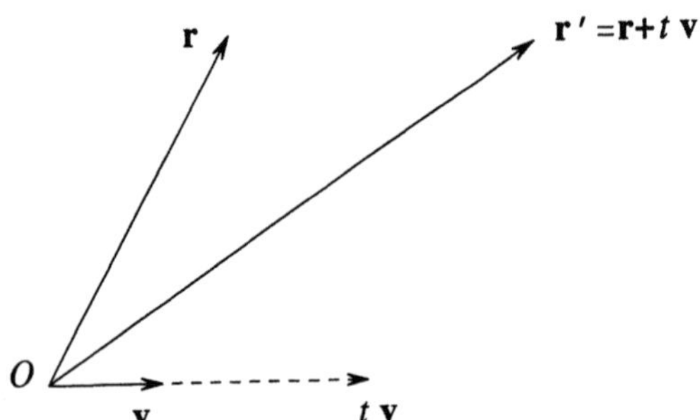


Fig. 1.2.1. (Galilean) acceleration.

In quantum mechanics we represent this by operators,

$$\Psi(\mathbf{r},t) \to U(\mathbf{v})\Psi(\mathbf{r},t); \tag{1.2.2}$$

we consider here spinless particles to simplify the discussion. If we want (1.2.2) to represent (1.2.1), we will have to assume that $U(\mathbf{v})\Psi(\mathbf{r},t)$ is equivalent to $\Psi(\mathbf{r}-\mathbf{v}t,t)$, so they should be equal up to a phase:

$$U(\mathbf{v})\Psi(\mathbf{r},t) = e^{-\frac{i}{\hbar}\lambda(\mathbf{v},\mathbf{r},t)}\Psi(\mathbf{r}-\mathbf{v}t,t). \tag{1.2.3}$$

As will be seen in a moment, the phase is essential to maintain invariance of the Schrödinger equation.

Consider $\mathbf{v}$ infinitesimal; writing

$$U(\mathbf{v}) = 1 - \frac{i}{\hbar}\mathbf{v}\frac{\partial}{\partial\mathbf{v}}\lambda(\mathbf{v},\mathbf{r},t)|_{\mathbf{v}=0} - \frac{i}{\hbar}t\frac{\hbar}{i}\mathbf{v}\boldsymbol{\nabla}_{\mathbf{r}} + O(v^2),$$

and defining the operator

$$\mathbf{B} \equiv \frac{\partial}{\partial\mathbf{v}}\lambda(\mathbf{v},\mathbf{r},t)|_{\mathbf{v}=0},$$

we have

$$U(\mathbf{v}) \simeq 1 - \frac{i}{\hbar}\mathbf{v}(\mathbf{B} + t\mathbf{P}), \tag{1.2.4}$$

and we have identified $-i\hbar\boldsymbol{\nabla} = \mathbf{P}$.

We can write the Schrödinger equation (for a free particle) as

$$\left(i\hbar\partial_t - \frac{1}{2m}\mathbf{P}^2\right)\Psi(\mathbf{r},t) = 0, \quad \partial_t \equiv \partial/\partial t.$$

The invariance of this equation under (1.2.3) or, in infinitesimal form, (1.2.4) is equivalent to demanding that the commutator of the Schrödinger operator, $i\hbar\partial_t - \mathbf{P}^2/2m$, and the generators of boosts, $\mathbf{B} + t\mathbf{P}$, commute, i.e., we must have

$$\left[i\hbar\partial_t - \frac{1}{2m}\mathbf{P}^2, B_j + tP_j\right] = 0$$

for every component $j = 1,2,3$. This implies a condition on the $\mathbf{B}$:

$$i\hbar P_j = \frac{1}{2m}[\mathbf{P}^2, B_j];$$

the simplest solution is to choose $B_j = mQ_j$, where $\mathbf{Q}$ is the nonrelativistic position operator,

$$\mathbf{Q}\Psi(\mathbf{r},t) \equiv \mathbf{r}\Psi(\mathbf{r},t).$$

We have thus obtained that (infinitesimal) Galilean accelerations may be written as

$$U(\mathbf{v}) = 1 - (i/\hbar)\mathbf{v}(-m\mathbf{Q} + t\mathbf{P}) + O(v^2), \tag{1.2.5}$$

that is to say, the operators

$$-m\mathbf{Q} + t\mathbf{P}$$

are the generators of Galilean boosts.

A full analysis of Galilean transformations is more complicated than would appear from what we have seen. The reason is that, because $\mathbf{Q}$ and $\mathbf{P}$ do not commute, it is not trivial to go from (1.2.5) to (1.2.3), and the connection with rotations and translations is not trivial, either. These topics may be found in the texts of Gottfried (1966) or Galindo and Pascual (1978); we will leave the subject here, since we have covered all that we require.

## 1.3 Lorentz Transformations. Normal Parameters

In relativity theory the passage from one inertial system to another one, moving with respect to it with speed $\mathbf{v}$, is given by the *Lorentz boosts* (or *accelerations*). Starting with the case where $\mathbf{v}$ is parallel to the $OZ$ axis, these boosts are given by

$$
\begin{aligned}
x &\longrightarrow x, \quad y \longrightarrow y, \\
z &\longrightarrow \frac{1}{\sqrt{1 - v^2/c^2}}(z + vt), \\
t &\longrightarrow \frac{1}{\sqrt{1 - v^2/c^2}}\left(t + \frac{v}{c^2}z\right).
\end{aligned}
\tag{1.3.1a}
$$

Here and henceforth $c$ will denote the speed of light.

We also write this with shorthand notation

$$
\mathbf{r} \longrightarrow L(\mathbf{v}_z)\mathbf{r}, \quad t \longrightarrow L(\mathbf{v}_z)t.
\tag{1.3.1b}
$$

(This really *is* shorthand: $L(\mathbf{v})\mathbf{r}$ depends also on $t$, and not only on $\mathbf{v}, \mathbf{r}$; likewise, $L(\mathbf{v})t$ depends also on $\mathbf{r}$.) For $\mathbf{v}$ directed in an arbitrary way, we use the following trick. Let $R(\mathbf{z} \to \mathbf{v})$ be a rotation carrying the $OZ$ axis over $\mathbf{v}$. For example, we may choose

$$
R(\mathbf{z} \to \mathbf{v}) = R(\boldsymbol{\alpha}), \quad R(\boldsymbol{\alpha})\mathbf{z} = \mathbf{v}/|\mathbf{v}|,
$$

with $\mathbf{z}$ the unit vector along $OZ$ and

$$
\cos\alpha = v_3/v, \quad \boldsymbol{\alpha} = (\alpha/v\sin\alpha)\mathbf{z} \times \mathbf{v}.
$$

Denoting by $L(\mathbf{v})$ the Lorentz boost with velocity $\mathbf{v}$, we define

$$
L(\mathbf{v}) = R(\mathbf{z} \to \mathbf{v})L(\mathbf{v}_z)R^{-1}(\mathbf{z} \to \mathbf{v}),
$$

where $\mathbf{v}_z$ is a vector of length $v$ along $OZ$. Using the explicit formulas (1.3.1), (1.1.1) for $L(\mathbf{v}_z)$ and $R$ we find that

$$
\begin{aligned}
\mathbf{r} \to L(\mathbf{v})\mathbf{r} &= \mathbf{r} - \frac{\mathbf{v}\mathbf{r}}{v^2}\mathbf{v} + \left(1 - \frac{\mathbf{v}^2}{c^2}\right)^{-1/2}\left(\frac{1}{v^2}\mathbf{r}\mathbf{v} + t\right)\mathbf{v}, \\
t \to L(\mathbf{v})t &= \left(1 - \frac{\mathbf{v}^2}{c^2}\right)^{-1/2}\left(t + \frac{\mathbf{v}\mathbf{r}}{c^2}\right).
\end{aligned}
\tag{1.3.2}
$$

**Exercise.** Verify that, for $t, t', \mathbf{r}, \mathbf{r}', \mathbf{v}$ arbitrary,

$$
c^2(L(\mathbf{v})t)(L(\mathbf{v})t') - (L(\mathbf{v})\mathbf{r})(L(\mathbf{v})\mathbf{r}') = c^2 tt' - \mathbf{r}\mathbf{r}',
\tag{1.3.3a}
$$

i.e., that the quadratic form

$$
c^2 tt' - \mathbf{r}\mathbf{r}' = \text{invariant}
\tag{1.3.3b}
$$

under Lorentz boosts ∎

Unlike what happens in the nonrelativistic case, the parameters $\mathbf{v}$ are now *not* normal; it is not true that the product of boosts by $\mathbf{v}$, $\mathbf{v}'$ is the boost by $\mathbf{v} + \mathbf{v}'$ (which may not even exist if $|\mathbf{v} + \mathbf{v}'| \geq c$ ). It is then convenient to use other parameters, which will be denoted by $\boldsymbol{\xi}, \boldsymbol{\eta}, \ldots$ such that, whenever $\boldsymbol{\xi}$ and $\boldsymbol{\eta}$ are parallel,

$$L(\boldsymbol{\xi})L(\boldsymbol{\eta}) = L(\boldsymbol{\xi} + \boldsymbol{\eta}). \tag{1.3.4}$$

Note that we use the same notation for $L(\mathbf{v})$ and $L(\boldsymbol{\xi})$; the context, and the latin/greek characters should be enough to indicate whether we are using velocities or the new normal parameters.

Let us choose $\boldsymbol{\xi}$ along $OZ$. If we write

$$L(\boldsymbol{\xi})z = A(\xi)z + B(\xi)ct, \quad L(\boldsymbol{\xi})t = \frac{1}{c}C(\xi)z + D(\xi)t,$$

where $A$, $B$, $C$, $D$ are functions to be determined, condition (1.3.3) tells us that

$$AB = CD, \quad A^2 - C^2 = D^2 - B^2 = 1,$$

so that there exists a function $\varphi(\boldsymbol{\xi})$ such that

$$A = D = \cosh \varphi(\boldsymbol{\xi}), \quad B = C = \sinh \varphi(\boldsymbol{\xi}).$$

The relation (1.3.4) implies that

$$\cos(\varphi(\boldsymbol{\xi}) + \varphi(\boldsymbol{\eta})) = \cosh \varphi(\boldsymbol{\xi}) \cosh \varphi(\boldsymbol{\eta}) + \sinh \varphi(\boldsymbol{\xi}) \sinh \varphi(\boldsymbol{\eta})$$

$$\sinh(\varphi(\boldsymbol{\xi}) + \varphi(\boldsymbol{\eta})) = \cosh \varphi(\boldsymbol{\xi}) \sinh \varphi(\boldsymbol{\eta}) + \sinh \varphi(\boldsymbol{\xi}) \cosh \varphi(\boldsymbol{\eta}),$$

and we can thus choose $\varphi(\boldsymbol{\xi}) = \xi$. Finally

$$x \longrightarrow x, \quad y \longrightarrow y,$$
$$z \longrightarrow (\cosh \xi)z + (\sinh \xi)ct, \tag{1.3.5a}$$
$$t \longrightarrow \frac{1}{c}(\sinh \xi)z + (\cosh \xi)t, \quad \boldsymbol{\xi} \parallel OZ.$$

The relation between the $\boldsymbol{\xi}$ and $\mathbf{v}$ is found by comparison with (1.3.1a):

$$\cos \xi = \frac{1}{\sqrt{1 - \mathbf{v}^2/c^2}}, \quad \sinh \xi = \frac{|\mathbf{v}|}{c} \frac{1}{\sqrt{1 - \mathbf{v}^2/c^2}}, \quad \boldsymbol{\xi} \parallel \mathbf{v}. \tag{1.3.5b}$$

$\boldsymbol{\xi}$ is sometimes called the *rapidity*. For a boost along $\boldsymbol{\xi}$, (1.3.5a) becomes

$$\mathbf{r} \to L(\boldsymbol{\xi})\mathbf{r} = \mathbf{r} - \frac{\boldsymbol{\xi}\mathbf{r}}{\xi^2}\boldsymbol{\xi} + \frac{1}{\xi}\left\{ (\cosh \xi)\frac{\boldsymbol{\xi}\mathbf{r}}{\xi}\boldsymbol{\xi} + c(\sinh \xi)t\boldsymbol{\xi} \right\},$$
$$t \to L(\boldsymbol{\xi})t = (\cosh \xi)t + \frac{1}{c}\frac{\sinh \xi}{\xi}\boldsymbol{\xi}\mathbf{r}; \tag{1.3.5c}$$

(1.3.4) still holds, as the reader may easily verify.

For speeds small compared with $c$,

$$\boldsymbol{\xi} \simeq \mathbf{v}/c,$$

and (1.3.2), or (1.3.5c), coincides with a Galilean boost, (1.2.1).

The transformations $\Lambda$ of the set $(\mathbf{r}, t)$ obtained by applying rotations and Lorentz boosts as a product,

$$\Lambda = LR, \tag{1.3.6}$$

are called *Lorentz transformations*. As we will see in the next sections, they form a group, called the *Lorentz group*, or, sometimes, and for reasons that will be apparent presently, the *orthochronous, proper Lorentz group*.

If we include possible products by space, $I_s$, and time, $I_t$, reversals,

$$I_s \ : \ \mathbf{r} \to -\mathbf{r}, \ t \to t; \ I_t \ : \ \mathbf{r} \to \mathbf{r}, \ t \to -t, \tag{1.3.7}$$

we obtain a set (which is also a group) called the *full Lorentz group*. Its elements are of one of the following forms:

$$LR, \ I_s LR, \ I_t LR, \ I_s I_t LR. \tag{1.3.8}$$

## 1.4 Minkowski Space. The Full Lorentz Group

As we saw in the previous section, Lorentz boosts mix space and time. A unified treatment of relativistic transformations demands that we work in a set that contains both. This is *Minkowskian spacetime* (or just *Minkowski space*). Its elements, or points, which will be denoted[3] by letters $x, y, \ldots$, are called *four-vectors*, and are determined by *four* coordinates, $x_\mu, \mu = 0, 1, 2, 3$,

$$x \sim \begin{pmatrix} x_0 \\ x_1 \\ x_2 \\ x_3 \end{pmatrix}, \tag{1.4.1}$$

where $x_0 = ct$ corresponds to a time coordinate and $x_j = r_j, j = 1, 2, 3$ are purely spatial coordinates.

We will consistently tag Minkowskian coordinates with Greek indices $\mu, \nu, \ldots$ varying from 0 to 3; italic indices $i, j, \ldots$ will be restricted to varying from 1 to 3. We will also denote by $\mathbf{r}$ the spatial part of $x$, and (1.4.1) may thus also be written as

$$x \sim \begin{pmatrix} ct \\ \mathbf{r} \end{pmatrix}.$$

At times a horizontal notation is convenient, and we write $x \sim (\mathbf{r}, ct)$, or $x \sim (\mathbf{r}, t)$ omitting the $c$.

Lorentz boosts may be represented by $4 \times 4$ matrices $L$, $x \to Lx$, with elements $L_{\mu\nu}$, so that

---

[3] Our conventions are not universal, although they are certainly quite common. See Sect. 1.9 for comparison with others.

$$(Lx)_\mu = \sum_{\nu=0}^{3} L_{\mu\nu} x_\nu;$$

explicitly, using (1.3.5c), we have

$$(Lx)_0 = (\cosh\xi)x_0 + \frac{\sinh\xi}{\xi} \sum_{j=1}^{3} \xi_j x_j,$$

$$(Lx)_i = x_i - \frac{1}{\xi^2}\left(\sum_j \xi_j x_j\right)\xi_i + \frac{1}{\xi}\left(\frac{\cosh\xi}{\xi}\sum_j \xi_j x_j + x_0\sinh\xi\right)\xi_i. \tag{1.4.2}$$

Rotations can also be defined as transformations in Minkowski space: $x \to Rx$, with

$$(Rx)_\mu = \sum_\nu R_{\mu\nu} x_\nu,$$

and (cf. (1.1.1))

$$(Rx)_0 = x_0,$$

$$(Rx)_i = (\cos\theta)x_i + \frac{1-\cos\theta}{\theta^2}\left(\sum_j \theta_j x_j\right)\theta_i + \frac{\sin\theta}{\theta}\sum_{kl}\epsilon_{ikl}\theta_k x_l. \tag{1.4.3}$$

Here $\epsilon_{ikl}$ is the Levi-Cività symbol, $+1$ for $ikl$ an even permutation of 123, $-1$ for an odd one, and zero if two indices are equal.

The transformations $L$, $R$ leave invariant the quadratic form $x \cdot y$ defined by

$$x \cdot y \equiv x_0 y_0 - \sum_{j=1}^{3} x_j y_j. \tag{1.4.4}$$

This form is known as the *Minkowski* (pseudo) *scalar product*, and can be also written in terms of the (pseudo) *metric tensor $G$*, with components $g_{\mu\nu}$,

$$g_{\mu\nu} = \begin{cases} 0, & \mu \neq \nu, \\ 1, & \mu = \nu = 0, \\ -1, & \mu = \nu \neq 0. \end{cases} \tag{1.4.5}$$

Indeed,

$$x \cdot y = \sum_{\mu\nu} g_{\mu\nu} x_\mu y_\nu = \sum_\mu g_{\mu\mu} x_\mu y_\mu = x^T G y. \tag{1.4.6}$$

In the last expression, $x$, $y$ are taken to be matrices (as in (1.4.1)), and the index $T$ means the transpose of a matrix.[4]

---

[4] Likewise, $M^*$ will be the complex conjugate of $M$ and $M^+ = M^{*T}$ the adjoint (or hermitean conjugate) of $M$.

The *Minkowski square*, denoted by $x^2$ if there is no danger of confusion, is defined as $x^2 \equiv x \cdot x$.

By direct computation we can verify that, when $\Lambda = LR$ for any $L$, $R$, then, for every pair $x$, $y$,

$$(\Lambda x) \cdot (\Lambda y) = x \cdot y. \tag{1.4.7}$$

In terms of the metric tensor, (1.4.7) tells us that

$$\Lambda^T G \Lambda = G. \tag{1.4.8}$$

These relations suggest that we define a group, called the *full Lorentz group*, and denoted by $\overline{\mathcal{L}}$, to be the set of all matrices $\overline{\Lambda}$ such that

$$\overline{\Lambda}^T G \overline{\Lambda} = G. \tag{1.4.9a}$$

It is obvious that such $\overline{\Lambda}$ form a group, and it is easy to verify that (1.4.9a) implies that

$$\overline{\Lambda} G \overline{\Lambda}^T = G. \tag{1.4.9b}$$

Let us take determinants in (1.4.9). We find that $(\det \overline{\Lambda})^2 = 1$, and hence $\det \overline{\Lambda} = \pm 1$. Consider space reversal, acting in Minkowski space by $(I_s x)_0 = x_0$, $(I_s x)_i = -x_i$. Clearly, $I_s$ is in $\overline{\mathcal{L}}$ and moreover $\det I_s = -1$. If $\overline{\Lambda}$ belongs to $\overline{\mathcal{L}}$ and $\det \overline{\Lambda} = -1$, then we can write identically

$$\overline{\Lambda} = I_s(I_s \overline{\Lambda}), \tag{1.4.10a}$$

and now $\det(I_s \overline{\Lambda}) = +1$. If we denote by $\mathcal{L}_+$ the subgroup of $\overline{\mathcal{L}}$ consisting of matrices with determinant unity, we have just shown that $\overline{\mathcal{L}}$ consists of matrices either in $\mathcal{L}_+$ or products of $I_s$ time matrices in $\mathcal{L}_+$.

Consider next the four-vector $n_t$, a unit vector along the time axis, with components $n_{t\mu} = \delta_{\mu 0}$. Given $\overline{\Lambda}$ in $\overline{\mathcal{L}}$, we may have either $(\overline{\Lambda} n_t)_0 > 0$ or $(\overline{\Lambda} n_t)_0 < 0$; it is not possible to have $(\overline{\Lambda} n_t)_0 = 0$. Moreover, if $(\overline{\Lambda} n_t)_0 > 0$ and $(\overline{\Lambda}' n_t)_0 > 0$, then $(\overline{\Lambda}^{-1} \overline{\Lambda}' n_t)_0 > 0$. (The proof of these statements may be found in Problems 1.1, 1.2, 1.3.) It then follows that the subset of $\overline{\mathcal{L}}$ consisting of transformations $\overline{\Lambda}$ with $(\overline{\Lambda} n_t)_0 > 0$ forms a group, called the *orthochronous Lorentz group*, and denoted by $\mathcal{L}^{\uparrow}$; the corresponding transformations preserve the arrow of time. If the matrix $\overline{\Lambda}$ in $\overline{\mathcal{L}}$ is such that $(\overline{\Lambda} n_t)_0 < 0$, then we can write identically

$$\overline{\Lambda} = I(I \overline{\Lambda}), \tag{1.4.10b}$$

where $I$ is the *total reversal*, $I = I_t I_s$: $Ix \equiv -x$. Clearly, $(I \overline{\Lambda} n_t)_0$ is now positive. We have proved that any element of $\overline{\mathcal{L}}$ is either an element of $\mathcal{L}^{\uparrow}$ or a product $I\Lambda$ with $\Lambda$ in $\mathcal{L}^{\uparrow}$.

Finally, the *proper, orthochronous Lorentz group* $\mathcal{L}^{\uparrow}_+$ (which we simply call, if there is no danger of confusion, the *Lorentz group*, $\mathcal{L}$) is the group of matrices $\Lambda$ such that

$$\Lambda^T G \Lambda = G, \quad \det \Lambda = 1, \quad (\Lambda n_t)_0 > 0. \tag{1.4.11}$$

As shown before, cf. (1.4.10), we have that any element in $\overline{\mathcal{L}}, \overline{\Lambda}$ is of one of the forms

$$I_s\Lambda, \; I_t\Lambda, \; I_s I_t\Lambda, \; \Lambda \tag{1.4.12}$$

with $\Lambda$ in $\mathcal{L}_+^\uparrow$.

The transformations $I_s$, $I_t$, $I$ are at times called *improper* transformations.

## 1.5 The Lorentz Group

In this section we further characterize the (orthochronous, proper) Lorentz group. We start by proving a simple, but basic, theorem.

**Theorem 1.** *If $R$ is in $\mathcal{L}$ and $Rn_t = n_t$, then $R$ is a rotation.*

To prove this, we note that the condition $Rn_t = n_t$ implies that $R$ is of the form

$$R = \begin{pmatrix} 1 & 0 & 0 & 0 \\ 0 & & & \\ 0 & & \hat{R} & \\ 0 & & & \end{pmatrix},$$

with $\hat{R}$ a $3 \times 3$ matrix. The condition $R^T G R = G$ implies that $\hat{R}^T \hat{R} = 1$; and $\det R = +1$ implies that also $\det \hat{R} = +1$. Therefore, $\hat{R}$ is a three-dimensional orthogonal matrix with unit determinant, i.e., a three-dimensional rotation. From now on we will denote by the same symbol $R$ the Minkowski space transformation and the restriction $(\hat{R})$ to ordinary three-space.

Now let $\Lambda$ be an arbitrary transformation in $\mathcal{L}$, and let $u \equiv \Lambda n_t$. We have $u_0 > 0$ and $u \cdot u = 1$. Consider the vector $\boldsymbol{\xi}$ such that $u_0 = \cosh|\boldsymbol{\xi}|$, $|\mathbf{u}| = \sinh|\boldsymbol{\xi}|$; this is possible because

$$1 = u \cdot u = (u_0)^2 - |\mathbf{u}|^2 = \cosh^2\xi - \sinh^2\xi.$$

We choose $\boldsymbol{\xi}$ directed along $\mathbf{u}$,

$$\boldsymbol{\xi}/|\boldsymbol{\xi}| = \mathbf{u}/|\mathbf{u}|,$$

so that

$$u_0 = \cosh\xi, \quad u_i = \frac{1}{\xi}(\sinh\xi)\xi_i. \tag{1.5.1}$$

Using the explicit expression for $L(\boldsymbol{\xi})$, (1.3.5c) or (1.4.2), we see that $L(\boldsymbol{\xi})n_t = u$. It follows that the transformation $L^{-1}(\boldsymbol{\xi})\Lambda$ is such that

$$L^{-1}(\boldsymbol{\xi})\Lambda n_t = n_t,$$

so by Theorem 1, $L^{-1}(\boldsymbol{\xi})\Lambda \equiv R$ has to be a rotation, characterized by some $\boldsymbol{\theta}$. We have therefore proved the following theorem (that was anticipated in Sect. 1.3):

**Theorem 2.** *Any (proper, orthochronous) Lorentz transformation, $\Lambda$, can be written as*

$$\Lambda = L(\boldsymbol{\xi})R(\boldsymbol{\theta}), \tag{1.5.2}$$

*where $R$ is a rotation and $L$ a Lorentz boost (the decomposition is not unique). In particular it follows that the Lorentz group is a six-dimensional group (three parameters from $\boldsymbol{\theta}$ and three from $\boldsymbol{\xi}$).*

We may recall that the Lorentz boost $L(\boldsymbol{\xi})$ can be written as

$$R'L(\boldsymbol{\xi}_z)R'',$$

with $R', R'' = R'^{-1}$ rotations and $L(\boldsymbol{\xi}_z)$ an acceleration along the $OZ$ axis. Thus, the general study of Lorentz transformations is reduced to that of rotations and pure accelerations, that may be taken to be along the $OZ$ axis.

**Exercise.** Given two pure boosts $L(\boldsymbol{\xi})$, $L(\boldsymbol{\eta})$, find $L(\boldsymbol{\zeta})$, $R(\boldsymbol{\theta})$ such that

$$L(\boldsymbol{\xi})L(\boldsymbol{\eta}) = L(\boldsymbol{\zeta})R(\boldsymbol{\theta}).$$

Note that in general (unless $\boldsymbol{\xi}, \boldsymbol{\eta}$ are parallel) the product of two boosts is *not* a pure boost ∎

We finish the characterization by presenting two more theorems, and a covariant parametrization of $\Lambda$.

**Theorem 3.** *A Lorentz transformation $\Lambda$ such that $\Lambda n_t = u$ is a pure boost, times a rotation around $\boldsymbol{\xi}$, where $\boldsymbol{\xi}$ is given in terms of $u$ by (1.5.1), if, and only if, $\Lambda$ commutes with all rotations around $\boldsymbol{\xi}$.*

To prove this, we use (1.4.2); because a rotation around $\boldsymbol{\xi}$, which we denote by $R_{\boldsymbol{\xi}}$, leaves $\boldsymbol{\xi}$ invariant, it follows that $L(\boldsymbol{\xi})$ and $R_{\boldsymbol{\xi}}$ commute. (Use that $\boldsymbol{\xi}(R_{\boldsymbol{\xi}}\mathbf{r}) = (R_{\boldsymbol{\xi}}^{-1}\boldsymbol{\xi})\mathbf{r} = \boldsymbol{\xi}\mathbf{r}$ for any $\mathbf{r}$.) The reciprocal is also easy. Given that $u = \Lambda n_t$, we construct via (1.5.1) $\boldsymbol{\xi}$ and then $L(\boldsymbol{\xi})$. Now, $L^{-1}(\boldsymbol{\xi})\Lambda = R$ is a rotation. As we have just seen, $L(\boldsymbol{\xi})$ commutes with rotations $R_{\boldsymbol{\xi}}$; so does $\Lambda$, and hence $R$. But a rotation that commutes with all rotations around an axis $\boldsymbol{\xi}$ is itself a rotation around that axis, so $\Lambda = L(\boldsymbol{\xi})R_{\boldsymbol{\xi}}$, finishing the proof.

**Theorem 4.** *We have, for any $\boldsymbol{\xi}$ and any rotation $R$,*

$$RL(\boldsymbol{\xi})R^{-1} = L(R\boldsymbol{\xi}), \tag{1.5.3}$$

*where $L(R\boldsymbol{\xi})$ is the boost characterized by $R\boldsymbol{\xi}$.*

The proof is straightforward using (1.4.2) and is left as an exercise.

Instead of parametrizing a Lorentz transformation $\Lambda = L(\boldsymbol{\xi})R(\boldsymbol{\theta})$ by the parameters $\boldsymbol{\xi}$, $\boldsymbol{\theta}$, it is at times convenient to use what is called a *covariant parametrization*. We define the set of parameters $\omega_{\mu\nu}$ in terms of $\boldsymbol{\xi}$, $\boldsymbol{\theta}$ by

$$\sum_{jk} \epsilon_{jkl}\omega_{jk} = \theta_l, \quad \omega_{j0} = \frac{1}{2}\xi_j; \quad \omega_{\alpha\beta} = -\omega_{\beta\alpha}. \tag{1.5.4}$$

For $\omega$ infinitesimal we write a Lorentz transformation as

$$\Lambda = 1 - \sum \omega_{\alpha\beta} X^{(\alpha\beta)} + O(\omega^2). \tag{1.5.5}$$

Then, the matrices $X^{(\alpha\beta)}$ have components

$$X_{\mu\nu}^{(\alpha\beta)} = -(\delta_{\mu\alpha} g_{\nu\beta} - \delta_{\mu\beta} g_{\nu\alpha}). \tag{1.5.6}$$

To prove this, we note that, on the one hand, and from the definition of $X$,

$$(\Lambda(\omega)x)_\mu \simeq x_\mu - \sum_{\alpha\beta} \sum_\nu \omega_{\alpha\beta} X_{\mu\nu}^{(\alpha\beta)} x_\nu; \tag{1.5.7}$$

on the other, from the explicit formulas for $R$, $L$ and using (1.5.4),

$$(R(\boldsymbol{\theta})x)_0 = x_0, \ \ (R(\boldsymbol{\theta})x)_i = x_i - \sum 2\omega_{ik} x_k;$$

$$(L(\boldsymbol{\xi})x)_0 \simeq x_0 + \sum 2\omega_{j0} x_j, \ \ (L(\boldsymbol{\xi})x)_i \simeq x_i + 2\omega_{i0} x_0,$$

so that letting $\Lambda = LR$, we get

$$(\Lambda x)_0 \simeq x_0 - \sum 2\omega_{0j} x_j, \ \ (\Lambda x)_i \simeq x_i + 2\omega_{i0} x_0 - \sum 2\omega_{ik} x_k;$$

on comparing this with (1.5.7), the expression (1.5.6) follows.

The invariance group of relativity, beyond $\mathcal{L}_+^\uparrow$, also includes space translations,

$$\mathbf{r} \longrightarrow \mathbf{r} + \mathbf{a},$$

and time translations,

$$ct \longrightarrow ct + a_0;$$

in four-vector notation,

$$x_\mu \longrightarrow x_\mu + a_\mu. \tag{1.5.8}$$

The group obtained by adjoining to $\mathcal{L}$ the translations will be called the *Poincaré*, or *inhomogeneous Lorentz group*, written $\mathcal{JL}$. Its elements are pairs $(a, \Lambda)$ with $a$ a four-vector and $\Lambda$ in $\mathcal{L}$. They act on an arbitrary vector $x$ by

$$(a, \Lambda)x = a + \Lambda x, \tag{1.5.9}$$

and satisfy the ensuing product and inverse law:

$$(a, \Lambda)(a', \Lambda') = (a + \Lambda a', \Lambda\Lambda'),$$

$$(a, \Lambda)^{-1} = (-\Lambda^{-1} a, \Lambda^{-1}). \tag{1.5.10}$$

The unit element of the group is the transformation $(0, 1)$.

## 1.6 Geometry of Minkowski Space

The geometrical properties of spacetime present some peculiarities owing to the indefinite character of the metric. A first peculiarity is that we can classify vectors $v$ of a Minkowskian space, *in a relativistically invariant way*, in the following classes: timelike, lightlike, and spacelike vectors. *Timelike vectors* $v$ are such that $v \cdot v > 0$. If $v_0 > 0$, we say they are *positive timelike*; if $v_0 < 0$, *negative* ($v_0 = 0$ is impossible). *Lightlike* vectors $v$, which satisfy $v \cdot v = 0$, are *positive lightlike* if $v_0 > 0$, *negative* if $v_0 < 0$. $v_0 = 0$ is only possible for the null vector, $v = 0$. Finally, we say that $v$ is *spacelike* if $v \cdot v < 0$; the sign of $v_0$ is *not* invariant now.

**Exercise.** (A) Prove that this classification is invariant under transformations in $\mathcal{L}_+^\uparrow$; in particular check invariance of sign $v_0$ if $v^2 \geq 0$. (B) Show that the trajectory of a particle with mass is given by a positive timelike vector, and that of a light ray by a positive lightlike vector.

*Hint.* Let $\mathbf{r}$ be the location of a particle (or signal) at time $t$. Form the four-vector $x, x_0 = ct, \mathbf{x} = \mathbf{r}$. The velocity of the particle (assuming uniform motion) is $\mathbf{V} = \mathbf{r}/t$ ∎

The following lemma is very useful:

**Lemma 1.** *(i) If $v$ is positive (negative) timelike, then there exists a vector $v^{(0)}$ and a Lorentz transformation $\Lambda$ such that $v = \Lambda v^{(0)}$, and $v_0^{(0)} = \pm m$, $\mathbf{v}^{(0)} = 0$, $m > 0$. (ii) If $v$ is positive (negative) lightlike there exists a $\overline{v}$ and $\Lambda$ with $v = \Lambda\overline{v}$ and $\overline{v}_0 = \pm 1$, $\overline{v}_1 = \overline{v}_2 = 0$, $\overline{v}_3 = 1$. (Here and before the signs $(\pm)$ are correlated to positive–negative.) (iii) If $v$ is spacelike, there exist a $v^{(3)}$ and $\Lambda$ with $v = \Lambda v^{(3)}$, $v_\mu^{(3)} = \delta_{\mu 3} v_3^{(3)}$, $v_3^{(3)} > 0$.*

This means that, in an appropriate reference system, a positive lightlike vector (say) can be chosen to be of the form $\overline{v}$,

$$\overline{v} = \begin{pmatrix} 1 \\ 0 \\ 0 \\ 1 \end{pmatrix}.$$

The clumsy but simple proof of this lemma uses the explicit expression for the Lorentz transformations to build explicit constructions.

The difference between a Euclidean space and Minkowski space is also apparent in the two following results:

**Theorem 1.** *If both $v$ and $v'$ are lightlike and they are orthogonal, that is, $v \cdot v' = 0$, then they are parallel: $v' = \alpha v$.*

For the proof, which uses the previous lemma, see Problem P.1.4.

**Theorem 2.** *If $v \cdot v \geq 0$, and $v \cdot u = 0$, then either $v$ and $u$ are proportional or necessarily $u$ is spacelike.*

The proof is again left as an exercise, using Lemma 1.

**Theorem 3.** *The only invariant numerical tensors in Minkowski space are combinations of the metric tensor, $g_{\mu\nu}$, and the* Levi-Cività *tensor $\epsilon_{\mu\nu\rho\sigma}$,*

$$
\epsilon_{\mu\nu\rho\sigma} = \begin{cases} \quad 1, & \text{if } \mu\nu\rho\sigma \text{ is an even permutation of } 1230, \\ -1, & \text{if } \mu\nu\rho\sigma \text{ is an odd permutation of } 1230, \\ \quad 0, & \text{if two indices are equal.} \end{cases}
$$

Note that $\epsilon_{ijk0} = \epsilon_{ijk}$, where $\epsilon_{ijk}$ is the Levi-Cività tensor in ordinary three-space.

**Theorem 4.** *Given a set of Minkowski vectors $v^{(a)}$, the only invariants that are continuous and that can be formed with them are functions of the scalar products $v^{(a)} \cdot v^{(b)}$ and, if there are four or more vectors, of the quantities*

$$
\sum g_{\mu\mu} g_{\nu\nu} g_{\rho\rho} g_{\sigma\sigma} \epsilon_{\mu\nu\rho\sigma} v_\mu^{(a)} v_\nu^{(b)} v_\rho^{(c)} v_\sigma^{(d)}.
$$

In spite of the fact that these theorems are similar to their analogues in Euclidean space and also in spite of their apparent simplicity, proofs are very complicated. For example, Theorem 4 fails if we remove the requisite of continuity: the functions $(\text{sign } v_0)\theta(v^2)$ or $\delta_4(v) \equiv \delta(v_0)\delta(\mathbf{v})$ are invariant: yet they *cannot* be written in terms of invariants. Proofs of Theorems 3 and 4 can be found in, for example, the treatise of Bogoliubov, Logunov and Todorov (1975).

Given a Minkowski vector, $v$, the set of Lorentz transformations $\Gamma$ that leave it invariant is called its *little group*[5] (or *stabilizer*), $\mathcal{W}(v)$. The little group of a vector $v$ depends only upon the sign of $v \cdot v$, in the sense that if, for example, $v \cdot v > 0$ and $u \cdot u > 0$, then the little groups $\mathcal{W}(v)$, $\mathcal{W}(u)$ are isomorphic. To prove this, we first note that $\mathcal{W}(v)$ and $\mathcal{W}(\Lambda v)$ are isomorphic for any $\Lambda$. Indeed, if $\Gamma v = v$, then $\Lambda\Gamma\Lambda^{-1}$ is in $\mathcal{W}(\Lambda v)$, and vice versa. Moreover, $\mathcal{W}(v)$ is identical with $\mathcal{W}(\alpha v)$ for any number $\alpha \neq 0$. Using this in conjunction with Lemma 1, we find that there are essentially only three little groups. To be precise, we have that, if $v \cdot v > 0$, the little group is isomorphic to $\mathcal{W}(n_t)$; if $v \cdot v = 0$, the little group is isomorphic to $\mathcal{W}(\bar{v})$, $\bar{v}_0 = \bar{v}_3$, $\bar{v}_1 = \bar{v}_2 = 0$; and if $v \cdot v < 0$, the little group is isomorphic to $\mathcal{W}(n^{(3)})$, $n_\mu^{(3)} = \delta_{\mu3}$. This greatly simplifies the study of the little groups.

**Theorem 5.** *(A) $\mathcal{W}(n_t) = SO(3)$, where by $SO(3)$ we denote the group of ordinary rotations. (B) $\mathcal{W}(\bar{v}) = SO(2) \times T_2$, where $SO_z(2)$ is the group of rotations around $OZ$, and $T_2$ is defined below. (C) $\mathcal{W}(n^{(3)}) = \mathcal{L}_+^\uparrow(3)$, where $\mathcal{L}_+^\uparrow(3)$ is identical to a Lorentz group that acts only on time and the spatial plane $XOY$, but leaves $OZ$ invariant.*

---

The result (A) is already known to us (Theorem 1 of Sect. 1.5). Result (C) is left as a simple exercise. We turn to the lightlike case (B). Let $\Gamma$ be an element of $\mathcal{W}(\bar{v})$, and let $N$ be the subspace of Minkowski space orthogonal to $\bar{v}$, that is, if $u$ is in $N$, then $u \cdot \bar{v} = 0$.

Clearly, the subspace $N$ is also invariant under $\Gamma$. A basis of $N$ is formed by the three vectors $v^{(a)}, a = 1, 2, 3$ with $v^{(1)} = n^{(1)}$, $v^{(2)} = n^{(2)}$, $n_\mu^{(a)} = \delta_{a\mu}$, and $v^{(3)} = \bar{v}$ : because $\bar{v}$ is lightlike the subspace orthogonal to $\bar{v}$ contains $\bar{v}$ itself. If $u$ is in $N$, we write $u = \sum_a \alpha_a v^{(a)}$. Because $\Gamma u$ is also in $N$, we can write

$$\Gamma u = \Sigma_{ab} \gamma_{ab} \alpha_b v^{(a)};$$

thus the matrix elements $\gamma_{ab}$ determine $\Gamma$, and vice versa. The conditions $\Gamma u \cdot \Gamma u' = u \cdot u'$ and $\Gamma \bar{v} = \bar{v}$ imply that

$$(\gamma_{ab}) = \begin{pmatrix} \cos\theta & \sin\theta & 0 \\ -\sin\theta & \cos\theta & 0 \\ \gamma_{31} & \gamma_{32} & 1 \end{pmatrix}, \tag{1.6.1}$$

with $\gamma_{31}, \gamma_{32}$ arbitrary. The set of matrices (1.6.1) has a mathematical structure like that of the Euclidean group of the plane, $SO_z(2) \times \mathcal{T}_2$ of rotations $SO_z(2)$ around $OZ$,

$$\begin{pmatrix} \cos\theta & \sin\theta & 0 \\ -\sin\theta & \cos\theta & 0 \\ 0 & 0 & 1 \end{pmatrix},$$

and "translations" $\mathcal{T}_2$,

$$\begin{pmatrix} 1 & 0 & 0 \\ 0 & 1 & 0 \\ \gamma_{31} & \gamma_{32} & 1 \end{pmatrix}.$$

To finish this section we present a few more definitions (see Fig. 1.6.1). The *light cone* is the set of vectors $v$ with $v^2 = 0$. If, moreover, $v_0 > 0$ $(v_0 < 0)$, we speak of the *future, forward or positive* (*past, backward or negative*) *light cone*, denoted by $V^+$ $(V^-)$. The set of vectors $u$ with $u^2 = m^2 > 0$ is denoted by $\Omega^\pm(m)$, $(\pm)$ according to the sign of $u_0$, and is called the *future, forward or positive* (*past, backward or negative*) *mass hyperboloid*, for $u_0 > 0$ $(u_0 < 0)$. This name derives from (*momentum*) *Minkowski space* (see below). The set of $w$ with $w \cdot w = -\mu^2$, $\mu^2 > 0$ is called the *imaginary mass hyperboloid*, $\Omega(i\mu)$.

**Exercise.** Verify that the sets $V^+$, $V^-$, $\Omega^+(m)$, $\Omega^-(m)$, $\Omega(i\mu)$ are invariant under $\mathcal{L}_+^\uparrow$, and that each vector in one of them can be reached by an appropriate transformation from any other one in the same set ∎

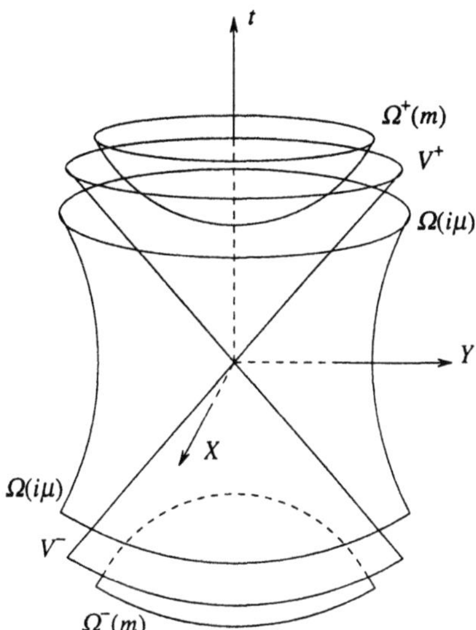



**Fig. 1.6.1.** Various regions in Minkowski space.

## 1.7 Transformation Properties of Physical Quantities Under the Lorentz Group

When effecting a Lorentz transformation not only space and time change; other quantities such as momentum and energy, or the electromagnetic potentials, are modified as well.

Consider first spacetime derivatives. Forming a four-component object,

$$\overline{\partial} \sim \begin{pmatrix} \partial/\partial x_0 \\ \partial/\partial x_1 \\ \partial/\partial x_2 \\ \partial/\partial x_3 \end{pmatrix},$$

and recalling that derivatives transform contragradiently to coordinates, we get

$$\Lambda \; : \; \overline{\partial} \longrightarrow (\Lambda^{-1T})\overline{\partial}.$$

A transformation law like that of the coordinates themselves is obtained by defining the *four-derivative* (or *derivative four-vector*) $\partial$:

$$\partial \sim \begin{pmatrix} c^{-1}\,\partial_t \\ -\boldsymbol{\nabla} \end{pmatrix}, \tag{1.7.1a}$$

i.e., writing

$$\partial_0 \equiv \partial/\partial x_0, \; \partial_i \equiv -\partial/\partial x_i. \tag{1.7.1b}$$

Indeed, one has $\partial = G\overline{\partial}$, so that

$$\partial \longrightarrow G(\Lambda^{-1T}\overline{\partial}) = \Lambda G\overline{\partial} = \Lambda\partial,$$

because $G\Lambda^{-1T} = \Lambda G$ (cf. (1.4.8)). Therefore,

$$\partial_\mu \longrightarrow \sum_j \Lambda_{\mu\nu}\partial_\nu; \ \partial \longrightarrow \Lambda\partial. \tag{1.7.1c}$$

The transformation laws for momentum and energy may be deduced in classical (i.e., nonquantum) relativistic mechanics[6]. We will infer them from the correspondence principle. In quantum mechanics momentum $\mathbf{p}$ is represented by the operator $-i\hbar\boldsymbol{\nabla}$, and energy by $i\hbar\partial_t$, with $\partial_t$ short for $\partial/\partial t$. Therefore we can form the *energy–momentum four-vector* (or just the *four-momentum*), $p$, with components $p_\mu$,

$$p \sim \begin{pmatrix} c^{-1}E \\ \mathbf{p} \end{pmatrix} = \begin{pmatrix} p_0 \\ p_1 \\ p_2 \\ p_3 \end{pmatrix}, \tag{1.7.2a}$$

which in quantum mechanics corresponds to $i\hbar\partial$; hence we can demand the transformation properties

$$\Lambda \ : \ p \to \Lambda p, \ (\Lambda p)_\mu = \sum_\nu \Lambda_{\mu\nu}p_\nu. \tag{1.7.2b}$$

For a free particle $E$ and $\mathbf{p}$ are connected by

$$E = +\sqrt{m^2c^4 + c^2\mathbf{p}^2},$$

so we have $p_0 > 0$ and

$$p^2 = p \cdot p = m^2c^2 : \tag{1.7.3}$$

the mass is a relativistic invariant. Equation (1.7.4) is the reason for the name "mass hyperboloid" of the previous section; the four-momenta that satisfy (1.7.4) fill $\Omega^+(m)$.

For electromagnetic potentials the transformation properties can be obtained from the principle of minimal replacement. Under this we have

$$-i\hbar\boldsymbol{\nabla} \to -i\hbar\boldsymbol{\nabla} - \frac{e}{c}\mathbf{A}, \ i\hbar\partial_t \to i\hbar\partial_t - e\phi,$$

with $\mathbf{A}$ the *vector potential* and $\phi$ the *scalar* one. We thus define the electromagnetic *four-potential* $A$, with components $A_{\mu}$, by

$$A \sim \begin{pmatrix} \phi \\ \mathbf{A} \end{pmatrix} = \begin{pmatrix} A_0 \\ A_1 \\ A_2 \\ A_3 \end{pmatrix}, \tag{1.7.4a}$$

and for consistency we postulate that

$$\Lambda \ : \ A \to \Lambda A, \ (\Lambda A)_\mu = \sum_\mu \Lambda_{\mu\nu}A_\nu. \tag{1.7.4b}$$

---

[6] See, for example, Landau and Lifshitz (1975); Goldstein (1965)

The minimum replacement principle can be unified thus:

$$i\hbar\partial_\mu \rightarrow i\hbar\partial_\mu - \frac{e}{c}A_\mu. \tag{1.7.5}$$

## 1.8 Covariant Form of the Maxwell Equations

Relativity was devised to make mechanics compatible with Maxwell's equations; therefore, to write the latter equations in a *manifestly covariant* form merely requires setting up some appropriate notation. The equations, in terms of the *electric field* $\boldsymbol{\mathcal{E}}$ and *magnetic field* $\boldsymbol{\mathcal{B}}$, are

$$\boldsymbol{\nabla} \times \boldsymbol{\mathcal{E}} + \frac{1}{c}\partial_t\boldsymbol{\mathcal{B}} = 0, \ \ \boldsymbol{\nabla}\boldsymbol{\mathcal{B}} = 0; \tag{1.8.1a}$$

$$\boldsymbol{\nabla}\boldsymbol{\mathcal{E}} = 4\pi\rho; \ \ \boldsymbol{\nabla} \times \boldsymbol{\mathcal{B}} - \frac{1}{c}\partial_t\boldsymbol{\mathcal{E}} = \frac{4\pi}{c}\mathbf{j}; \tag{1.8.1b}$$

$$\partial_t\rho + \boldsymbol{\nabla}\mathbf{j} = 0. \tag{1.8.2}$$

We use Gauss's system of units, and $\mathbf{j}$, $\rho$ are, respectively, the *current* and *charge density*. In terms of $\mathbf{A}$, $\phi$,

$$\boldsymbol{\mathcal{E}} = -\frac{1}{c}\partial_t\mathbf{A} - \boldsymbol{\nabla}\phi, \ \ \boldsymbol{\mathcal{B}} = \boldsymbol{\nabla} \times \mathbf{A}. \tag{1.8.3}$$

We start by writing the last in covariant form. To do this, we define the antisymmetric *field strength tensor*, $F_{\mu\nu}$, as

$$F_{\mu\nu} = \partial_\mu A_\nu - \partial_\nu A_\mu, \tag{1.8.4}$$

so that, comparing this with (1.8.3), we get

$$\mathcal{E}_i = F_{i0}; \ \ \mathcal{B}_k = \frac{-1}{2}\sum_{ij}\epsilon_{kij}F_{ij}; \ \ F_{ij} = -\sum_k\epsilon_{ijk}\mathcal{B}_k. \tag{1.8.5}$$

Substituting in (1.8.1b), we thus obtain

$$\sum_\mu g_{\mu\mu}\partial_\mu F_{\mu\nu} = 4\pi J_\nu, \tag{1.8.6}$$

and we have defined the *four-current J*,

$$J_0 = \rho, \ \ J_i = \frac{1}{c}j_i \ : \ J \sim \begin{pmatrix} \rho \\ c^{-1}\mathbf{j} \end{pmatrix}. \tag{1.8.7}$$

With this last definition, (1.8.2) can be written as $\partial \cdot J = 0$, or explicitly,

$$\sum_\mu g_{\mu\mu}\partial_\mu J_\mu = 0. \tag{1.8.8}$$

The conditions (1.8.1a) can also be expressed covariantly as

$$\sum g_{\mu\mu}g_{\sigma\sigma}g_{\rho\rho}\partial_\mu\epsilon_{\mu\nu\rho\sigma}F_{\rho\sigma} = 0, \tag{1.8.9}$$

and are identically satisfied if $F_{\mu\nu}$ is given in terms of $A_\mu$ by (1.8.4).

Equation (1.8.6) can be written in terms of $A$:

$$(\partial \cdot \partial)A_\mu - \partial_\mu(\partial \cdot A) = 4\pi J_\mu, \qquad (1.8.10)$$

where the *d'Alembertian operator* is defined as

$$\Box \equiv \partial^2 \equiv \partial \cdot \partial \equiv \frac{1}{c^2}\partial_t^2 - \Delta,$$

and $\partial \cdot A = c^{-1}\partial_t\phi + \text{div } \mathbf{A}$.

*Gauge transformations* can also be written in a manifestly covariant manner. For the four-potential $A_\mu(x) = A_\mu(\mathbf{r}, t)$, we define them as the transformations

$$A_\mu \rightarrow A'_\mu(x) = A_\mu(x) - \partial_\mu f(x), \qquad (1.8.11)$$

with $f(x)$ an arbitrary function of $\mathbf{r}$, $t$. The transformation (1.8.11) leaves invariant the Maxwell equations and the tensor $F_{\mu\nu}$, which is the object that contains the observable electric and magnetic fields, $\boldsymbol{\mathcal{E}}, \boldsymbol{\mathcal{B}}$.

The four-vector *density of energy–momentum* for the (classical) electromagnetic field, $\Pi_\mu$, is

$$\Pi_0(\mathbf{r}, t) = \frac{1}{8\pi}(\boldsymbol{\mathcal{E}}^2 + \boldsymbol{\mathcal{B}}^2), \quad \boldsymbol{\Pi}(\mathbf{r}, t) = \frac{1}{4\pi}\boldsymbol{\mathcal{E}} \times \boldsymbol{\mathcal{B}}, \qquad (1.8.12)$$

so that the energy and momentum of the free electromagnetic radiation field are

$$\begin{aligned}
E_{\text{rad}} &= \int d^3r\, \Pi_0(\mathbf{r}, t), \\
\mathbf{P}_{\text{rad}} &= \frac{1}{c}\int d^3r\, \boldsymbol{\Pi}(\mathbf{r}, t).
\end{aligned} \qquad (1.8.13)$$

We can derive them from a *density of energy–momentum tensor*, $\Theta_{\mu\nu}$, with

$$\Pi_\mu = \Theta_{\mu 0}, \quad \Theta_{\mu\nu} = \frac{1}{4\pi}\left(\frac{1}{4}g_{\mu\nu}\sum g_{\alpha\alpha}g_{\beta\beta}F_{\alpha\beta}^2 + \sum g_{\rho\rho}F_{\mu\rho}F_{\rho\nu}\right). \quad (1.8.14)$$

## 1.9 Minkowski Space: Metric, Conventions

Recall that we defined

$$g_{\mu\nu} = \begin{cases} 1, & \mu = \nu = 0, \\ -1, & \mu = \nu \neq 0, \\ 0, & \mu \neq \nu, \end{cases}$$

and the scalar product

$$x \cdot y = x_0 y_0 - \sum_{i=1}^{3} x_i y_i = \sum_{\mu\nu} g_{\mu\nu} x_\mu y_\nu.$$

Also,

$$F_{\mu\nu} = \partial_\mu A_\nu - \partial_\nu A_\mu.$$

These conventions are far from being unanimously accepted. The ones we have chosen here are Feynman's. They also agree with the ones that are currently the most popular, with one proviso: it is becoming fashionable to denote components by superscripts, rather than subscripts. This is the case, in particular, in the treatises of Bjorken and Drell (1964, 1965), Bogoliubov and Shirkov (1959), Schweber (1961), Itzykson and Zuber (1980) etc. Therefore, one goes from the notations of these authors to ours[7] by just exchanging $x_\mu \leftrightarrow x^\mu$, $\epsilon_{\mu\nu\rho\sigma} \leftrightarrow \epsilon^{\mu\nu\rho\sigma}$, etc. Bjorken and Drell (1964, 1965) also define $F_{\mu\nu}$ with the opposite sign to ours:

$$F_{\mu\nu}^{\mathrm{BD}} = -F_{\mu\nu}^{\mathrm{here}}.$$

Sometimes one uses contravariant indices, $x^\mu \equiv \sum_\nu g_{\mu\nu} x_\nu$, and Einstein's convention of summing over repeated indices:

$$x \cdot y \equiv x^\mu y_\mu \equiv g^{\mu\nu} x_\nu y_\mu \equiv \sum g_{\mu\nu} x_\nu y_\mu.$$

Jauch and Rohrlich (1959) follow our convention for the components *but* define a metric tensor opposite to ours:

$$g_{\mu\nu}^{\mathrm{JR}} = -g_{\mu\nu}.$$

Therefore, $(x \cdot y)^{\mathrm{JR}} = -x \cdot y$, etc.

Some authors (starting with Minkowski himself) define an *imaginary* fourth (time) coordinate, $x_4 = ict$, and the scalar product with reversed sign so that

$$(x \cdot y)^{\mathrm{M}} = \sum_{\alpha=1}^{4} x_\alpha y_\alpha \ :$$

the scalar product apparently has the same structure as an ordinary Euclidean one.

This simplification is, I believe, fictitious. The price to pay is that, first, Lorentz transformations become *complex* (the $i$ has to go somewhere!) and, what is worse, the $i$ in $x_4 = ict$ is peculiar. If we write a wave function as $\Psi$, then the operation of complex conjugation $\Psi \to \Psi^*$ has to be defined so as *not* to affect the $i$ of $x_4 = ict$. Some authors, like Rose (1961), introduce a new operation of complex conjugation, $\Psi \to \Psi^x$, with these properties. Perhaps not surprisingly, Minkowski's convention is falling out of use.

We end up with two useful relations between the tensors $\epsilon$, $g$ (valid with *our* conventions):

---

[7] We have kept Feynman notation with subscripts because it is more convenient for studying the nonrelativistic and semirelativistic limits.

$$\sum_{\alpha\beta} g_{\alpha\beta}\epsilon_{\alpha\mu\rho\sigma}\epsilon_{\beta\nu\tau\lambda} = g_{\mu\nu}(g_{\rho\lambda}g_{\sigma\tau} - g_{\rho\tau}g_{\sigma\lambda})$$
$$+ g_{\mu\lambda}(g_{\rho\tau}g_{\sigma\nu} - g_{\rho\nu}g_{\sigma\tau})$$
$$+ g_{\mu\tau}(g_{\rho\nu}g_{\sigma\lambda} - g_{\rho\lambda}g_{\sigma\nu}); \tag{1.9.1}$$

$$\sum_{\alpha\beta} g_{\alpha\alpha}g_{\beta\beta}\epsilon_{\mu\nu\alpha\beta}\epsilon_{\alpha\beta\rho\sigma} = 2(g_{\nu\rho}g_{\mu\sigma} - g_{\mu\rho}g_{\nu\sigma}).$$

## Problems

**P.1.1.** Prove that $\Lambda^T G\Lambda = G$ implies that $\Lambda n_t \neq 0$.

*Solution.* Consider the 00 components of $\Lambda^T G\Lambda = G$, and $\Lambda G\Lambda^T = G$; then,

$$\Lambda_{00}^2 - \sum_i \Lambda_{i0}^2 = 1; \quad \Lambda_{00}^2 - \sum_i \Lambda_{0i}^2 = 1.$$

From any of these, $|\Lambda_{00}| \geq 1$ so $|(\Lambda n_t)_0| \geq 1$.

**P.1.2.** Show that $\Lambda_{00} > 0$, $\Lambda'_{00} > 0$ imply that $(\Lambda\Lambda')_{00} > 0$.

*Solution.* Using the evaluations of the previous problem and Schwartz's inequality,

$$\left|\sum_i \Lambda_{0i}\Lambda'_{i0}\right| \leq \sqrt{\sum \Lambda_{0i}\Lambda_{0i}}\sqrt{\sum \Lambda'_{i0}\Lambda'_{i0}} < \Lambda_{00}\Lambda'_{00}.$$

Hence,

$$(\Lambda\Lambda')_{00} = \Lambda_{00}\Lambda'_{00} + \sum_i \Lambda_{0i}\Lambda'_{i0} > \Lambda_{00}\Lambda'_{00} - \left|\sum_i \Lambda_{0i}\Lambda'_{i0}\right| > 0.$$

**P.1.3.** Show that $\Lambda_{00} > 0$ implies that $(\Lambda^{-1})_{00} > 0$.

**P.1.4.** Prove that $u^2 = v^2 = u \cdot v = 0$ implies that $u = \alpha v$ (Theorem 1 in Sect. 1.6).

*Solution.* Suppose, for example, that $v$ is positive lightlike. Choose a reference system where $v_0 = v_3 = 1$, $v_1 = v_2 = 0$. The condition $u \cdot v = 0$ tells us that $u_0 = u_3 = \alpha$; the condition $u^2 = 0$ then forces $u_1$, $u_2$ to vanish: in this system of reference, hence in all, $u = \alpha v$.

**P.1.5.** Prove that one cannot fix the gauge (i.e., one cannot introduce a subsidiary condition making $A_\mu$ unique) in a form which is manifestly relativistically invariant. (This, of course, does not invalidate the formulation of Maxwell's equations in terms of $A_\mu$; what it means is that gauge and relativistic invariance are intertwined in a nontrivial manner. See, for example, Sect. 5.2.)

# 2. The Klein–Gordon Equation.
## Relativistic Equation for Spinless Particles

## 2.1 The Klein–Gordon Equation. Generalities

When Schrödinger developed his nonrelativistic wave equation, he also proposed a relativistic generalization. This equation, studied in greater detail by Klein and Gordon (whose name it now bears) can be immediately obtained from the relation (1.7.4) for a free particle,

$$p \cdot p = m^2 c^2,$$

with the substitutions suggested by the correspondence principle,

$$p_0 = \frac{1}{c} E \rightarrow i\hbar \partial_t, \ \mathbf{p} \rightarrow -i\hbar \nabla.$$

For a wave function $\Psi(\mathbf{r}, t)$ representing free, spinning or spinless particles we then get the *Klein–Gordon equation*,

$$\left\{ \left( \frac{i\hbar}{c} \partial_t \right)^2 - (i\hbar \nabla)^2 - m^2 c^2 \right\} \Psi(\mathbf{r}, t) = 0. \tag{2.1.1a}$$

This can be written in a manifestly covariant manner as

$$(\hbar^2 \partial \cdot \partial + m^2 c^2) \Psi(x) = 0, \tag{2.1.1b}$$

or in the form of a time evolution equation:

$$(i\hbar \partial_t)^2 \Psi(\mathbf{r}, t) = (-c^2 \hbar^2 \triangle + m^2 c^4) \Psi(\mathbf{r}, t). \tag{2.1.1c}$$

In any of these forms, (2.1.1) is really an inevitable equation for the relativistic wave function of a free particle; all you have to assume is that the eigenvalues of energy and momentum satisfy (1.7.4), i.e., that $E(\mathbf{p}) = (m^2 c^4 + \mathbf{p}^2 c^2)^{1/2}$. Any other equation we may want to use must necessarily imply (2.1.1), indeed, for *free particles* be essentially equivalent to it. This, of course, will no longer be the case when interactions are introduced.

Before considering explicit situations, let us add a few general words. First of all, a full quantum treatment requires more than just a wave equation; a Lorentz-invariant scalar product, and a particle spinsity $\rho(\mathbf{r}, t)$ and a density of probability current $\mathbf{j}(\mathbf{r}, t)$ with appropriate properties also need to be given. In the NR (= nonrelativistic) case,

$$\rho(\mathbf{r},t) = |\Psi(\mathbf{r},t)|^2, \ \mathbf{j}(\mathbf{r},t) = \frac{i\hbar}{2m}(\Psi\nabla\Psi^* - \Psi^*\nabla\Psi), \tag{2.1.2a}$$

and these quantities satisfy a continuity equation,

$$\partial_t\rho + \nabla\mathbf{j} = 0. \tag{2.1.2b}$$

This last equation may be written as

$$\partial_0\rho - \frac{1}{c}\sum\partial_i j_i = 0, \tag{2.1.2c}$$

so we define a *four-current* (which, for charged particles, will be related to the electromagnetic one) by setting

$$J_0 = \rho, \ \ J_i = \frac{1}{c}j_i, \tag{2.1.3}$$

and (2.1.2c) now reads

$$\partial \cdot J = 0. \tag{2.1.4}$$

This equation will be Lorentz invariant provided $J$ transforms like a vector:

$$\Lambda \ : \ J_\mu \to (\Lambda J)_\mu = \sum_\nu \Lambda_{\mu\nu} J_\nu. \tag{2.1.5}$$

In particular, it follows that $\rho$ cannot be relativistically invariant, which was to be expected: $\rho$ equals the number of particles (certainly invariant) per volume, which changes with Lorentz boosts.

## 2.2 Wave Equation for Free Spinless Particles

Equation (2.1.1), as it stands, is not a satisfactory equation even for spinless particles for a number of reasons. An important one is that, being of second order in the time derivative, the values of the wave function $\Phi(\mathbf{r},t_0)$ at time $t_0$ do *not* uniquely determine $\Phi(\mathbf{r},t)$; knowledge of $\partial_t\Phi(\mathbf{r},t_0)$ is also necessary. To try and solve the problems, let us start by going over to momentum space. We define

$$\Phi(\mathbf{p},t) = \frac{1}{(2\pi\hbar)^{3/2}} \int d^3r \ e^{-i\mathbf{pr}/\hbar}\Phi(\mathbf{r},t), \tag{2.2.1a}$$

with inverse

$$\Phi(\mathbf{r},t) = \frac{1}{(2\pi\hbar)^{3/2}} \int d^3p \ e^{i\mathbf{pr}}\Phi(\mathbf{r},t); \tag{2.2.1b}$$

the same letter will be used for $\Phi$ and its Fourier transform: we shall rely on the argument to remove the ambiguity. Then, (2.1.1c) becomes

$$(i\hbar\partial_t)^2\Phi(\mathbf{p},t) = (c^2\mathbf{p}^2 + m^2c^4)\Phi(\mathbf{p},t). \tag{2.2.2}$$

We can take the square root of this and postulate the evolution equation for spinless particles,

$$i\hbar\partial_t\Phi(\mathbf{p},t) = +\sqrt{c^2\mathbf{p}^2 + m^2c^4}\,\Phi(\mathbf{p},t).\tag{2.2.3}$$

We have chosen the positive square root to guarantee positive energies. Equation (2.2.3) is a Schrödinger-like evolution equation,

$$i\hbar\partial_t\Phi(\mathbf{p},t) = H_0\Phi(\mathbf{p},t),\tag{2.2.4a}$$

where the free Klein–Gordon Hamiltonian is

$$H_0 = +\sqrt{m^2c^2 + c^2\mathbf{p}^2}.\tag{2.2.4b}$$

We will refer to (2.2.4) as the Klein–Gordon Schrödinger equation, *KGS equation* for short. In position space,

$$i\hbar\partial_t\Phi(\mathbf{r},t) = H_0\Phi(\mathbf{r},t),\tag{2.2.5a}$$

$$H_0\Phi(\mathbf{r},t) \equiv \int d^3r'\, h_0(\mathbf{r}-\mathbf{r}\,')\Phi(\mathbf{r}\,',t),\tag{2.2.5b}$$

where the kernel $h_0$ is given by

$$h_0(\mathbf{r}) = \frac{1}{(2\pi\hbar)^3}\int d^3p\; e^{i\mathbf{pr}/\hbar}\sqrt{m^2c^4 + c^2\mathbf{p}^2}.\tag{2.2.5c}$$

Expanding this in powers of $1/c^2$, we get

$$h_0(\mathbf{r}) \simeq \frac{mc^2}{(2\pi\hbar)^3}\int d^3p\; e^{i\mathbf{pr}/\hbar}\left(1 + \frac{\mathbf{p}^2}{2m^2c^2} + \dots\right)$$

$$= mc^2\left[\delta(\mathbf{r}) - \frac{1}{2m^2c^2}\hbar^2\triangle\delta(\mathbf{r}) + \dots\right],$$

so that, to an error $O(1/c^2)$,

$$H_0\Phi(\mathbf{r},t) \simeq mc^2\Phi(\mathbf{r},t) - \frac{\hbar^2}{2m}\triangle\Phi(\mathbf{r},t) + \dots,$$

i.e., $H_0$ tends to the NR (= nonrelativistic) Hamiltonian. The wave function $\Phi$ will coincide, in the NR limit, with the ordinary one up to the phase $\exp(-i/\hbar)mc^2t$ due to the displacement of energies $mc^2$ between the relativistic expression and the usual NR convention, where one takes $E = 0$ for $\mathbf{p} = 0$.

We also write (2.2.5b) formally as

$$i\hbar\partial_t\Phi(\mathbf{r},t) = +c\sqrt{mc^2 - \hbar^2\triangle}\;\Phi(\mathbf{r},t),\tag{2.2.5d}$$

and the meaning of the square root is provided by the integral expression (2.2.5a,b).

Because (2.2.4), (2.2.5) are the square root of the Klein–Gordon equation, one would expect them to be relativistically invariant. A formal proof follows. Given a Lorentz transformation, $\Lambda$, we define its action as

$$\Lambda\;:\;\Phi(x) \to \Phi_\Lambda(x) \equiv U(\Lambda)\Phi(x) \equiv \Phi(\Lambda^{-1}x).\tag{2.2.6}$$

Because we can decompose $\Lambda = LR$, we may consider rotations and boosts separately. Since $H_0$ only depends upon $\mathbf{P}^2$ (and not on $\mathbf{P}$), rotation invariance is manifest. For boosts $L(\boldsymbol{\xi})$ we consider the infinitesimal case and, since we already have invariance under rotations, we may, if we so wish, consider the case where $\boldsymbol{\xi}$ is directed along $OZ$.

Let us write the KGS equation as

$$(i\hbar\partial_t - H_0)\Phi(\mathbf{r}, t) = 0;$$

we will have invariance if the transformed $\Phi_L$ satisfies the same equation:

$$(i\hbar\partial_t - H_0)\Phi_L(\mathbf{r}, t) = 0.$$

Here, using (1.3.5a), (2.2.6) we have that, for infinitesimal $\xi$,

$$\Phi_L(\mathbf{r}, t) \simeq \Phi(x, y, z - \xi ct, t - c^{-1}\xi z)$$

$$\simeq \Phi(\mathbf{r}, t) + \left(-\xi ct\frac{\partial}{\partial z} - \frac{1}{c}\xi z\frac{\partial}{\partial t}\right)\Phi(\mathbf{r}, t).$$

Therefore we will have invariance provided

$$\left[i\hbar\partial_t - H_0, -ct\frac{\partial}{\partial z} - \frac{1}{c}z\frac{\partial}{\partial t}\right]\Phi = 0$$

for all $\Phi$ satisfying the KGS equation.

Now,

$$\left[i\hbar\partial_t, -ct\frac{\partial}{\partial z}\right] = -i\hbar c\frac{\partial}{\partial z},$$

and

$$\left[-H_0, -\frac{1}{c}z\partial_t\right] = \left[\sqrt{m^2c^4 + c\mathbf{P}^2}, z\right]\frac{1}{c}\partial_t$$

$$= -\frac{i\hbar}{c}\frac{c^2}{H_0}P_z\partial_t = -cP_z\frac{1}{H_0}i\hbar\partial_t;$$

other commutators vanish. Acting on $\Phi$, and because $i\hbar\partial_t\Phi = H_0\Phi$,

$$\left[i\hbar\partial_t - H_0, -ct\frac{\partial}{\partial z} - \frac{1}{c}z\partial_t\right]\Phi = c\left(-i\hbar\frac{\partial}{\partial z} - P_z\right)\Phi = 0.$$

Invariance under finite transformations follows from invariance under infinitesimal ones since the $\xi$ parameters are normal. For a finite transformation along the $OZ$ axis,

$$U(L(\boldsymbol{\xi_z})) = \exp\left(-\xi ct\frac{\partial}{\partial z} - \frac{1}{c}\xi z\partial_t\right);$$

along an arbitrary axis,

$$U(L(\boldsymbol{\xi})) = \exp\frac{-i}{\hbar}\boldsymbol{\xi}\mathbf{N}, \tag{2.2.7a}$$

where the acceleration generators $\mathbf{N}$ are

$$\mathbf{N} = -cti\hbar\boldsymbol{\nabla} - \frac{1}{c}ri\hbar\partial_t = ct\mathbf{P} - \frac{1}{c}\mathbf{r}H_0 = x_0\mathbf{P} - \mathbf{r}P_0;$$

$$N_i = x_0 P_i - x_i P_0, \tag{2.2.7b}$$

and we have defined $P_0 = H_0/c$. This can be compared with the known case of rotations,

$$L_i = \sum \epsilon_{ijk} x_j P_k; \quad U(R(\boldsymbol{\theta})) = \exp\frac{-i}{\hbar}\boldsymbol{\theta}\mathbf{L}.$$

In the NR limit, $H_0 \simeq mc^2$, $\xi \simeq \mathbf{v}/c$ and

$$\boldsymbol{\xi}\mathbf{N} \simeq \mathbf{v}(-m\mathbf{r} + t\mathbf{P}),$$

which, as was to be expected, agrees with the generators of Galilean accelerations, (1.2.5).

In many applications it is useful to have a manifestly covariant expression for Lorentz transformations. Let us write an infinitesimal one as

$$U(\Lambda) \simeq 1 - \frac{i}{\hbar}\sum g_{\mu\mu}g_{\nu\nu}\omega_{\mu\nu}M_{\mu\nu}, \tag{2.2.8a}$$

where the $\omega_{\mu\nu}$ were defined in (1.5.4) and are antisymmetric and the $M_{\mu\nu}$, also taken to be antisymmetric, can be considered the representatives of the $X^{(\alpha\beta)}$ in (1.5.5). Comparing this with pure boosts, (2.2.7), and with rotations, we get

$$\tfrac{1}{2}\xi_i = \omega_{i0}, \ \tfrac{1}{2}\sum \theta_i\epsilon_{ijk} = \omega_{jk},$$

$$M_{i0} = x_i P_0 - x_0 P_i, \ M_{jk} = x_j P_k - x_k P_j, \tag{2.2.8b}$$

or, together,

$$M_{\mu\nu} = x_\mu P_\nu - x_\nu P_\mu. \tag{2.2.8c}$$

## 2.3 Plane Waves. Current. Scalar Product

We now look for plane waves, solutions to the KGS equation with well-defined momentum:

$$\mathbf{P}\Phi^{(\mathbf{P})}(\mathbf{r}, t) = \mathbf{p}\Phi^{(\mathbf{P})}(\mathbf{r}, t), \tag{2.3.1a}$$

so that

$$\Phi^{(\mathbf{P})}(\mathbf{r}, t) = C(\mathbf{p}, t)e^{i\mathbf{p}\mathbf{r}/\hbar}. \tag{2.3.1b}$$

Substituting this into (2.2.5a), we find an equation for $C$,

$$i\hbar\partial_t C(\mathbf{p}, t) = \sqrt{m^2c^4 + c^2\mathbf{p}^2}\, C(\mathbf{p}, t),$$

with solution

$$C(\mathbf{p},t) = A(\mathbf{p})e^{-iE(\mathbf{p})t/\hbar}, \tag{2.3.1c}$$

$$E(\mathbf{p}) = +\sqrt{m^2c^4 + c^2\mathbf{p}^2},$$

where $A(\mathbf{p})$ is arbitrary. Note that we can write $\Phi^{(\mathbf{P})}$ in an invariant manner:

$$\Phi^{(p)}(x) \equiv \Phi^{(\mathbf{P})}(\mathbf{r},t) = A(p)e^{-ip\cdot x/\hbar},$$
$$p_0 \equiv \frac{1}{c}\sqrt{m^2c^4 + c^2\mathbf{p}^2}. \tag{2.3.2}$$

The more general solution of the KGS evolution equation can be written as a superposition of plane waves:

$$\Phi(\mathbf{r},t) = \frac{1}{(2\pi\hbar)^{3/2}} \int d^3p f(\mathbf{p})e^{-ip\cdot x/\hbar}. \tag{2.3.3}$$

Let us now evaluate the current. Because a relativistic particle can be massless, the definition (2.1.2a),

$$\mathbf{j} = \frac{i\hbar}{2m}(\Phi\boldsymbol{\nabla}\Phi^* - \Phi\boldsymbol{\nabla}\Phi),$$

is not very convenient. We define instead the current three-vector by

$$\mathbf{J} = i\hbar(\Phi\boldsymbol{\nabla}\Phi^* - \Phi^*\boldsymbol{\nabla}\Phi). \tag{2.3.4}$$

To find the fourth component, $J_0$, we postulate the continuity equation

$$\frac{1}{c}\partial_t J_0 + \boldsymbol{\nabla}\mathbf{J} = 0, \tag{2.3.5a}$$

or, with covariant notation,

$$\partial \cdot J = 0. \tag{2.3.5b}$$

Using now (2.3.4) and (2.1.1c), we get

$$\frac{1}{c}\partial_t J_0 = -\boldsymbol{\nabla}\mathbf{J} = -i\hbar\{\Phi(\triangle\Phi^*) - \Phi^*(\triangle\Phi)\}$$
$$= \frac{-i\hbar}{c^2}\{\Phi(\partial_t^2\Phi)^* - \Phi^*(\partial_t^2\Phi)\}$$
$$= \frac{-i\hbar}{c^2}\partial_t\{\Phi\partial_t\Phi^* - \Phi^*\partial_t\Phi\},$$

so an acceptable solution is to take

$$J_0 = \frac{i\hbar}{c}(\Phi^*\partial_t\Phi - \Phi\partial_t\Phi^*). \tag{2.3.6}$$

We can introduce with a general character the notation $\overleftrightarrow{\partial}$,

$$f\overleftrightarrow{\partial}g = f\partial g - (\partial f)g;$$

with it we can unify (2.3.4), (2.3.6) as

$$J_\mu(x) = i\hbar\Phi^*(x)\overleftrightarrow{\partial}_\mu \Phi(x). \tag{2.3.7}$$

The covariance of $J_\mu$ is obvious from (2.3.7):

$$\Lambda \; : \; J_\mu(x) \to \sum \Lambda_{\mu\nu} J_\nu(\Lambda^{-1}x).$$

What is not obvious is that $J_0$ is positive definite, as befits a probability density. It *is* so if $\Phi$ satisfies (2.2.4); the equation quadratic in $\partial_t$, say (2.1.1), is *not* sufficient (cf. Problems P.2.1, P.2.2).

Let us now turn to the scalar product, starting in **p**-space. The obvious generalization of the nonrelativistic definition is to take

$$\langle \Phi_1 | \Phi_2 \rangle = \int d^3p\, C(\mathbf{p}) \Phi_1^*(\mathbf{p}, t) \Phi_2(\mathbf{p}, t), \tag{2.3.8}$$

and try to fix $C$ by requiring Lorentz invariance of (2.3.8). Invariance under rotations is achieved if $C(\mathbf{p}) = C(|\mathbf{p}|)$. For boosts along $OZ$, the four-momentum behaves like

$$p_x \to p_x' = p_x,$$
$$p_y \to p_y' = p_y,$$
$$p_z \to p_z' = (\cosh\xi)p_z + (\sinh\xi)p_0,$$
$$p_0 \to p_0' = (\sinh\xi)p_z + (\cosh\xi)p_0,$$

where

$$p_0 = \sqrt{m^2c^2 + \mathbf{p}^2}, \;\; p_0' = \sqrt{m^2c^2 + \mathbf{p}'^2}.$$

The Jacobian of the transformation is thus such that

$$d^3p \to d^3p' = (dp_x dp_y dp_z)\left( \cosh\xi + \sinh\xi\frac{p_z}{\sqrt{m^2c^2 + \mathbf{p}^2}} \right),$$

i.e.,

$$d^3p' = (d^3p)\frac{p_0'}{p_0}.$$

It follows that $C(|\mathbf{p}|)d^3p$ will be invariant provided $C(|\mathbf{p}|) = (1/n)p_0$, with $n$ an arbitrary number. It is customary to choose $n = 2c$, so that we obtain the *relativistically invariant scalar product*, in momentum space,

$$\langle \Phi_1 | \Phi_2 \rangle = c^{-1} \int \frac{d^3p}{2p_0} \Phi_1^*(\mathbf{p}, t) \Phi_2(\mathbf{p}, t),$$
$$p_0 = \frac{1}{c}E(\mathbf{p}) = \sqrt{m^2c^2 + \mathbf{p}^2}. \tag{2.3.9}$$

In the NR limit,

$$\langle \Phi_1 | \Phi_2 \rangle \underset{NR}{\simeq} \frac{1}{2mc^2} \int d^3p\, \Phi_1^*(\mathbf{p}, t) \Phi_2(\mathbf{p}, t), \tag{2.3.10}$$

which coincides with the ordinary definition up to the factor $1/2mc^2$. Because of this, some authors define the scalar product as

$$\langle \Phi_1 | \Phi_2 \rangle = (2mc^2) \int \frac{d^3p}{2E(\mathbf{p})} \Phi_1^*(\mathbf{p}, t) \Phi_2(\mathbf{p}, t),$$

a definition that presents the problem of not working for massless particles. In view of this, we prefer to stick to (2.3.9) and recognize a factor $\sqrt{2E(\mathbf{p})}$ between the normalizations of relativistic and nonrelativistic wave functions.

To go over to position space we start by redefining the transformation (2.2.1) to make it relativistically invariant. We thus define

$$\Phi(\mathbf{r},t) = \frac{1}{(2\pi\hbar)^{3/2}} \int \frac{d^3p}{2E(\mathbf{p})} e^{i\mathbf{pr}/\hbar}\Phi(\mathbf{p},t), \qquad (2.3.11a)$$

with inverse

$$\Phi(\mathbf{p},t) = \frac{2E(\mathbf{p})}{(2\pi\hbar)^{3/2}} \int d^3r e^{-i\mathbf{pr}/\hbar}\Phi(\mathbf{r},t). \qquad (2.3.11b)$$

Substituting this into (2.3.9), we get

$$\langle\Phi_1|\Phi_2\rangle = \frac{1}{(2\pi\hbar)^3} \int d^3p \frac{(2E(\mathbf{p}))^2}{2E(\mathbf{p})} \int d^3r' \int d^3r''$$

$$\times e^{i\mathbf{p}(\mathbf{r}'-\mathbf{r}'')/\hbar}\Phi_1^*(\mathbf{r}',t)\Phi_2(\mathbf{r}'',t)$$

$$= \frac{2}{(2\pi\hbar)^3} \int d^3r' \int d^3r'' \int d^3p E(\mathbf{p})e^{i\mathbf{p}(\mathbf{r}'-\mathbf{r}'')/\hbar}\Phi_1^*(\mathbf{r}',t)\Phi_2(\mathbf{r}'',t).$$

Using the property, valid for any function $f(\mathbf{r})$,

$$\int d^3r E(\mathbf{p})e^{i\mathbf{pr}/\hbar}f(\mathbf{r}) = \int d^3r e^{i\mathbf{pr}/\hbar}c\sqrt{m^2c^2 - \triangle\hbar^2}f(\mathbf{r}),$$

and the evolution equation (2.2.5d), we work this to

$$\langle\Phi_1|\Phi_2\rangle = \frac{c}{(2\pi\hbar)^3} \int d^3p \int d^3r' \int d^3r'' e^{i\mathbf{p}(\mathbf{r}'-\mathbf{r}'')/\hbar}$$

$$\times \left\{ \left( \sqrt{m^2c^2 - \hbar^2\triangle'}\,\Phi_1(\mathbf{r}',t) \right)^* \Phi_2(\mathbf{r}'',t) \right.$$

$$\left. + \Phi_1^*(\mathbf{r}',t)\sqrt{m^2c^2 - \hbar^2\triangle''}\,\Phi_2(\mathbf{r}'',t) \right\}$$

$$= \frac{1}{(2\pi\hbar)^3} \int d^3r' \int d^3r'' \int d^3p e^{i\mathbf{p}(\mathbf{r}'-\mathbf{r}'')/\hbar}$$

$$\times \{-i\hbar(\partial_t\Phi_1^*(\mathbf{r}',t))\Phi_2(\mathbf{r}'',t) + \Phi_1^*(\mathbf{r}',t)i\hbar\partial_t\Phi_2(\mathbf{r}'',t)\}.$$

The $d^3p$ integration is now immediate and we end up with the explicit expression

$$\langle\Phi_1|\Phi_2\rangle = i\hbar \int d^3r\Phi_1^*(\mathbf{r},t) \overset{\leftrightarrow}{\partial_t} \Phi_2(\mathbf{r},t). \qquad (2.3.12)$$

Appearances notwithstanding, the equality of (2.3.12) and (2.3.9) proves that the right-hand side of (2.3.12) is time-independent and relativistically invariant (for $\Phi_i$ being solutions of the KGS equation!)

Let us finally standardize the plane waves. Just as we proved that $d^3p/2E(\mathbf{p})$ is invariant, so we would check the invariance of

$$2p_0\delta(\mathbf{p} - \mathbf{p}'), \; p_0 \equiv \sqrt{m^2c^2 + \mathbf{p}^2}.$$

We then define *plane-wave states* $|p\rangle$ normalized invariantly as

$$\langle p|p'\rangle = 2E(\mathbf{p})\delta(\mathbf{p} - \mathbf{p}') = 2cp_0\delta(\mathbf{p} - \mathbf{p}'). \tag{2.3.13}$$

The corresponding wave functions will be given by (2.3.1), with the constant $A(\mathbf{p})$ fixed by (2.3.13). Taking it positive, we thus write

$$\Phi^{(p)}(\mathbf{r}, t) = \frac{1}{(2\pi\hbar)^{3/2}} e^{-iE(p)t/\hbar} e^{i\mathbf{pr}/\hbar},$$
$$\langle \Phi^{(p)}|\Phi^{(p')}\rangle = 2E(p)\delta(\mathbf{p} - \mathbf{p}'). \tag{2.3.14}$$

The corresponding density of particles is

$$J_0 = \frac{i\hbar}{c}\Phi^{(p)*}\partial_t\Phi^{(p)} = 2p_0, \tag{2.3.15}$$

i.e., we have $2\sqrt{m^2c^2 + \mathbf{p}^2}$ particles per unit volume.

## 2.4 Interaction with the Classical Electromagnetic Field. Gauge Invariance

In principle we can introduce the interactions of a charged particle with the electromagnetic field via the *minimal replacement*,

$$i\hbar\partial_\mu \rightarrow i\hbar\partial_\mu - \frac{e}{c}A_\mu, \tag{2.4.1}$$

so that the Klein–Gordon equation will become

$$\left\{\hbar^2\left(\partial + \frac{ie}{\hbar c}A\right) \cdot \left(\partial + \frac{ie}{\hbar c}A\right) + m^2c^2\right\}\Phi(x) = 0. \tag{2.4.2}$$

The situation, however, is less straightforward than it may seem. First of all, and although it may be reasonable to postulate (2.4.2) for *elementary* spinless particles, there is no reason why we could not add other interaction terms, such as

$$\lambda \sum g_{\mu\mu}g_{\nu\nu}F_{\mu\nu}F_{\mu\nu}\Phi, \tag{2.4.3}$$

for *composite* particles; this is known to be the case for spin $1/2$ particles. On dimensional grounds we expect terms like (2.4.3) to be suppressed with respect to (2.4.2) by powers of $1/c$, and an explicit calculation showing that this is indeed the case will be presented later.

To discuss the second question, let us rewrite (2.4.2) as a KGS-type equation:

$$i\hbar\partial_t\Phi = eA_0\Phi + c\sqrt{m^2c^2 + \left(\mathbf{P} - \frac{e}{c}\mathbf{A}\right)^2}\,\Phi,\qquad(2.4.4\text{a})$$

i.e., the corresponding Hamiltonian is

$$H = eA_0 + c\sqrt{m^2c^2 + \left(\mathbf{P} - \frac{e}{c}\mathbf{A}\right)^2}.\qquad(2.4.4\text{b})$$

Now, for strong attractive fields, $eA_0 < 0$, $|eA_0| > mc^2$, the spectrum of $H$ will contain negative energies, which makes no physical sense.

Nevertheless, we will use (2.4.4), remembering that it will only be valid for weak fields and velocities not too close to that of light; the last requisite holding for nonelementary particles. For elementary spinless particles (of which none has been found in nature[1]) this restriction can be lifted.

Let us now discuss gauge invariance. Under gauge transformations,

$$A_\mu(x) \to A_\mu(x) - \partial_\mu f(x),\qquad(2.4.5\text{a})$$

with $f$ arbitrary. We then postulate that (2.4.5a) has to be accompanied by a transformation of the wave function, which, just as for the NR ($=$ nonrelativistic) case, we take to be[2]

$$\Phi(x) \to e^{ief(x)/c\hbar}\Phi(x).\qquad(2.4.5\text{b})$$

We define the *gauge covariant derivative*, $D_\mu$, by

$$D_\mu \equiv \partial_\mu + \frac{ie}{\hbar c}A_\mu(x).\qquad(2.4.6)$$

Then, it follows that, under (2.4.5), $D_\mu\Phi$ transforms like $\Phi$ itself, i.e., just with a phase:

$$D_\mu\Phi(x) \to e^{ief(x)/c\hbar}D_\mu\Phi(x).\qquad(2.4.7)$$

Indeed, from (2.4.5),

$$D_\mu\Phi(x) \to \left\{\partial_\mu + \frac{ie}{\hbar c}(A_\mu - \partial_\mu f)\right\}e^{ief(x)/c\hbar}\Phi(x)$$

$$= e^{ief(x)/c\hbar}\left(\partial_\mu + \frac{ie}{\hbar c}A_\mu\right)\Phi(x) = e^{ief(x)/c\hbar}D_\mu\Phi(x).$$

## 2.5 Particle in a Coulomb Field

We will now consider the case where a charged scalar particle interacts with a classical, static Coulomb field. This is important, from a practical point of

---

[1] Our formulation will, however, be valid for particles like charged pions ($\pi$) or kaons ($K$) up to energies where their structure is apparent.

[2] We denote by $e$ the charge of the particle. We trust that no confusion will arise with the number $e$ of the exponential function.

view, for the study of so-called mesic atoms, where a particle $\pi^-$ or $K^-$ is bound electromagnetically to a nucleus that replaces one of the electrons of the atom.

It is in principle unclear how to introduce the interaction. One may start with a wave function

$$\Phi(\mathbf{r}, t) = e^{-iEt/\hbar}\varphi_E(\mathbf{r}), \qquad (2.5.1)$$

and replace this in the Klein–Gordon equation (including the Coulomb field),

$$\left\{\hbar^2\left(\partial + \frac{ie}{\hbar c}A_C\right)\left(\partial + \frac{ie}{\hbar c}A_C\right) + m^2c^2\right\}\Phi(x) = 0, \qquad (2.5.2)$$

$$A_{C0} = e_1 e_2/r; \quad \mathbf{A}_C = 0.$$

Here $e_1$, $e_2$ are the charges of the particle and of the source of the field. We then find the equation

$$\left(E + \frac{\hbar c\alpha_0}{r}\right)^2 \varphi_E(\mathbf{r}) = (m^2c^4 + c^2\mathbf{P}^2)\varphi_E(\mathbf{r})$$
$$= (m^2c^4 - \hbar^2c^2\triangle)\varphi_E(\mathbf{r}), \qquad (2.5.3)$$

where $\alpha_0 \equiv -e_1 e_2/\hbar c$. The solution to (2.5.3) is rather trivial since it becomes identical with the ordinary (nonrelativistic) Schrödinger Coulombic problem with the replacements

$$l_{NR} \to \lambda \equiv \sqrt{(l + 1/2)^2 - \alpha_0^2} - 1/2,$$

$$E_{NR} \to \frac{E^2 - m^2c^4}{2mc^2},$$

$$-e_1 e_2 \to E\hbar\alpha_0/mc.$$

The details may be found in, for example, the treatise of Schiff (1968). Unfortunately (2.5.3) is not acceptable: it is not even compatible with the superposition principle.

A second possibility (as it turns out, the correct one) is to start from (2.4.4) so that we have the equation

$$\left\{c\sqrt{m^2c^2 - \hbar^2\triangle} + \frac{e_1 e_2}{r}\right\}\varphi_E(\mathbf{r}) = E\varphi_E(\mathbf{r}). \qquad (2.5.4)$$

An exact solution of (2.5.4) is not known; but is is possible to solve it in a power series in $1/c^2$. The expansion is less straightforward than it may seem; indeed, not only has one to take into account the $c$ dependence of (2.5.4), but it is also necessary to remember that the scalar product (2.3.12) is different in the relativistic and nonrelativistic cases. One finds that the first corrections to the ordinary nonrelivistic equation,

$$\left(mc^2 - \frac{\hbar^2}{2m}\triangle + \frac{e_1 e_2}{r}\right)\varphi_E^{NR}(\mathbf{r}) = (E_{NR} + mc^2)\varphi_E^{NR}(\mathbf{r}), \qquad (2.5.5)$$

do not start at order $1/c^2$, but that the first nonzero correction is actually of order $1/c^4$ (apart from the merely kinetic term):

$$\left\{ mc^2 - \frac{\hbar^2}{2m}\Delta - \frac{\hbar^4\Delta^2}{8m^3c^2} - \frac{\hbar^6\Delta^3}{16m^5c^4} + \frac{e_1e_2}{r} \right.$$
$$\left. + \frac{e_1e_2}{32m^4c^4}\left[\Delta,\left[\Delta,\frac{1}{r}\right]\right] \right\} \varphi_E(\mathbf{r}) = E\varphi_E(\mathbf{r}) + 0(c^{-6}).$$

$$(2.5.6)$$

This equation may be solved easily by considering the terms

$$\frac{-\hbar^4\Delta^2}{8m^3c^2}, \frac{-\hbar^6\Delta^3}{16m^5c^4}, \frac{e_1e_2}{32m^4c^4}\left[\Delta,\left[\Delta,\frac{1}{r}\right]\right]$$

to be perturbations of the exactly known solutions of (2.5.5). This is left to the reader.

The proof that (2.5.4) is indeed the correct equation, as well as the deduction of the approximation (2.5.6), and higher-order ones, may be found in Bjorken and Drell (1964). The proofs will also be given in the present text after we have developed field-theoretic interactions in Chap. 10; see, in particular, Problem P.10.3.


## Problems

**P.2.1.** Prove that the density $J_0$ is positive definite for wave functions $\Phi_E$ that are solutions of the KGS equation.

**P.2.2.** Consider the NR limit of the Klein–Gordon equation (*not* the KGS equation). What happens to the negative part of $J_0$?

**P.2.3.** Evaluate the commutation relations of $N_j$, $L_k$.

*Solution.*

$$[L_k, L_j] = i\hbar \sum \epsilon_{kjl} L_l$$

$$[L_k, N_j] = i\hbar \sum \epsilon_{kjl} N_l$$

$$[N_k, N_l] = -i\hbar c \sum \epsilon_{kjl} L_l .$$

# 3. Spin 1/2 Particles

## 3.1 The Dirac Equation

Given a positive-definite operator, such as $(m^2c^4 + \mathbf{P}^2c^2)$, there is a mathematical theorem that guarantees that there is one, and only one, square root that is also positive definite, denoted by $+(m^2c^4 + \mathbf{P}^2c^2)^{1/2}$. Other square roots become possible if we give up positive definiteness. This may appear to spoil the theory by allowing negative energies; but, if the operator is Hermitean, states corresponding to negative energies will be orthogonal to positive-energy states and a sensible physical theory is obtained if we restrict ourselves to the latter. We can, moreover, ensure manifest covariance by looking for an equation not only linear in $\partial_t$, but also linear in the space derivatives; that equation, we expect, will describe relativistic spin 1/2 particles, such as the electron[1]. We then use a multicomponent wave function[2], $\underset{\sim}{\Psi}$, and look for an equation *linear* in the $P_\mu$, the *Dirac equation*,

$$i\hbar\partial_t\underset{\sim}{\Psi}(\mathbf{r},t) = \underset{\sim}{H}_0\underset{\sim}{\Psi} = -i\hbar c\underset{\sim}{\boldsymbol{\alpha}}\boldsymbol{\nabla}\underset{\sim}{\Psi}(\mathbf{r},t) + mc^2\underset{\sim}{\beta}\underset{\sim}{\Psi}(\mathbf{r},t), \tag{3.1.1}$$

where the *free Dirac Hamiltonian* $\underset{\sim}{H}_0$ satisfies

$$\underset{\sim}{H}_0^2 = m^2c^4 + c^2\mathbf{P}^2; \tag{3.1.2}$$

$$\underset{\sim}{H}_0 = -i\hbar c\underset{\sim}{\boldsymbol{\alpha}}\boldsymbol{\nabla} + mc^2\underset{\sim}{\beta}. \tag{3.1.3}$$

We have introduced $m$'s, $\hbar$'s and $c$'s so that the quantities $\underset{\sim}{\alpha}_i$, $\underset{\sim}{\beta}$ are dimensionless.

If we want (3.1.3) to imply (3.1.2) we have to admit that the $\underset{\sim}{\alpha}_i$, $\underset{\sim}{\beta}$ are *matrices*. In fact, squaring $\underset{\sim}{H}_0$ we find that (3.1.2) will be satisfied provided

$$m^2c^4 + c^2\hbar^2\Delta = m^2c^4\underset{\sim}{\beta}^2 - \hbar^2c^2\sum\underset{\sim}{\alpha}_j\underset{\sim}{\alpha}_k\partial_j\partial_k + i\hbar mc^3\sum(\underset{\sim}{\alpha}_l\underset{\sim}{\beta} + \underset{\sim}{\beta}\underset{\sim}{\alpha}_l)\partial_l,$$

which holds if, and only if, the $\underset{\sim}{\alpha}_i$, $\underset{\sim}{\beta}$ satisfy the *anticommutation relations*,

---

[1] We will refer many times specifically to electrons, for definiteness, when indeed the considerations apply to any other (elementary) spin 1/2 particle.

[2] We temporarily denote matrices by putting a tilde under the symbol. That $\underset{\sim}{\Psi}$ is a vertical matrix is indeed what we should have for a particle with spin. For spin 1/2, we would expect *two* components; we find *four*. This may be connected to the existence of unphysical, negative energy states.

$$\{\alpha_j, \alpha_k\} = 2\delta_{jk}, \quad \beta^2 = 1, \quad \{\alpha_l, \beta\} = 0, \tag{3.1.4}$$

for all $j$, $k$, $l$. (The *anticommutator* of $A$ and $B$ is defined as $\{A, B\} = AB + BA$ for any matrices or operators $A$, $B$.)

If $T$ is an arbitrary matrix with nonzero determinant, we may form the matrices

$$\alpha'_j = T\alpha_j T^{-1}, \quad \beta' = T\beta T^{-1},$$

and these matrices will satisfy the same relations (3.1.4) as the $\alpha_j$, $\beta$. One can interpret these transformations as a change of basis, $\Psi \to T\Psi$. Up to this arbitrariness, one can prove that the matrices $\alpha_j$, $\beta$ are essentially unique (see Appendix A.4). An explicit realization is the so-called *Pauli realization*[3], where

$$\alpha = \begin{pmatrix} 0 & \sigma \\ \sigma & 0 \end{pmatrix}, \beta = \begin{pmatrix} 1 & 0 \\ 0 & -1 \end{pmatrix}, \tag{3.1.5a}$$

with $\sigma$ the Pauli matrices and $0$, $1$ are the $2 \times 2$ zero and unit matrices, respectively:

$$\sigma_1 = \begin{pmatrix} 0 & 1 \\ 1 & 0 \end{pmatrix}, \sigma_2 = \begin{pmatrix} 0 & -i \\ i & 0 \end{pmatrix}, \sigma_3 = \begin{pmatrix} 1 & 0 \\ 0 & -1 \end{pmatrix},$$

$$0 = \begin{pmatrix} 0 & 0 \\ 0 & 0 \end{pmatrix}, 1 = \begin{pmatrix} 1 & 0 \\ 0 & 1 \end{pmatrix}. \tag{3.1.5b}$$

The Dirac equation can be rewritten in a manner which makes apparent its relativistic covariance (of course, we will later give a formal proof of this). To do so, we left multiply (3.1.1) by $\beta$ and rearrange. We find that

$$\frac{i\hbar}{c}\beta\partial_t\Psi + i\hbar\beta\alpha\nabla\Psi = mc\Psi;$$

here and henceforth we suppress the tilde under matrix objects, whose character should be clear from the context. Defining next the *Dirac $\gamma$ matrices* $\gamma_\mu$ by (in any representation)

$$\gamma_0 \equiv \beta, \quad \gamma_i \equiv \beta\alpha_i, \tag{3.1.6a}$$

we note that they satisfy

$$\{\gamma_\mu, \gamma_\nu\} = 2g_{\mu\nu}. \tag{3.1.6b}$$

---

[3] Another useful realization is *Weyl*'s. Here, defining

$$\gamma_0 = \beta^{\text{Weyl}}, \quad \gamma = \beta^{\text{Weyl}}\alpha^{\text{Weyl}},$$

one has

$$\gamma_\mu = \begin{pmatrix} 0 & \tilde{\sigma}_\mu \\ \sigma_\mu & 0 \end{pmatrix}, \tilde{\sigma}_0 = \sigma_0 = 1, \tilde{\sigma}_j = -\sigma_j.$$

Remembering that $\partial_i \equiv -\partial/\partial r_i$, we finally are able to write the Dirac equation as

$$i\hbar\gamma \cdot \partial\Psi(x) = mc\,\Psi(x). \tag{3.1.7}$$

In the relativistic theory of spin $1/2$ particles the $\gamma$ matrices appear many times in the combination $\gamma \cdot v$, where $v$ is a Minkowski four-vector whose components may be numbers or operators. We therefore introduce the so-called (Feynman) *slash notation*:

$$\not{v} \equiv \gamma \cdot v = \sum_{\mu\nu} g_{\mu\nu}\gamma_\mu v_\nu = \sum_\mu g_{\mu\mu}\gamma_\mu v_\mu. \tag{3.1.8}$$

We thus write, for example,

$$\not{\partial} \equiv \gamma \cdot \partial, \quad \not{A} \equiv \gamma \cdot A, \text{ etc.,}$$

and the Dirac equation (3.1.7) as

$$(i\hbar\not{\partial} - mc)\Psi(x) = 0. \tag{3.1.9}$$


## 3.2 Invariance Properties of the Dirac Equation

Given a Lorentz transformation $\Lambda$, we postulate a corresponding transformation of the Dirac wave function,

$$\Lambda \; : \; \Psi(x) \rightarrow U(\Lambda)\Psi(x) = D(\Lambda)\Psi(\Lambda^{-1}x), \tag{3.2.1}$$

where the matrix $D$ will be determined so that the Dirac equation is left unaltered by (3.2.1). We will consider rotations and boosts separately.


### 3.2.1 Rotations

For an infinitesimal rotation $R(\boldsymbol{\theta})$ we define

$$\begin{aligned}
U(R(\boldsymbol{\theta})) &= 1 - \frac{i}{\hbar}\boldsymbol{\theta}\mathbf{J} + 0(\theta^2), \\
D(R(\boldsymbol{\theta})) &= 1 - \frac{i}{\hbar}\boldsymbol{\theta}\mathbf{S} + 0(\theta^2);
\end{aligned} \tag{3.2.2a}$$

$\mathbf{J}$ will be the *total angular momentum* operators, and $\mathbf{S}$ the *spin* ones, for a Dirac particle. Furthermore, we can also write

$$\Psi(R^{-1}x) \equiv \Psi(R^{-1}\mathbf{r}, t) = \left(1 - \frac{i}{\hbar}\boldsymbol{\theta}\mathbf{L}\right)\Psi(\mathbf{r}, t) + 0(\theta^2), \tag{3.2.2b}$$

where $\mathbf{L}$ is the *orbital angular momentum* operator,

$$\mathbf{L} = \mathbf{r} \times \mathbf{P} = -i\hbar\mathbf{r} \times \boldsymbol{\nabla},$$

the last expression being valid in the position representation.

**Exercise.** Check (3.2.2b) by using (1.1.1) ∎

Consider now (3.2.1) for an infinitesimal rotation,

$$U(R)\Psi(x) = D(R)\Psi(R^{-1}x);$$

using (3.2.2) we get

$$
\begin{aligned}
U(R)\Psi(x) &= \left(1 - \frac{i}{\hbar}\boldsymbol{\theta}\mathbf{J}\right)\Psi(\mathbf{r}, t) + 0(\theta^2) \\
&= \left\{1 - \frac{i}{\hbar}(\mathbf{L} + \mathbf{S})\right\}\Psi(\mathbf{r}, t) + 0(\theta^2).
\end{aligned}
\tag{3.2.3}
$$

Because **J** does not involve the time, Dirac's equation will be invariant under rotations provided its components commute with the Dirac Hamiltonian:

$$[H_0, J_k] = 0, \quad k = 1, 2, 3.$$

Remembering the commutation relations $[L_k, P_j] = i\hbar \sum \epsilon_{kjl} P_l$, we find that

$$
\begin{aligned}
[H_0, J_k] &= [c\boldsymbol{\alpha}\mathbf{P} + mc^2\beta, L_k + S_k] \\
&= -i\hbar c \sum \epsilon_{klj}\alpha_l P_j + \sum cP_l[\alpha_l, S_k] + mc^2[\beta, S_k];
\end{aligned}
\tag{3.2.4}
$$

we have also made use of the fact that the commutators $[P_j, S_l]$, $[\alpha_l, L_j]$, $[\alpha_l, L_j]$ and $[\beta, L_j]$ vanish because the operators involved act upon independent variables. Expression (3.2.4) will vanish provided one has

$$[\beta, S_k] = 0, \quad [\alpha_l, S_k] = i\hbar \sum \epsilon_{lkj}\alpha_j,$$

the solution of which is to take

$$
\mathbf{S} = \frac{\hbar}{2}\boldsymbol{\Sigma}, \quad \boldsymbol{\Sigma} = \begin{pmatrix} \boldsymbol{\sigma} & 0 \\ 0 & \boldsymbol{\sigma} \end{pmatrix},
\tag{3.2.5}
$$

an expression clearly related to the usual nonrelativistic one.

Exponentiating, we get the expressions

$$
\begin{aligned}
U(R(\theta))\Psi(\mathbf{r}, t) &= e^{-\frac{i}{\hbar}\boldsymbol{\theta}\mathbf{S}}\Psi(R^{-1}(\boldsymbol{\theta})\mathbf{r}, t) \\
&= e^{-\frac{i}{\hbar}\boldsymbol{\theta}(\mathbf{S}+\mathbf{L})}\Psi(\mathbf{r}, t).
\end{aligned}
\tag{3.2.6}
$$

One can also write

$$
e^{-\frac{i}{\hbar}\boldsymbol{\theta}\mathbf{S}} = D(R(\boldsymbol{\theta})) = \cos\frac{\theta}{2} - i\frac{\sin\theta/2}{\theta}\boldsymbol{\theta}\boldsymbol{\Sigma}.
\tag{3.2.7}
$$

### 3.2.2 Boosts

Under a Lorentz boost, $L(\boldsymbol{\xi})$,

$$\Psi(\mathbf{r}, t) \rightarrow U(L(\boldsymbol{\xi}))\Psi(\mathbf{r}, t) = D(L(\boldsymbol{\xi}))\Psi(L^{-1}(\xi)x).
\tag{3.2.8a}$$

For infinitesimal $\boldsymbol{\xi}$, we define

$$U(L(\boldsymbol{\xi})) = 1 - \frac{i}{\hbar}\boldsymbol{\xi}\mathbf{N} + O(\xi^2),$$
$$D(L(\boldsymbol{\xi})) = 1 - \frac{i}{\hbar}\boldsymbol{\xi}\boldsymbol{\lambda} + O(\xi^2). \tag{3.2.8b}$$

(Note that $\mathbf{N}$ will be *different* from the spin-zero expression (2.2.7), since it will now contain a spin part.) We also have

$$\Psi(L^{-1}(\boldsymbol{\xi})x) = \left\{ 1 - \frac{i\boldsymbol{\xi}}{\hbar}(ct\mathbf{P} - \frac{i\hbar}{c}\mathbf{r}\partial_t) \right\} \Psi(\mathbf{r},t) + 0(\xi^2), \tag{3.2.8c}$$

an expression identical to that for spinless particles.

Because (3.2.8) now involves the time variable, the invariance of the Dirac equation will be equivalent to the statement that the Dirac operator,

$$\mathcal{D} = i\hbar\beta\partial_t + i\hbar c\beta\boldsymbol{\alpha}\boldsymbol{\nabla} = i\hbar c\slashed{\partial},$$

commutes with the generators of accelerations, $\mathbf{N}$. In fact, we can write the Dirac equation as

$$\mathcal{D}\Psi = mc^2\Psi.$$

Applying a Lorentz boost, we get

$$U(L)\mathcal{D}\Psi = mc^2 U(L)\Psi;$$

if $U$ and $\mathcal{D}$ commute, this implies that

$$\mathcal{D}U\Psi = mc^2 U\Psi,$$

so the transformed function $U(L)\Psi$ also satisfies the Dirac equation. Therefore we require

$$[\mathcal{D}, \mathbf{N}] = 0, \quad \mathbf{N} = \boldsymbol{\lambda} + ct\mathbf{P} - \frac{i\hbar}{c}\mathbf{r}\partial_t.$$

This will fix the $\lambda_j$. Choose first the $z$ component of $\mathbf{N}$. The above condition reads

$$\left[ i\hbar\beta\partial_t + i\hbar c\beta\boldsymbol{\alpha}\boldsymbol{\nabla}, \lambda_z - i\hbar ct\frac{\partial}{\partial_z} - \frac{i\hbar}{c}z\partial_t \right] = 0.$$

Evaluating the commutator we find the conditions

$$[\gamma_a, \lambda_z] = 0, \quad a = 1,2;$$

$$\frac{1}{i\hbar}[\beta, \lambda_z] = \gamma_3; \quad \frac{1}{i\hbar}[\gamma_3, \lambda_z] = \beta,$$

whose solution is (we give it for all three components)

$$\boldsymbol{\lambda} = \frac{i\hbar}{2}\gamma_0\boldsymbol{\gamma} = \frac{i\hbar}{2}\boldsymbol{\alpha}, \tag{3.2.9a}$$

so that the generators of accelerations for Dirac particles are

$$\mathbf{N} = ct\mathbf{P} - \frac{i\hbar}{c}\mathbf{r}\partial_t + \frac{i\hbar}{2}\boldsymbol{\alpha}. \tag{3.2.9b}$$

Exponentiating, and using the fact that for any numerical vector $\mathbf{n}$, $(\mathbf{n}\boldsymbol{\alpha})^2 = \mathbf{n}^2$, we obtain

$$U(L(\boldsymbol{\xi}))\Psi(x) = e^{-\frac{i}{\hbar}\boldsymbol{\xi}\mathbf{N}}\Psi(x) = e^{\frac{1}{2}\boldsymbol{\xi}\boldsymbol{\alpha}}\Psi(L^{-1}(\boldsymbol{\xi})x);$$

$$D(L(\boldsymbol{\xi})) = e^{\frac{1}{2}\boldsymbol{\xi}\boldsymbol{\alpha}} = \cosh\frac{\xi}{2} + \frac{\sinh\xi/2}{\xi}\boldsymbol{\xi}\boldsymbol{\alpha}. \tag{3.2.10}$$

It is remarkable that, while the matrices $D(R(\boldsymbol{\theta}))$ are unitary, the $D(L(\boldsymbol{\xi}))$ are *not*.

It is possible to produce a manifestly covariant expression for arbitrary Lorentz transformations. If we write

$$U(\Lambda(\omega)) = 1 - \frac{i}{\hbar}\sum g_{\mu\mu}g_{\nu\nu}\omega_{\mu\nu}(M_{\mu\nu} + \rho_{\mu\nu}) + 0(\omega^2), \tag{3.2.11a}$$

where the parameter $\omega$ and the $M$ are those given in (2.2.8), then use of (3.2.10), (3.2.5) allows us to find the $\rho_{\mu\nu}$. For a boost,

$$-\sum_i(\rho_{i0}\omega_{i0} + \rho_{0i}\omega_{0i}) = \boldsymbol{\xi}\boldsymbol{\lambda} = 2\sum\omega_{i0}\lambda_i,$$

so that

$$\rho_{j0} = -\lambda_j = \frac{-i\hbar}{2}\gamma_0\gamma_j = \frac{\hbar}{2}\frac{i}{2}[\gamma_j, \gamma_0];$$

for a rotation

$$\sum\omega_{ij}\rho_{ij} = \boldsymbol{\theta}\mathbf{S} = \sum\epsilon_{kij}\omega_{ij}S_k = \sum\omega_{ij}\frac{\hbar}{2}\epsilon_{kij}\Sigma_k$$

so that

$$\rho_{lj} = \frac{\hbar}{2}\sum_k\epsilon_{klj}\Sigma_k = \frac{\hbar}{2}\frac{i}{2}[\gamma_l, \gamma_j];$$

here we have used the equality

$$\frac{i}{2}[\gamma_l, \gamma_j] = \sum_k\epsilon_{klj}\begin{pmatrix}\sigma_k & 0 \\ 0 & \sigma_k\end{pmatrix} = \sum_k\epsilon_{klj}\Sigma_k.$$

Putting the results together, we have

$$\rho_{\mu\nu} = \frac{\hbar}{2}\sigma_{\mu\nu}, \sigma_{\mu\nu} \equiv \frac{i}{2}[\gamma_\mu, \gamma_\nu],$$

$$M_{\mu\nu} = x_\mu P_\nu - x_\nu P_\mu. \tag{3.2.11b}$$

**Exercise.** Check Lorentz invariance of the Dirac equation "covariantly", directly using (3.2.11). That is to say, verify that

$$[i\hbar\slashed{\partial}, M_{\mu\nu} + \rho_{\mu\nu}] = 0.$$

*Hint.* Use the commutation relations

$$[\partial_\mu, x_\nu] = g_{\mu\nu}, \quad [\gamma_\mu, \sigma_{\alpha\beta}] = 2i(g_{\mu\alpha}\gamma_\beta - g_{\mu\beta}\gamma_\alpha) \;\blacksquare \tag{3.2.12}$$

For finite Lorentz transformations, one has

$$D^{-1}(\Lambda)\gamma_\mu D(\Lambda) = \sum_\nu \Lambda_{\mu\nu}\gamma_\nu, \tag{3.2.13a}$$

i.e., the $\gamma_\mu$ transform as a Minkowski vector (justifying the notation). With covariant notation we have the explicit expression

$$D(\Lambda) = \exp\frac{-i}{\hbar}\sum g_{\mu\mu}g_{\nu\nu}\omega_{\mu\nu}\rho_{\mu\nu} = \exp\frac{-i}{2}\sum g_{\mu\mu}g_{\nu\nu}\omega_{\mu\nu}\sigma_{\mu\nu}. \tag{3.2.13b}$$

For the proof, see Problem P.3.2.

### 3.2.3 Parity

To represent space reversal we introduce the parity operator

$$\mathcal{P} \; : \; \Psi(\mathbf{r},t) \to \mathcal{P}\Psi(\mathbf{r},t) \equiv C_P\Psi(-\mathbf{r},t), \tag{3.2.14}$$

where $C_P$ is a $4 \times 4$ matrix that must be determined so that $\mathcal{P}$ has the properties desirable of a parity operator: one thus requires

$$\mathcal{P}\mathbf{P} = -\mathbf{P}\mathcal{P}, \; \mathcal{P}\mathbf{L} = \mathbf{L}\mathcal{P}, \; \mathcal{P}\mathbf{S} = \mathbf{S}\mathcal{P} \tag{3.2.15a}$$

and parity invariance of the free Dirac equation,

$$[\mathcal{P}, H_0] = 0. \tag{3.2.15b}$$

Moreover we demand that $\mathcal{P}^2 = \eta_P^2$ with $\eta_P^2$ a phase so that $\mathcal{P}^2$ is equivalent to the identity.

The first two relations of (3.2.15a) are automatic from the definition (3.2.14) of $\mathcal{P}$. The third relation implies that

$$[C_P, \boldsymbol{\Sigma}] = 0,$$

whereas (3.2.15b) requires

$$[C_P, \beta] = 0, \; \{C_P, \boldsymbol{\alpha}\} = 0.$$

The solution is unique up to a constant of unit modulus, $\eta_P$; we write $C_P = \eta_P\overline{C}_P$, choosing $\overline{C}_P$ to satisfy $\overline{C}_P^2 = 1$; then

$$\overline{C}_P = \beta = \gamma_0,$$

so that

$$\mathcal{P}\Psi(\mathbf{r},t) = \eta_P\beta\Psi(-\mathbf{r},t). \tag{3.2.16}$$

The constant $\eta_P$, which can be chosen real and thus equal to $(+1)$ or $(-1)$, is called the *intrinsic parity* of the particle.

### 3.2.4 Time Reversal

We know that, in the nonrelativistic case, time reversal is represented by an *antiunitary* operator (cf. (1.1.13)). We then try

$$\mathcal{T}\Psi(\mathbf{r},t) = C_T\Psi^*(\mathbf{r},-t);$$

$C_T$ is a numerical matrix that will be determined, as we did for parity, requiring

$$\mathcal{T}\mathbf{P} = -\mathbf{P}\mathcal{T}, \ \mathcal{T}\mathbf{L} = -\mathbf{L}\mathcal{T}, \ \mathcal{T}\mathbf{S} = -\mathbf{S}\mathcal{T} \tag{3.2.17}$$

and invariance of the Dirac equation. These requirements give the conditions

$$C_T\alpha_j^* + \alpha_j C_T = 0, \ C_T\beta^* = \beta C_T, \tag{3.2.18a}$$

or, in terms of the $\gamma$ matrices, and in the Pauli or Weyl representations,

$$C_T\gamma_0^* = \gamma_0 C_T, \ C_T\gamma_j^* C_T^{-1} = -\gamma_j. \tag{3.2.18b}$$

The expression of $C_P$ is independent of the realization of the $\gamma$ matrices; but $C_T$ *does* depend on it. In the Pauli or Weyl realizations, $C_T = \eta_T\overline{C}_T$, with $\overline{C}_T^2 = 1$ and

$$\overline{C}_T = -i\alpha_1\alpha_3 = i\gamma_1\gamma_3, \tag{3.2.19}$$

so that

$$\mathcal{T}\Psi(\mathbf{r},t) = \eta_T i\gamma_1\gamma_3\Psi^*(\mathbf{r},t). \tag{3.2.20}$$

## 3.3 Density of Particles. Current. Scalar Product

As we have defined it, the Dirac wave function is a vertical four-component matrix[4]:

$$\Psi = \begin{pmatrix} \Psi_1 \\ \Psi_2 \\ \Psi_3 \\ \Psi_4 \end{pmatrix}. \tag{3.3.1a}$$

It is convenient to define a related horizontal four-component matrix[5],

---

[4] The fact that the Dirac matrices have four components is only *indirectly* linked to the dimension of spacetime. If spacetime were $n$-dimensional, one can prove that the Dirac matrices should be $2^{n/2}$-dimensional. (For odd dimensions there is no solution for the $\gamma$ matrices).

[5] We may call $\overline{\Psi}$ the *Dirac adjoint* of $\Psi$. The expression $\overline{\Psi} = \Psi^+\gamma_0$ (with $\overline{\Psi}$ possessing the properties described below, in particular that $\overline{\Psi}\Psi$ be scalar) is not valid in all representations; it is valid, nevertheless, in the ones we use most, the Pauli and Weyl realizations. In general, $\overline{\Psi} = \Psi^+ M$, with $M$ a matrix such that $MD^+(\Lambda)M^{-1} = D^{-1}(\Lambda)$.

$$\overline{\Psi} \equiv \Psi^+ \gamma_0, \tag{3.3.1b}$$

which in the Pauli realization has the explicit form

$$\overline{\Psi} = (\Psi_1^*, \Psi_2^*, -\Psi_3^*, -\Psi_4^*). \tag{3.3.1c}$$

The importance of $\overline{\Psi}$ lies in the fact that it transforms contragradiently to $\Psi$ so that, for example, the quantity $\overline{\Psi}\Psi$ is a scalar (invariant) under Lorentz transformations. In fact, from (3.2.1), and under a Lorentz transformation $\Lambda$,

$$\Psi(x) \to D(\Lambda)\Psi(\Lambda^{-1}x),$$

so that

$$\begin{aligned}
\overline{\Psi}(x) &\to \Psi^+(\Lambda^{-1}x)D^+(\Lambda)\gamma_0 = \Psi^+(\Lambda^{-1}x)\gamma_0\gamma_0 D^+(\Lambda)\gamma_0 \\
&= \overline{\Psi}(\Lambda^{-1}x)\gamma_0 D^+(\Lambda)\gamma_0.
\end{aligned}$$

Therefore,

$$\overline{\Psi}(x)\Psi(x) \to \overline{\Psi}(\Lambda^{-1}x)\gamma_0 D^+(\Lambda)\gamma_0 D(\Lambda)\Psi(\Lambda^{-1}x);$$

to prove that $\overline{\Psi}\Psi$ is a scalar it only remains to show that

$$\gamma_0 D^+(\Lambda)\gamma_0 = D^{-1}(\Lambda). \tag{3.3.2}$$

To do so, we use (3.2.11):

$$D(\Lambda) = \exp\left(\frac{-i}{2}\sum g_{\mu\mu}g_{\nu\nu}\omega_{\mu\nu}\sigma_{\mu\nu}\right);$$

then, in the Pauli or Weyl realizations,

$$\gamma_0\gamma_\mu^+\gamma_0 = \gamma_\mu,$$

so that

$$\gamma_0\sigma_{\mu\nu}^+\gamma_0 = \sigma_{\mu\nu}$$

and, finally,

$$\begin{aligned}
\gamma_0 D^+(\Lambda)\gamma_0 &= \gamma_0\left(\exp\left(-\frac{i}{2}\sum g_{\mu\mu}g_{\nu\nu}\omega_{\mu\nu}\sigma_{\mu\nu}\right)^+\right)\gamma_0 \\
&= \exp\left(\frac{i}{2}\sum g_{\mu\mu}g_{\nu\nu}\omega_{\mu\nu}\sigma_{\mu\nu}\right) = D^{-1}(\Lambda),
\end{aligned}$$

as required. We have therefore checked that $\overline{\Psi}\Psi$ is a scalar:

$$\Lambda \;:\; \overline{\Psi}(x)\Psi(x) \to \overline{\Psi}(\Lambda^{-1}x)\Psi(\Lambda^{-1}x). \tag{3.3.3}$$

We proved in (3.2.13) that the $\gamma_\mu$ transform as a Minkowski vector; since $\overline{\Psi}\Psi$ is a scalar, we see that a candidate for the *density of particles-current four-vector* is[6]

---

[6] We hope that, in spite of the identitical notation, $J$ (four-current) will not be mistaken for $\mathbf{J}$ (generator of rotations).

$$J_\mu(x) = \overline{\Psi}(x)\gamma_\mu\Psi(x). \tag{3.3.4}$$

$J$ is indeed a good candidate. It satisfies a continuity equation,

$$\partial \cdot J(x) = 0, \tag{3.3.5a}$$

or, in components,

$$\frac{1}{c}\partial_t J_0(\mathbf{r}, t) + \boldsymbol{\nabla}\mathbf{J}(\mathbf{r}, t) = 0. \tag{3.3.5b}$$

Moreover, the density of particles is manifestly positive definite:

$$\rho(\mathbf{r}, t) = J_0(\mathbf{r}, t) = \Psi^+(\mathbf{r}, t)\Psi(\mathbf{r}, t) \geq 0. \tag{3.3.5c}$$

**Exercise.** Verify (3.3.5a) using the Dirac equation ∎

Because of these properties, we will be able to define a satisfactory scalar product; we set

$$\begin{aligned}
\langle \Psi | \Psi' \rangle &= \int d^3r \overline{\Psi}(\mathbf{r}, t)\gamma_0\Psi'(\mathbf{r}, t) \\
&= \int d^3r \Psi^+(\mathbf{r}, t)\Psi'(\mathbf{r}, t) \\
&= \sum_{a=1}^{4} \int d^3r \Psi_a^*(\mathbf{r}, t)\Psi_a'(\mathbf{r}, t).
\end{aligned} \tag{3.3.6}$$

This is certainly positive definite, and it is also Lorentz invariant.

**Exercise.** Check explicitly the Lorentz invariance of (3.3.6).

*Hint.* Realize that $d^3r = dx\, dy\, dz$ behaves as the product of three space coordinates, and $\overline{\Psi}\gamma_0\Psi'$ as a time coordinate ∎

Equation (3.3.6) is also time-independent, which is verified by straightforward calculation using Dirac's equation:

$$\begin{aligned}
\partial_t\langle \Psi | \Psi' \rangle &= \int d^3r\, \partial_t(\overline{\Psi}(\mathbf{r}, t)\gamma_0\Psi'(\mathbf{r}, t)) \\
&= \int d^3r\{(\partial_t\overline{\Psi}(\mathbf{r}, t))\gamma_0\Psi'(\mathbf{r}, t) + \overline{\Psi}(\mathbf{r}, t)\gamma_0\partial_t\Psi'(\mathbf{r}, t)\} \\
&= \int d^3r \left\{ \frac{1}{-i\hbar}(-i\hbar c\boldsymbol{\alpha}\boldsymbol{\nabla}\Psi + mc^2\beta\Psi)^+\gamma_0\gamma_0\Psi' \right. \\
&\quad \left. + \Psi^+\gamma_0\gamma_0\frac{1}{i\hbar}(-i\hbar c\boldsymbol{\alpha}\boldsymbol{\nabla}\Psi' + mc^2\beta\Psi') \right\}.
\end{aligned}$$

The terms with $mc^2$ cancel one against the other, so also using the fact that $\alpha^+ = \alpha$,

$$\partial_t\langle \Psi | \Psi' \rangle = -c \int d^3r\, \mathrm{div}(\Psi^+\boldsymbol{\alpha}\Psi) = 0.$$

**Exercise.** Show that $\overline{\Psi}(x)$ obeys the equation

$$\overline{\Psi}(x)(i\hbar\,\overleftarrow{\partial\!\!\!/}+mc^2)=0, \tag{3.3.7}$$

where the arrow means that the derivative acts to the left: for any $f$,

$$f(x)\,\overleftarrow{\partial_\mu}\equiv(\partial_\mu f(x))\;\blacksquare$$

**Exercise.** Check that $J_\mu$ transforms as a Minkowski four-vector,

$$\Lambda\;:\;J_\mu(x)\to\sum_\nu\Lambda_{\mu\nu}J_\nu(\Lambda^{-1}x). \tag{3.3.8}$$

*Hint.* Use (3.2.1), (3.2.13) $\blacksquare$

## 3.4 Minimal Replacement. Gauge Invariance. Large and Small Components: Nonrelativistic Limit of the Dirac Equation

To clarify the interpretation of the Dirac equation we will now include the interaction with a (classical) electromagnetic field, considering also the nonrelativistic (NR) limit. Interactions with the electromagnetic field are implemented by postulating, for particles with charge $e$, the minimal replacement

$$-i\hbar\nabla\to-i\hbar\nabla-\frac{e}{c}\mathbf{A},\;i\hbar\partial_t\to i\hbar\partial_t-e\phi, \tag{3.4.1a}$$

or, in covariant notation,

$$i\hbar\partial_\mu\to i\hbar D_\mu=i\hbar\left(\partial_\mu+\frac{ie}{\hbar c}A_\mu\right),\;A_0=\phi. \tag{3.4.1b}$$

Gauge invariance is formulated as in the scalar or the nonrelativistic case. A gauge transformation will consist of a modification of the four-potential,

$$A_\mu(x)\to A_\mu(x)-\partial_\mu f(x), \tag{3.4.2a}$$

together with a phase change of the wave function,

$$\Psi(x)\to e^{ief(x)/c\hbar}\Psi(x). \tag{3.4.2b}$$

Under (3.4.2) the combination

$$i\hbar D_\mu\Psi=\left(i\hbar\partial_\mu-\frac{e}{c}A_\mu(x)\right)\Psi(x),$$

only gets altered by a phase.

In view of this, we postulate that the Dirac equation in the presence of a prescribed (classical) electromagnetic field may be written as

$$i\hbar D\!\!\!/\,\Psi=\left(i\hbar\gamma\cdot\partial-\frac{e}{c}\gamma\cdot A\right)\Psi(x)=mc\Psi(x), \tag{3.4.3}$$

which makes it gauge invariant.

To study the NR limit it is advantegeous to work in the Pauli realization, which we will do in the rest of this section. We also write $\Psi(x)$ as a pair of two-component matrices,

$$\Psi = \left(\begin{array}{c} \Psi_b \\ \Psi_s \end{array}\right); \ \Psi_b = \left(\begin{array}{c} \Psi_1 \\ \Psi_2 \end{array}\right), \ \Psi_s = \left(\begin{array}{c} \Psi_3 \\ \Psi_4 \end{array}\right). \tag{3.4.4}$$

The $\Psi_b$ and $\Psi_s$ are called respectively the *large* (or *big*) and *small* components, for reasons that will be obvious presently. In terms of these one can write the Dirac equation (3.4.3) as a pair of coupled two-component equations:

$$i\hbar\partial_t\Psi_b = c\boldsymbol{\sigma}\left(\mathbf{P} - \frac{e}{c}\mathbf{A}\right)\Psi_s + mc^2\Psi_b + e\phi\Psi_b,$$
$$i\hbar\partial_t\Psi_t = c\boldsymbol{\sigma}\left(\mathbf{P} - \frac{e}{c}\mathbf{A}\right)\Psi_b - mc^2\Psi_s + e\phi\Psi_s. \tag{3.4.5}$$

In the NR limit, the energy will be $E \simeq mc^2 + E_{NR}$: to get equations like the ordinary NR ones we have to substract the rest energy $mc^2$. This, in particular, means that we have to separate off a phase from the wave functions, $\exp mc^2 t/i\hbar$ (for stationary states). We thus define the functions $\Psi_{b,s}^{NR}$,

$$\Psi_{b,s}(\mathbf{r},t) = e^{-imc^2 t/\hbar}\Psi_{b,s}^{NR}(\mathbf{r},t), \tag{3.4.6}$$

in terms of which (3.4.5) become

$$i\hbar\partial_t\Psi_b^{NR} = c\boldsymbol{\sigma}\left(\mathbf{P} - \frac{e}{c}\mathbf{A}\right)\Psi_s^{NR} + e\phi\Psi_b^{NR},$$

$$i\hbar\partial_t\Psi_s^{NR} = c\boldsymbol{\sigma}\left(\mathbf{P} - \frac{e}{c}\mathbf{A}\right)\Psi_b^{NR} + e\phi\Psi_s^{NR}$$

$$-2mc^2\Psi_s^{NR}.$$

From the second equation we get that[7]

$$\Psi_s^{NR} = \boldsymbol{\sigma}\left(\mathbf{P} - \frac{e}{c}\mathbf{A}\right)\frac{1}{2mc}\Psi_b^{NR} + 0\left(\frac{1}{c^2}\right)\Psi_s, \tag{3.4.7}$$

and substituting this into the first we also find the equation

$$i\hbar\partial_t\Psi_b^{NR} \underset{NR}{\simeq} \frac{1}{2m}\left\{\boldsymbol{\sigma}\left(\mathbf{P} - \frac{e}{c}\mathbf{A}\right)\right\}^2 \Psi_b^{NR} + e\phi\Psi_b^{NR}. \tag{3.4.8}$$

The latter is *Pauli's equation*. After some rearrangements of the $\boldsymbol{\sigma}$, using

$$\sigma_j\sigma_k = \delta_{jk} + i\sum \epsilon_{jks}\sigma_s,$$

we can rewrite it as

---

[7] When taking the NR limit some care has to be exercised with the size of the potentials $|A_\mu|$. In (3.4.7) the term $O(1/c^2)$ contains factors $e\phi/mc^2$; likewise we will neglect terms $O(|e\mathbf{A}|/mc^2)$. Therefore the NR limit is only consistent for weak fields, $|e\phi| \ll mc^2$, $|e\mathbf{A}| \ll mc^2$.

$$i\hbar\partial_t\Psi_b^{NR} = \left\{\frac{1}{2m}\left(\mathbf{P}-\frac{e}{c}\mathbf{A}\right)^2 + e\phi - \frac{2\mu_B}{\hbar}\frac{\hbar\boldsymbol{\sigma}}{2}\mathcal{B}\right\}\Psi_b^{NR}, \tag{3.4.9}$$

the standard nonrelativistic equation[8], with the correct value of the *magnetic moment* $\mu$,

$$\mu = \mu_B = e\hbar/2mc,$$

*for elementary particles* such as the electron or muon (or, presumably, quarks). We have found that, in the NR limit,

$$\Psi(\mathbf{r},t) \underset{NR}{\simeq} e^{-imc^2t/\hbar}\begin{pmatrix} \Psi_b^{NR}(\mathbf{r},t) \\ 0 \end{pmatrix}; \tag{3.4.10}$$

note that (3.4.7) implies that $\Psi_s^{NR}$ tends to zero relative to $\Psi_b^{NR}$. Then $\Psi_b^{NR}$ satisfies the ordinary nonrelativistic Pauli equation *with the correct value* of $\mu$, which thus appears as a prediction of the relativistic theory.

For particles which are not elementary, such as the proton or neutron, these arguments show that the minimal substitution is not enough, or else we would have obtained the (wrong) results $\mu_p = e\hbar/2m_pc$, $\mu_n = 0$. We can obtain NR equations with the correct values of the magnetic moments by adding a (phenomenological) extra term to (3.4.3), writing

$$\left(i\hbar\gamma\cdot\partial - \frac{e}{c}\gamma\cdot A\right)\Psi + \frac{\hbar e\delta}{4mc}(\sigma\cdot F)\Psi = mc^2\Psi, \tag{3.4.11}$$

where we have defined

$$\sigma\cdot F \equiv \sum_{\mu\nu} g_{\mu\mu}g_{\nu\nu}\sigma_{\mu\nu}F_{\mu\nu}, \quad F_{\mu\nu} = \partial_\mu A_\nu - \partial_\nu A_\mu,$$

and $\delta$ is a constant that we adjust so that the predicted value of the magnetic moment coincides with the experimental one; from (3.4.11) we get (see, for example, Problem P.3.5),

$$\mu = \frac{e\hbar}{2mc} + \frac{e\hbar\delta}{2mc}.$$

## 3.5 Plane Waves. States with Well-Defined Spin

We look for solutions to the Dirac equation with well-defined momentum and energy:

$$\mathbf{P}\Psi_E^{(p)}(\mathbf{r},t) = \mathbf{p}\Psi_E^{(p)}(\mathbf{r},t), \tag{3.5.1a}$$

$$H_0\Psi_E^{(p)}(\mathbf{r},t) = E\Psi_E^{(P)}(\mathbf{r},t). \tag{3.5.1b}$$

Because of the latter, we can write

---

[8] See, for example, Landau and Lifshitz (1958); Galindo and Pascual (1978); Gottfried (1966); Ynduráin (1988); etc.

$$\Psi_E^{(\mathrm{P})}(\mathbf{r}, t) = e^{-iEt/\hbar}\psi_E^{(\mathrm{P})}(\mathbf{r}). \tag{3.5.2a}$$

Since $\Psi_E^{(\mathrm{P})}$ must also satisfy the Klein–Gordon equation, it follows that $E, \mathbf{p}$ are related by

$$E^2 = m^2c^4 + c^2\mathbf{p}^2,$$

so that, choosing only the solution with positive energy,

$$E = E(\mathbf{p}) = +c\sqrt{m^2c^2 + \mathbf{p}^2} \equiv cp_0. \tag{3.5.2b}$$

Equation (3.5.1a) determines the $\mathbf{r}$ dependence:

$$\psi_E^{(\mathrm{P})}(\mathbf{r}) = \frac{1}{(2\pi\hbar)^{3/2}}e^{i\mathbf{p}\mathbf{r}/\hbar}u(p). \tag{3.5.2c}$$

(The factor $(2\pi\hbar)^{-3/2}$ is introduced for convenience.) We can combine the space and time dependence and also write

$$\Psi^{(p)}(x) \equiv \Psi_E^{(\mathrm{P})}(\mathbf{r}, t) = \frac{1}{(2\pi\hbar)^{3/2}}e^{-ip\cdot x/\hbar}u(p). \tag{3.5.2d}$$

$u(p)$ is called a *Dirac spinor*; because $\Psi^{(p)}(x)$ satisfies the Dirac equation, it follows that $u$ must be a solution of the algebraic equation

$$(c\boldsymbol{\alpha}\mathbf{p} + mc^2\beta)u(p) = E(\mathbf{p})u(p), \tag{3.5.3a}$$

or, in covariant form,

$$(\gamma \cdot p - mc)u(p) = 0. \tag{3.5.3b}$$

Note that $u(p)$ depends only on $\mathbf{p}$, as $p_0$ is here a *dependent* quantity, given by (3.5.2b).

Let us write $u$ as a pair of two-element matrices, $u_{b/s}$ (*large/small* components):

$$u(p) = \begin{pmatrix} u_b \\ u_s \end{pmatrix}; \; u_b = \begin{pmatrix} u_1 \\ u_2 \end{pmatrix}, \; u_s = \begin{pmatrix} u_3 \\ u_4 \end{pmatrix}.$$

Equation (3.5.3a, b) tells us that

$$\begin{aligned} mc^2u_b + c\mathbf{p}\boldsymbol{\sigma}u_s &= Eu_b, \\ -mc^2u_s + c\mathbf{p}\boldsymbol{\sigma}u_b &= Eu_s, \end{aligned} \tag{3.5.3c}$$

so we get the most general form of $u$ satisfying (3.5.3):

$$u(p) = \sqrt{E(\mathbf{p}) + mc^2}\begin{pmatrix} \chi(\mathbf{p}) \\ c(E(\mathbf{p}) + mc^2)^{-1}\mathbf{p}\boldsymbol{\sigma}\chi(\mathbf{p}) \end{pmatrix}; \tag{3.5.4a}$$

$\chi(\mathbf{p})$ is an arbitrary Pauli (two-component) spinor. Equation (3.5.3c) also yields the consistency condition

$$\frac{c^2\mathbf{p}^2}{E^2 - m^2c^4} = 1,$$

which, not surprisingly, implies that $E = \pm E(\mathbf{p})$. Of course we reject the negative solution and fall back on (3.5.2b).

It is convenient (and so will be done here) to assume Pauli spinors normalized to

$$\chi^+(\mathbf{p})\chi(\mathbf{p}) = 1, \tag{3.5.4b}$$

so that

$$u^+(p)u(p) = 2E(\mathbf{p}), \quad \bar{u}(p)u(p) = 2mc^2,$$

$$\overline{\Psi}^{(p)}(x)\gamma_0\Psi^{(p)}(x) = \frac{2E(\mathbf{p})}{(2\pi\hbar)^3}, \tag{3.5.5}$$

i.e., a density of $2E(p)(2\pi\hbar)^{-3}$ particles per unit volume.

To specify $u$ completely we would have to require the corresponding state to be an eigenstate of another quantum number commuting with momentum and energy; we will thus consider states with well-defined spin, and construct the *covariant spin operators* in a direction given by the vector $\mathbf{n}$, $\mathbf{n}^2 = 1$. To do this, we start by sitting in the reference system in which the particle is at rest. Therefore, $\mathbf{p} = 0$ and, from (3.5.4a), the Dirac spinor $u(\mathbf{p} = 0)$ will be of the form

$$u(\mathbf{p} = 0) = \sqrt{2mc^2}\left(\begin{array}{c} \chi^{(0)} \\ 0 \end{array}\right);$$

only the large components will be nonzero. We can then choose $\chi_n^{(0)}(\pm) = \chi(\mathbf{p} = 0)$ such that $u(\mathbf{p} = 0)$ is an eigenstate of the operator for the projection of the spin along $\mathbf{n}$, $S(\mathbf{n}, 0)$,

$$S(\mathbf{n}, 0) = \frac{\hbar}{2}\left(\begin{array}{cc} \mathbf{n}\boldsymbol{\sigma} & 0 \\ 0 & \mathbf{n}\boldsymbol{\sigma} \end{array}\right). \tag{3.5.6}$$

Such $\chi_n^{(0)}$ can be written as

$$\chi_n^{(0)}(\pm) = \frac{1}{\sqrt{2(1 \pm n_z)}}\left(\begin{array}{c} n_z \pm 1 \\ n_x + in_y \end{array}\right),$$

$$\mathbf{n}\boldsymbol{\sigma}\chi_n^{(0)}(\pm) = \pm\chi_n^{(0)}(\pm); \quad \chi_n^{(0)}(\lambda')^+\chi_n^{(0)}(\lambda) = \delta_{\lambda\lambda'}. \tag{3.5.7}$$

In this system of reference, $\mathbf{p} = 0$, the wave functions with $\mathbf{n}$ component of spin equal to $\pm\hbar/2$ are thus

$$\Psi^{(\mathbf{P}=0,\pm,n)}(x) = u(0, \pm 1/2)e^{ip^{(0)}\cdot x/\hbar}/(2\pi\hbar)^{3/2}, \tag{3.5.8a}$$

where $p^{(0)}$ is the four-vector

$$p_0^{(0)} = mc, \quad \mathbf{p}^{(0)} = 0, \tag{3.5.8b}$$

and

$$u(0, \pm 1/2) = \sqrt{2mc^2}\left(\begin{array}{c} \chi_n^{(0)}(\pm) \\ 0 \end{array}\right). \tag{3.5.8c}$$

Note that the factor $\sqrt{2mc^2}$ is included here for consistency with (3.5.4).

Consider now the boost $L_{\mathbf{p}}$ such that $L_{\mathbf{p}}p^{(0)} = p$, where $p$ is the four-momentum of the particle in an arbitrary reference system. The rapidity parameters $\boldsymbol{\xi}$ that correspond to this, $L(\boldsymbol{\xi}) \equiv L_{\mathbf{p}}$, are obtained immediately using (1.5.1); we have

$$\cosh\xi = p_0/m, \quad \sinh\xi = |\mathbf{p}|/mc, \quad \boldsymbol{\xi}/|\boldsymbol{\xi}| = \mathbf{p}/|\mathbf{p}|. \tag{3.5.9a}$$

Under this Lorentz transformation the wave function becomes

$$L_{\mathbf{p}} \;:\; \Psi^{(\mathbf{p}=0,\pm,n)}(x) \to D(L_{\mathbf{p}})\Psi^{(\mathbf{p}=0,\pm,n)}(L_p^{-1}x) \equiv \Psi^{(p,\pm,n)}(x);$$

taking into account that

$$p^{(0)} \cdot L_p^{-1}x = (L_p p^{(0)}) \cdot x = p \cdot x,$$

and recalling (3.5.2d) and (3.5.8), we obtain the expression

$$\Psi^{(p,\pm,n)}(x) = \frac{1}{(2\pi\hbar)^{3/2}} e^{-ip\cdot x/\hbar} D(L_{\mathbf{p}}) u_n(0, \pm 1/2). \tag{3.5.9b}$$

To finish the evaluation of this wave function we next use (3.5.8), (3.5.9) and the expression (3.2.10) for $D(L(\boldsymbol{\xi})) = D(L_{\mathbf{p}})$, so that

$$u_n(p, \pm 1/2) = D(L_{\mathbf{p}}) u_n(0, \pm 1/2),$$

$$D(L_{\mathbf{p}}) = e^{\frac{1}{2}\boldsymbol{\xi}\boldsymbol{\alpha}} = \left(\cosh\xi + \frac{\sinh\xi}{\xi}\boldsymbol{\alpha}\boldsymbol{\xi}\right)^{1/2}$$

$$= \frac{1}{\sqrt{mc}}(p_0 + \mathbf{p}\boldsymbol{\alpha})^{1/2} = \frac{1}{\sqrt{2mc(p_0 + mc)}}(p_0 + mc + \boldsymbol{\alpha}\mathbf{p}); \tag{3.5.10}$$

thus, and as was to be expected from the general expression (3.5.4a),

$$u_n(p, \pm 1/2) = \sqrt{\frac{c}{p_0 + mc}} \left( \begin{array}{c} (p_0 + mc)\chi_n^{(0)}(\pm) \\ \mathbf{p}\boldsymbol{\sigma}\chi_n^{(0)}(\pm) \end{array} \right)$$

$$= \sqrt{E(\mathbf{p}) + mc^2} \left( \begin{array}{c} \chi_n^{(0)}(\pm) \\ \dfrac{c}{E(\mathbf{p}) + mc^2}\mathbf{p}\boldsymbol{\sigma}\chi_n^{(0)}(\pm) \end{array} \right); \tag{3.5.11a}$$

$$\Psi^{(p,\pm,n)}(x) = \frac{1}{(2\pi\hbar)^{3/2}} e^{-ip\cdot x/\hbar} u_n(p, \pm 1/2), \tag{3.5.11b}$$

and we recall that, for spin $\pm\hbar/2$ along $\mathbf{n}$,

$$\chi_n^{(0)}(\pm) = \frac{1}{\sqrt{2(1 \pm n_z)}} \left( \begin{array}{c} n_z \pm 1 \\ n_x + in_y \end{array} \right). \tag{3.5.11c}$$

When $\mathbf{n}$ is parallel to $OZ$, (3.5.11c) is not defined; we should take

$$\chi_z^{(0)}(+) = \left( \begin{array}{c} 1 \\ 0 \end{array} \right), \quad \chi_z^{(0)}(-) = \left( \begin{array}{c} 0 \\ 1 \end{array} \right). \tag{3.5.12}$$

The operator whose eigenspinors are the $u(p, \pm 1/2)$ is

$$S(\mathbf{n}, \mathbf{p}) = D(L_\mathbf{p})S(\mathbf{n}, 0)D^{-1}(L_\mathbf{p}). \tag{3.5.13}$$

**Exercise.** Verify that $S(\mathbf{n}, \mathbf{p})u(p, \pm 1/2) = \pm\dfrac{\hbar}{2}u(p, \pm 1/2)$ ∎

A particularly important case is that of the *helicity operator* and *helicity quantum number*. The latter is defined as the projection of the spin along the direction of motion. Given $\mathbf{p}$, we then chose $\mathbf{n} = \mathbf{p}/|\mathbf{p}| \equiv \hat{\mathbf{p}}$. By direct evaluation, or using Theorem 3 of Sect. 1.5, one can verify that $D(L_\mathbf{p})$ and $S(\hat{\mathbf{p}}, 0)$ commute. Moreover, $S(\hat{\mathbf{p}}, 0)$ is now

$$S(\hat{\mathbf{p}}, 0) = \mathbf{n}\mathbf{S} = \frac{1}{|\mathbf{p}|}\mathbf{p}\mathbf{S},$$

so that the helicity operator $S_p$ is given by the simple expression

$$S_p \equiv S(\mathbf{p}/|\mathbf{p}|, \mathbf{p}) = \frac{1}{|\mathbf{p}|}\mathbf{p}\mathbf{S} = \frac{\hbar}{2|\mathbf{p}|}\begin{pmatrix} \mathbf{p}\boldsymbol{\sigma} & 0 \\ 0 & \mathbf{p}\boldsymbol{\sigma} \end{pmatrix}. \tag{3.5.14a}$$

This operator is of course not defined for particles at rest. In terms of the $\gamma$ matrices,

$$S_p = \frac{\hbar}{2|\mathbf{p}|}\gamma_5\gamma_0\boldsymbol{\gamma}\mathbf{p}, \tag{3.5.14b}$$

$$\gamma_5 \equiv i\gamma_0\gamma_1\gamma_2\gamma_3,$$

an expression that is *independent* of the realization one chooses.

With the scalar product defined in (3.3.6), one has the orthogonality relations

$$\langle\Psi^{(p,\lambda,n)}|\Psi^{(p',\lambda',n)}\rangle = 2\delta_{\lambda\lambda'}E(\mathbf{p})\delta(\mathbf{p} - \mathbf{p}'). \tag{3.5.15}$$

## 3.6 Radial Form of the Dirac Equation. Free-Particle Solutions

In many problems we require solutions to the Dirac equation with well-defined total angular momentum and third component thereof, in addition to energy. (We *cannot* have at the same time well-defined values of orbital angular momentum and spin, since $\mathbf{L}$ and $\mathbf{S}$ do *not* separately commute with the Hamiltonian.) We thus seek wave functions $\Psi_{E\omega}^{j,M}$ with

$$H\Psi_{E\omega}^{(j,M)}(\mathbf{r}, t) = E\Psi_{E,\omega}^{(j,M)}(\mathbf{r}, t),$$

$$\mathbf{J}^2\Psi_{E\omega}^{(j,M)}(\mathbf{r}, t) = \hbar^2 j(j+1)\Psi_{E\omega}^{(j,M)}(\mathbf{r}, t),$$

$$J_3\Psi_{E\omega}^{(j,M)}(\mathbf{r}, t) = \hbar M\Psi_{E\omega}^{(j,M)}(\mathbf{r}, t). \tag{3.6.1}$$

$\omega$ is an extra quantum number that we have to specify to fix the wave function uniquely. $H$ is the Hamiltonian, which we assume to be of the general form

$$H = -ic\hbar\boldsymbol{\alpha}\boldsymbol{\nabla} + mc^2\beta + V_0(r) + \beta V_S(r), \tag{3.6.2}$$

with $V_a$, $a = 0$, $S$ scalar functions[9].

The first equation of (3.6.1) implies that we can write

$$\Psi_{E\omega}^{(j,M)}(\mathbf{r},t) = e^{-iEt/\hbar}\psi_{E\omega}^{(j,M)}(\mathbf{r}), \tag{3.6.3}$$

and $\psi_{E\omega}^{(j,M)}(\mathbf{r})$ still satisfies equations like (3.6.1).

To fix the functions completely we have to specify $\omega$, by considering an operator that commutes with $H$, $\mathbf{J}^2$, $J_3$. Two choices are commonly used: parity, so that

$$\mathcal{P}\psi_{E\omega}^{(j,M)}(\mathbf{r}) = \eta(j,\omega)\psi_{E\omega}^{(j,M)}(\mathbf{r}), \tag{3.6.4}$$

(the value of $\eta$ will be found below, (3.6.10)), and the $K$ operator,

$$K = \beta(2\mathbf{SL} + \hbar^2). \tag{3.6.5}$$

That $K$ commutes with $H$, $\mathbf{J}$ is verified by direct computation. The square of $K$ can be related to $\mathbf{J}$. Indeed, using the fact that

$$(\boldsymbol{\sigma}\mathbf{L})^2 = \mathbf{L}^2 - \hbar\boldsymbol{\sigma}\mathbf{L},$$

we find

$$K^2 = \hbar^2\left\{\begin{pmatrix} 1 & 0 \\ 0 & -1 \end{pmatrix}\begin{pmatrix} \boldsymbol{\sigma}\mathbf{L} + \hbar & 0 \\ 0 & \boldsymbol{\sigma}\mathbf{L} + \hbar \end{pmatrix}\right\}^2 = \hbar^4(j + 1/2)^2.$$

It thus follows that the possible eigenvalues of $K$ are[10] $\hbar^2\kappa$, $\kappa = \pm(j + 1/2)$:

$$K\psi_{E\omega}^{(j,M)}(\mathbf{r}) = \hbar^2\kappa\psi_{E\kappa}^{(j,M)}(\mathbf{r}) = \pm(j + 1/2). \tag{3.6.6}$$

The quantum number is now $\kappa/|\kappa|$ and, as one could expect and we will check, it is related to parity.

### 3.6.1 Radial Form of $H$

Let us now write $H$, as given by (3.6.2), separating radial and spin-angular parts. We start by defining

$$\partial_r \equiv \partial/\partial r, \; r \equiv |\mathbf{r}|; \; \alpha_r \equiv \frac{1}{|\mathbf{r}|}\mathbf{r}\boldsymbol{\alpha},$$

---

[9] The Coulomb potential, or the potential for quark and gluon interactions at short distances, is of the type $V_0$. A potential of the type $V_S$ appears in the theory of nuclear interactions.

[10] Many authors (such as Rose, 1961 or Akhiezer and Berestetskii, 1963) denote the eigenvalues of $K$ by $-\kappa$, so there is a difference of sign between their definition of $\kappa$ and ours: $\kappa^{\mathrm{R}} = -\kappa^{\mathrm{ours}}$.

and write the identities

$$\frac{1}{r}\mathbf{r}\boldsymbol{\nabla} = \partial_r,$$

$$\boldsymbol{\nabla} = \frac{\mathbf{r}}{r}\left(\frac{1}{r}\mathbf{r}\boldsymbol{\nabla}\right) - \frac{1}{r^2}\mathbf{r}\times(\mathbf{r}\times\boldsymbol{\nabla})$$

$$= \frac{\mathbf{r}}{r}\partial_r - \frac{\mathbf{r}}{r}\times\left(\frac{\mathbf{r}}{r}\times\boldsymbol{\nabla}\right).$$

From this we find that

$$-i\hbar c\boldsymbol{\alpha}\boldsymbol{\nabla} = -i\hbar c\alpha_r\partial_r - \frac{1}{r^2}c\boldsymbol{\alpha}(\mathbf{r}\times\mathbf{L}).$$

Taking into account now that $\mathbf{rL}=0$, we obtain

$$i\boldsymbol{\alpha}(\mathbf{r}\times\mathbf{L}) = r\alpha_r, \quad \boldsymbol{\Sigma}\mathbf{L} = \frac{2}{\hbar}r\alpha_r\mathbf{SL},$$

and so we finally get the desired form of $H$:

$$H = -i\hbar c\alpha_r\partial_r + \frac{2ic}{\hbar r}\alpha_r\mathbf{SL} + mc^2\beta + V(r), \tag{3.6.7}$$

$$V(r) = V_0 + \beta V_S.$$

Because $\mathbf{J} = \mathbf{L} + \mathbf{S}$, we have

$$\frac{2}{\hbar}\mathbf{SL} = \frac{1}{\hbar}(\mathbf{J}^2 - \mathbf{S}^2 - \mathbf{L}^2),$$

which can be used to rewrite the angular momentum–spin part of (3.6.7) in a form that will be useful soon.

To construct the wave functions $\psi_{E\omega}^{(j,M)}$ now, we start by composing Pauli spinors $\chi(s_3)$ corresponding to the $z$ component of spin $s_3$,

$$\frac{1}{2}\sigma_z\chi(s_3) = s_3\chi(s_3),$$

and the spherical harmonics (Appendix A.1),

$$Y_\lambda^l(\Omega) = Y_\lambda^l(\theta,\varphi)$$

corresponding to orbital angular momentum $l$, and $z$ component $\lambda$, to obtain the objects (sometimes called *spinor harmonics*, or *spherical harmonics with spin*),

$$\mathcal{Y}_M^{l\pm}(\theta,\varphi) = \sum_{s_3+\lambda=M}\left(l,\lambda;\frac{1}{2},s_3|l\pm 1/2, M\right)Y_\lambda^l(\theta,\varphi)\chi(s_3).$$

Here the $(\ldots|\ldots)$ are the usual Clebsch–Gordan coefficients. The $\mathcal{Y}_M^{l\pm}$, two-component objects just like the $\chi$, correspond to a well-defined value of $\mathbf{J}^2$, equal to $\hbar j(j+1)$ with $j = l\pm 1/2$, and to the eigenvalue $\hbar M$ of $J_3$. Under space reversal, $\mathcal{Y}_M^{l\pm}\to(-1)^l\mathcal{Y}_M^{l\pm}$, because that is how the $Y_M^l$ transform, and the $\chi(s_3)$ are invariant. Given the expression (3.2.16) of the parity operator acting on Dirac spinors,

$$\mathcal{P}\begin{pmatrix} \Psi_b(\mathbf{r},t) \\ \Psi_s(\mathbf{r},t) \end{pmatrix} = \eta_P \begin{pmatrix} \Psi_b(-\mathbf{r},t) \\ -\Psi_s(-\mathbf{r},t) \end{pmatrix}, \tag{3.6.8}$$

with $\eta_P$ the intrinsic parity of the particle, it follows that we will have to combine two $\mathcal{Y}^l$ differing in one unit of $l$ to obtain Dirac wave functions with well-defined parity. We thus write

$$\psi_{E+}^{(j,M)}(r) \equiv \begin{pmatrix} f_{l+}(r)\mathcal{Y}_M^{l+}(\theta,\varphi) \\ g_{l+}(r)\mathcal{Y}_M^{l+1,-}(\theta,\varphi) \end{pmatrix}, \quad j = l + 1/2, \tag{3.6.9a}$$

$$\psi_{E-}^{(j,M)}(r) \equiv \begin{pmatrix} f_{l-}(r)\mathcal{Y}_M^{l+1,-}(\theta,\varphi) \\ g_{l-}(r)\mathcal{Y}_M^{l+}(\theta,\varphi) \end{pmatrix}, \quad j = l + 1/2. \tag{3.6.9b}$$

($\theta,\varphi$ are polar angles of $\mathbf{r}$). In the nonrelativistic limit, the orbital angular momentum is $l$, for the $(+)$ case, and $l' = l + 1$, for the $(-)$ case, as will be shown in detail later.

These functions are eigenfunctions of the parity operator,

$$\mathcal{P}\psi_{E\omega}^{(j,M)}(\mathbf{r}) = \omega\eta_P(-1)^{j-1/2}\psi_{E\omega}^{(j,M)}(\mathbf{r}), \omega = \pm; \tag{3.6.10a}$$

the corresponding eigenvalue is $\eta(j,\omega)$,

$$\eta(j,\omega) = \eta_P\omega(-1)^{j-1/2}. \tag{3.6.10b}$$

Using the explicit form of the Clebsch–Gordan coefficients one can verify (see Problem P.3.3) the important relations

$$\begin{aligned} \frac{1}{r}\mathbf{r}\boldsymbol{\sigma}\mathcal{Y}_M^{l+}(\theta,\varphi) &= -\mathcal{Y}_M^{l+1,-}(\theta,\varphi), \\ \frac{1}{r}\mathbf{r}\boldsymbol{\sigma}\mathcal{Y}_M^{l-}(\theta,\varphi) &= -\mathcal{Y}_M^{l-1,+}(\theta,\varphi), \end{aligned} \tag{3.6.11}$$

and with them we can rewrite the eigenvalue equation

$$H\psi_{E\omega}^{(j,M)}(\mathbf{r}) = E\psi_\omega^{(j,M)}(\mathbf{r})$$

as two coupled equations for the (scalar) functions[11] $f$, $g$:

$$\begin{aligned} i\hbar c\partial_r g_{l+} + i\hbar c\frac{l+2}{r}g_{l+} + mc^2 f_{l+} + V_b f_{l+} &= E f_{l+}, \\ i\hbar c\partial_r f_{l+} + i\hbar c\frac{-l}{r}f_{l+} - mc^2 g_{l+} + V_p g_{l+} &= E g_{l+}; \end{aligned} \tag{3.6.12a}$$

---

[11] The notation for $f,g$ is not unanimously accepted. Many authors, such as Bethe and Salpeter (1957), Rosen (1961), Akhiezer and Berestetskii (1963), exchange $f$ and $g$: $f^{\mathrm{BS}} = g^{\mathrm{ours}}$; $g^{\mathrm{BS}} = f^{\mathrm{ours}}$.

$$i\hbar c\partial_r g_{l-} + i\hbar c\frac{-l}{r}g_{l-} + mc^2 f_{l-} + V_b f_{l-} = E f_{l-},$$

$$(3.6.12b)$$

$$i\hbar c\partial_r f_{l-} + i\hbar c\frac{l+2}{r}f_{l-} - mc^2 g_{l-} + V_p g_{l-} = E g_{l-}.$$

Both equations can be written simultaneously in terms of the quantum number $\kappa$, defined in (3.6.6). Setting $\omega = \pm$, $\kappa(\omega) \equiv \omega(j, +1/2)$ we have

$$i\hbar c\left(\partial_r g_\kappa + \frac{1}{r}g_\kappa + \frac{\kappa}{r}g_\kappa\right) + mc^2 f_\kappa + V_b f_\kappa = E f_\kappa,$$

$$i\hbar c\left(\partial_r f_\kappa + \frac{1}{r}f_\kappa - \frac{\kappa}{r}f_\kappa\right) - mc^2 g_\kappa + V_p g_\kappa = E g_\kappa.$$

For $\omega = +$, (3.6.12c) is equivalent to (3.6.12a) with $l = j - 1/2$; for $\omega = -$, (3.6.12c) becomes (3.6.12b), also with $l = j - 1/2$. Note that $g_\kappa$ is imaginary relative to $f_\kappa$, if the potentials are real.

**Exercise.** Verify that the $\kappa$ do indeed coincide with the eigenvalues of the $K$ operator defined in (3.6.5) ∎

### 3.6.2 Free Particles

For $V = 0$ the solution to (3.6.12) is straightforward. We denote functions corresponding to free particles by $f^{(0)}$, $g^{(0)}$. Solving then for these, we obtain

$$g_{l+}^{(0)}(r) = \frac{i\hbar c}{mc^2 + E}\left(\partial_r - \frac{l}{r}\right)f_{l+}^{(0)}(r),$$

$$(3.6.13)$$

$$g_{l-}^{(0)}(r) = \frac{i\hbar c}{mc^2 + E}\left(\partial_r + \frac{l+2}{r}\right)f_{l-}^{(0)}(r);$$

thus we get for the $f^{(0)}$ the equations

$$-\hbar^2 c^2\left(\partial_r^2 + \frac{2}{r}\partial_r \frac{l(l+1)}{r^2}\right)f_{l+}^{(0)}(r) = (E^2 - m^2 c^4)f_{l+}^{(0)}(r),$$

$$-\hbar^2 c^2\left(\partial_r^2 + \frac{2}{r}\partial_r - \frac{(l+1)(l+2)}{r^2}\right)f_{l-}^{(0)}(r) = (E^2 - m^2 c^4)f_{l-}^{(0)}(r).$$

The solution is similar to the nonrelativistic one. Defining the wave number $k$ by $E^2 - m^2 c^4 = \hbar^2 c^2 k^2$, we have

$$f_{l+}^{(0)}(r) = N_+^{(0)} j_l(kr),$$

$$(3.6.14)$$

$$f_{l-}^{(0)}(r) = N_-^{(0)} j_{l+1}(kr),$$

where the $j_l$ are the spherical Bessel functions (Appendix A.2),

$$j_l(kr) = \sqrt{\frac{\pi}{2kr}} J_{l+1/2}(kr)$$

$$= (-1)^l \left(\frac{r}{k}\right)^l \left(\frac{1}{r}\frac{d}{dr}\right)^l \frac{\sin kr}{kr} \underset{r\to\infty}{\simeq} \frac{\sin(kr - \pi l/2)}{kr}. \tag{3.6.15}$$

The $g^{(0)}$ are found using the recursion formulas

$$\frac{d}{dz}j_l(z) = j_{l-1}(z) - \frac{l+1}{z}; \quad \frac{d}{dz}j_l(z) = -j_{l+1}(z) + \frac{l}{z}j_l(z); \tag{3.6.16}$$

thus we obtain

$$g_{l+}^{(0)}(r) = \frac{-i\hbar ck}{mc^2 + E} f_{l+1,+}^{(0)}(r), \quad g_{l-}^{(0)}(r) = \frac{i\hbar ck}{mc^2 + E} f_{l-1,-}^{(0)}(r). \tag{3.6.17}$$

The factors $N_\pm$ are determined by considering the scalar product of the $\psi$. We have

$$\langle \psi_{E\omega}^{(j,M)} | \psi_{E'\omega'}^{(j',M')} \rangle = \int d^3r \, \psi_{E\omega}^{(j,M)}(\mathbf{r})^+ \psi_{E'\omega'}^{(j',M')}(\mathbf{r}); \tag{3.6.18a}$$

since the $Y_\lambda^l, \chi(s_3)$ are orthonormal, and the Clebsch–Gordan matrix unitary,

$$\langle \psi_{E\omega}^{(j,M)} | \psi_{E'\omega'}^{(j',M')} \rangle$$

$$= \delta_{jj'}\delta_{MM'}\delta_{\omega\omega'} \int_0^\infty dr \, r^2(f^*f' + g^*g'). \tag{3.6.18b}$$

According to this, we will obtain the normalization

$$\int_0^\infty dr \, r^2(f^{(0)*}f'^{(0)} + g^{(0)*}g'^{(0)}) = 2E\frac{\pi}{2k^2}\delta(k - k'), \tag{3.6.19}$$

if we choose (up to a phase)

$$N_\pm^{(0)} = \sqrt{E + mc^2}. \tag{3.6.20}$$

The behaviours of the $f^{(0)}, g^{(0)}$ both as $r \to 0$ and as $r \to \infty$ are found using the asymptotic expression (3.6.15) for the second, and that

$$j_l(r) \underset{r\to 0}{\simeq} r^l \pi^{1/2}/2^{l+1}\Gamma(l + 3/2)$$

for the first. We get

$$\begin{pmatrix} f_{l+}^{(0)}(r) \\ g_{l+}^{(0)}(r) \end{pmatrix} \underset{r\to\infty}{\simeq} \sqrt{E + mc^2} \begin{pmatrix} \dfrac{\sin(kr - \pi l/2)}{kr} \\ \dfrac{i\hbar ck}{E + mc^2} \dfrac{\cos(kr - \pi l/2)}{kr} \end{pmatrix}, \tag{3.6.21}$$

$$\begin{pmatrix} f_{l+}^{(0)}(r) \\ g_{l+}^{(0)}(r) \end{pmatrix} \underset{r\to 0}{\simeq} \sqrt{E + mc^2} \begin{pmatrix} 1 \\ 0 \end{pmatrix} \delta_{l0}; \tag{3.6.22}$$

$$\begin{pmatrix} f_{l-}^{(0)}(r) \\ g_{l-}^{(0)}(r) \end{pmatrix} \underset{r \to \infty}{\simeq} \sqrt{E + mc^2} \begin{pmatrix} -\dfrac{\cos(kr - \pi l/2)}{kr} \\ \dfrac{i\hbar ck}{E + mc^2} \dfrac{\sin(kr - \pi l/2)}{kr} \end{pmatrix}, \qquad (3.6.23)$$

$$\begin{pmatrix} f_{l-}^{(0)}(r) \\ g_{l-}^{(0)}(r) \end{pmatrix} \underset{r \to 0}{\simeq} \sqrt{E + mc^2} \begin{pmatrix} 0 \\ \dfrac{i\hbar ck}{E + mc^2} \end{pmatrix} \delta_{l0}. \qquad (3.6.24)$$

## 3.7 The Problem of Negative Energies in the Dirac Equation. The Dirac Sea. Hole Theory. Charge Conjugation

### 3.7.1 Negative Energies. The Dirac Sea. Holes

In all our discussion up till now we have neglected the negative-energy solutions, a pest that plagues in one form or another all relativistic wave equations. We will now consider them, starting with the Dirac equation. The Dirac equation, which we now write as

$$(i\hbar\gamma \cdot \partial - mc)\Psi_E(x) = 0, \qquad (3.7.1)$$

presents, whenever there is a solution of

$$(-i\hbar c\boldsymbol{\alpha}\boldsymbol{\nabla} + mc^2\beta)\Psi_E = E\Psi_E \qquad (3.7.2)$$

with $E > 0$, another solution $\Psi_{-E}$ corresponding to negative energy.

So long as we do not introduce interactions this fact presents no difficulties: we can legislate that only the states with $E > 0$ are physical and then, because $\langle \Psi_E | \Psi_{-E'} \rangle = 0$, they will not suffer interference from the negative-energy, unphysical ones. However, as soon as we introduce interaction with (say) the electromagnetic radiation, transitions $\Psi_E \to \Psi_{-E'} +$ photons become possible. A simple calculation for the states of the hydrogen atom, for example, tells us that the fundamental (positive) energy state decays into a negative energy state in a time of

$$\tau \sim \frac{\pi\hbar}{2\alpha^6 mc^2} \sim 10^{-8}\text{s},$$

which is plain nonsense.

To solve this difficulty Dirac postulated the hypothesis that *all the states with negative energy are occupied* (the *Dirac sea*). Since electrons are fermions, the Pauli principle prevents transitions $\Psi_E \to \Psi_{-E'} + \gamma$ because the state $\Psi_{-E'}$ is already occupied. In spite of its infinite charge and mass, this sea, being homogeneous, exerts no direct influence on us. There are, however, indirect effects. The Pauli principle forbids, as just mentioned, transitions in which an electron falls into the sea; but there is nothing against our *extracting* particles from the sea, that is to say, a reaction like

$$\gamma + \Psi_{-E'} \to \Psi_E; \ E, E' > 0 \tag{3.7.3}$$

should be possible. Through it, we have produced an electron with positive, hence physical, energy leaving a hole in the sea, in the place occupied by the state $\Psi_{-E'}$. The hole will appear as the *absence* of an electron of charge $-|e|$ and energy $-|E'|$, that is to say, as a particle of charge $+|e|$ and positive energy $+|E'|$: the theory suggests the existence of particles with properties identical to electrons except that they have the opposite charge. These particles, called *positrons*, were found experimentally by Anderson soon after being postulated by Dirac, so it would seem that, in spite of the incredible sea, we should take the theory seriously. In fact, a similar situation occurs in solids. Here all electron states are filled up to a certain energy, $E_F$ (the *Fermi level*). If we count energies from $E_F$, the situation is the same as for the Dirac sea.

Returning to our discussion, besides the extraction of a particle and creation of a hole, the converse process is possible: if we have a hole in the Dirac sea corresponding to energy $-E_1$ (and thus a positron with energy $E_1$), there is some chance that an electron will "fall" into the hole: thus, it may happen that an electron and a positron annihilate each other, the surplus energy materializing as photons. Both types of process have indeed been seen experimentally; we will calculate them in Chap. 11.

In spite of this the existence of the Dirac sea is hard to swallow and, as we will see later, it is possible (and desirable) to reformulate the theory without it.

### 3.7.2 Charge Conjugation

We may get rid of the sea: what we cannot dismiss is the existence of positrons. To deal with them it is convenient to introduce a new symmetry, *charge conjugation* (or more generally, *particle–antiparticle conjugation*), which we now describe. The Dirac equation for an electron in the presence of an electromagnetic field is

$$(i\gamma \cdot \partial - m + |e|\gamma \cdot A)\Psi(x) = 0; \tag{3.7.4}$$

in the rest of this section we use natural units, $\hbar = c = 1$, and denote the charge of the electron by $-|e|$. A negative-energy $(-|E|)$ solution of (3.7.4) is of the form

$$\Psi(x) = e^{i|E|t}\psi(\mathbf{r}).$$

We can put this into correspondence with a positive-energy one, with energy precisely equal to $-(-|E|) = |E|$, by replacing $\Psi$ by its complex conjugate, $\Psi(x) \to \Psi^*(x)$. If we want this new wave function to represent a positron, it will have to obey an equation like (3.7.4), with $|e|$ replaced by $-|e|$. That is to say, we have to find a *charge conjugation* operator $\mathcal{C}$ such that the charge-conjugate function[12],

$$\Psi^c(x) \equiv \mathcal{C}\Psi(x), \tag{3.7.5a}$$

$$\mathcal{C}\Psi(x) = \eta_C C\Psi^*(x) \tag{3.7.5b}$$

(where $C$ is a numerical matrix, and we have extracted for convenience the phase $\eta_C$), satisfies the equation

$$(i\gamma \cdot \partial - m - |e|\gamma \cdot A)\Psi^c(x) = 0, \tag{3.7.6}$$

which is identical to (3.7.4) with the change $|e| \to -|e|$. Note also that the energy of $\Psi^c$ is now *positive*.

To find $C$ we take complex conjugates in (3.7.4) and left multiply by $C$. We find that

$$\begin{aligned}
0 &= C(-i\gamma^* \cdot \partial - m + |e|\gamma^* \cdot A)\Psi^* \\
&= C(-i\gamma^* \cdot \partial - m + |e|\gamma^* \cdot A)C^{-1}C\Psi^* \\
&= (-i\gamma^c \cdot \partial - m + |e|\gamma^c \cdot A)\Psi^c,
\end{aligned}$$

where

$$\gamma_\mu^c = C\gamma_\mu^* C^{-1}.$$

We will obtain (3.7.6) if, and only if, we can find a matrix $C$ such that $\gamma_\mu^c = -\gamma_\mu$, that is to say, such that

$$C\gamma_\mu^* C^{-1} = -\gamma_\mu. \tag{3.7.7}$$

The solution is not difficult to find. Up to a constant, and in both the Pauli and Weyl representations,

$$C = i\gamma_2. \tag{3.7.8}$$

If we also want the transformation to preserve the norm of the scalar products, $\eta_C C$ must be unitary, so $\eta_C$, called the intrinsic charge conjugation parity, or $C$ parity, of the particle, must be a phase, which can be chosen to be $\pm 1$.

Together with the transformation $\mathcal{C}$ for wave functions, we may define an analogous one for spinors. Given a spinor $u(p, \lambda)$, we define the corresponding *conjugate spinor*, $v(p, \lambda)$, by

$$v(p, \lambda) \equiv Cu^*(p, \lambda). \tag{3.7.9a}$$

The conjugate spinor $v$ satisfies the equation

$$(\not{p} + m)v(p, \lambda) = 0,$$

---

[12] On *wave functions* the charge conjugation operator is antiunitary. However, we will see in Chap. 6 *et seq.* that it is *unitary* when defined on *states* and quantum fields.

i.e., similar to (3.5.3b), but with the sign of the mass term reversed.

Alert readers will have noticed that the existence of the charge conjugation symmetry of the Dirac equation is really independent of such dubious concepts as sea and holes, and will have realized that corresponding considerations also apply to the Klein–Gordon equation. In fact, if a scalar particle has charge $e$, we write its wave equation as

$$(\partial + ieA) \cdot (\partial + ieA)\Phi(x) = -m^2\Phi(x). \tag{3.7.10}$$

If $\Phi$ is a solution of (3.7.10) corresponding to negative energy, the conjugate function,

$$\Phi^c(x) = \eta_C\Phi^*(x), \tag{3.7.11}$$

is a positive-energy wave function and satisfies the equation

$$(\partial - ieA) \cdot (\partial - ieA)\Phi^c(x) = -m^2\Phi^c(x); \tag{3.7.12}$$

the particles it describes have charge $-e$. If we call the charge conjugate of a given particle the *antiparticle*, it would seem that we come to the conclusion that relativistic quantum theory implies the existence of antiparticles for every kind of particle, and not only spin 1/2 ones: something that has been amply verified experimentally. (In some cases, like that of the photon, the antiparticle is *identical* to the particle. Obviously such particles cannot have charge.) This is true also for bosons, for which the interpretation of the sea fails: one more indication that the theory contains important elements of truth, but has to be reformulated. This we will do in Chap. 6, *et seq.*

The relation between particle and antiparticle is symmetric. If we give the name "particle" to the electron and "antiparticle" to the positron, it is because we are more familiar with the first and, at least in our vicinity, there are more electrons than positrons. This asymmetry of nature, in spite of the symmetry of the equations, is one of the more fascinating problems in cosmology.

If a particle is denoted by the symbol $P$, then the corresponding antiparticle is denoted by putting a bar over it: $\overline{P}$. Thus the antiproton is $\bar{p}$, and the antineutron is $\bar{n}$. Electrons and positrons are an exception, being denoted by $e^-$, $e^+$ for historical reasons.

## 3.8 Covariants and Projectors

### 3.8.1 Covariants

Using the transformation properties of the Dirac wave functions and $\gamma_\mu$ matrices one can easily check the transformation properties of the following bilinears:

$$\overline{\Psi}(x)\Psi(x) = \text{scalar}, \quad \overline{\Psi}(x)\gamma_5\Psi(x) = \text{pseudoscalar};$$

$\overline{\Psi}\gamma_\mu\Psi(x) = \text{vector}, \; \overline{\Psi}(x)\gamma_\mu\gamma_5\Psi(x) = \text{axial vector (pseudovector)};$

$$\overline{\Psi}(x)\sigma_{\mu\nu}\Psi(x) = \text{antisymmetric tensor.} \tag{3.8.1}$$

Here we have defined

$$\sigma_{\mu\nu} = \frac{i}{2}[\gamma_\mu, \gamma_\nu],$$

and the matrix $\gamma_5$,

$$\gamma_5 = i\gamma_0\gamma_1\gamma_2\gamma_3; \; \gamma_5^2 = 1; \; \{\gamma_5, \gamma_\mu\} = 0; \; \gamma_5 = \begin{pmatrix} 0 & 1 \\ 1 & 0 \end{pmatrix},$$

the last explicit expression being valid in the Pauli realization.

Because of Pauli's theorem (Appendix A.4), which guarantees that any $4 \times 4$ matrix is a combination of the unit matrix and of $\gamma_5$, $\gamma_\mu$, $\gamma_\mu\gamma_5$ and $\sigma_{\mu\nu}$, it follows that (3.8.1) exhaust the independent bilinears in $\Psi^*, \Psi$.

### 3.8.2 Projectors

In this subsection we will construct projectors such that, when they are applied to an arbitrary wave function, they will give solutions of the Dirac equation with specified values of the spin component. We will work in momentum space and will also consider functions with well-defined energy so that the Dirac equation reads

$$(\gamma \cdot p - mc)\psi(p) = 0, \; p_0 = \sqrt{\mathbf{p}^2 + m^2c^2}.$$

Given an *arbitrary* function $\psi(p)$, we can use the commutation relations of the $\gamma_\mu$ to check that $(\gamma \cdot p - mc)(\gamma \cdot p + mc) = p \cdot p - m^2c^2 = 0$, so that the function

$$\frac{1}{2mc}(\gamma \cdot p + mc)\psi(p)$$

is automatically a solution of the Dirac equation. It thus follows that

$$\Lambda(p) \equiv \frac{1}{2mc}(\not{p} + mc) \tag{3.8.2}$$

is the projector over solutions of the Dirac equation.

**Exercise.** Check that $\Lambda(p)^2 = \Lambda(p)$ and that, in the Pauli or Weyl realizations, $\gamma_0\Lambda^+(p)\gamma_0 = \Lambda(p)$ ∎

Let us now construct the projector over the spin states defined in Sect. 3.5. To be definite, we seek projectors $\Sigma_+(n, p)$ such that, for any $\Psi$, $\Sigma_+(n, p)\Psi = (\text{const})u_n(p, +1/2)$, where the $u_n(p, +1/2)$ are those constructed in (3.5.11). If we wanted spinors with the third spin component along $\mathbf{n}$ equal to $-1/2$, we simply realize that they coincide with the spinors with the third component $+1/2$ along $-\mathbf{n}$, so our construction is quite general.

To obtain $\Sigma_+$ we start in the reference system in which the particle is at rest, with momentum $p^{(0)}$, $p_0^{(0)} = mc$, $\mathbf{p}^{(0)} = 0$. Defining a four-vector $n^{(0)}$ by $n_0^{(0)} = 0$, $\mathbf{n}^{(0)} = \mathbf{n}$, we may verify by explicit calculation that

$$\Sigma_+(n^{(0)}, p^{(0)}) = \frac{1 + \gamma_5 n^{(0)} \cdot \gamma}{2} \frac{1 + \gamma_0}{2},$$

is the desired projector. To go over to an arbitrary reference system we replace

$$\frac{1 + \gamma_0}{2} \to \frac{mc + \gamma \cdot p}{2mc},$$

$$n^{(0)} \to n \equiv L_p n^{(0)},$$

with $L_p$ the Lorentz boost such that $L_p p^{(0)} = p$. Then,

$$\Sigma_+(n, p) = \frac{1 + \gamma_5 \gamma \cdot n}{2} \cdot \frac{mc + \gamma \cdot p}{2mc}. \tag{3.8.3a}$$

If this acts on solutions to the Dirac equation, we may replace the last term in (3.8.3a) by unity, so that we simply have

$$\Sigma_+(n, p) = \frac{1 + \gamma_5 \gamma \cdot n}{2}$$

**Exercise.** Check that, indeed, for an arbitrary $\psi$,

$$\Sigma_+(n, p)\psi(p) = (\text{const})u_n(p, +1/2),$$

with $u_n$ given by (3.5.11). Check that $\Sigma_+^2 = \Sigma_+$, and that $\gamma_0 \Sigma_+^\dagger \gamma_0 = \Sigma_+$ (the latter relation being valid in the Pauli or Weyl realizations) ∎


## 3.9 Massless Spin 1/2 Particles

Relativity allows for the existence of massless particles. The best known such particle is the photon, which has spin 1 and shall be studied later. Now we will develop the theory of massless spin 1/2 particles, which is appropriate for describing neutrinos[13] and is also useful for approximately describing particles with mass at energies much larger than their rest energy, $E \gg mc^2$.

Although the theory is, of course, independent of the realization we use for the $\gamma$ matrices, it is convenient, for explicit calculations in our case, to use Weyl's representation:

$$\gamma_\mu^{\mathrm{W}} = \begin{pmatrix} 0 & \tilde{\sigma}_\mu \\ \sigma_\mu & 0 \end{pmatrix}, \quad \gamma_5^{\mathrm{W}} = \begin{pmatrix} 1 & 0 \\ 0 & -1 \end{pmatrix}, \tag{3.9.1a}$$

with $\sigma_i$ being the Pauli matrices and

---

[13] It is of course impossible to make sure that a particle is exactly massless. For the neutrinos, and considering the so-called electron neutrino (the most common one), we know that $m_\nu < 10^{-4} m_e$, so even if its mass were not exactly zero we are not committing a large error by neglecting it.

$$\tilde{\sigma}_i = -\sigma_i, \ \tilde{\sigma}_0 = \sigma_0 = 1. \tag{3.9.1b}$$

The equations found in the previous sections possess a smooth limit as $m \to 0$, although some peculiarities appear then. In fact, Dirac's equation now becomes

$$i\partial_t \Psi(\mathbf{r}, t) = -ic\boldsymbol{\alpha}\boldsymbol{\nabla}\Psi(\mathbf{r}, t), \tag{3.9.2}$$

and it does *not* contain the matrix $\beta$. Because of this, and since $\boldsymbol{\alpha} = \gamma_0\boldsymbol{\gamma}$ commutes with $\gamma_5$, it follows that, if $\Psi$ satisfies (3.9.2), the functions

$$\Psi_L(x) = \frac{1 - \gamma_5}{2}\Psi(x)$$

$$\Psi_R(x) = \frac{1 + \gamma_5}{2}\Psi(x) \tag{3.9.3}$$

are also solutions of (3.9.2). Particles with wave functions of the type $\Psi_R$ are called *right handed*, and those of type $\Psi_L$ *left handed*, for reasons that will be apparent presently.

**Exercise.** (A) Check that $(1 \pm \gamma_5)/2$ are projectors. (B) Check that the decomposition (3.9.3) is Lorentz invariant.

*Hint.* Use the expression (3.2.13) for $D(\Lambda)$ in terms of $\sigma_{\mu\nu}$, and the fact that $\sigma_{\mu\nu}$ commutes with $\gamma_5$ ∎

In the Weyl representation, (3.9.1),

$$\boldsymbol{\alpha}^{\mathrm{W}} = \gamma_0^{\mathrm{W}}\boldsymbol{\gamma}^{\mathrm{W}} = \begin{pmatrix} \boldsymbol{\sigma} & 0 \\ 0 & -\boldsymbol{\sigma} \end{pmatrix},$$

$$\frac{1 - \gamma_5^{\mathrm{W}}}{2} = \begin{pmatrix} 0 & 0 \\ 0 & 1 \end{pmatrix}, \ \frac{1 + \gamma_5^{\mathrm{W}}}{2} = \begin{pmatrix} 1 & 0 \\ 0 & 0 \end{pmatrix},$$

and we can write (3.9.2) as two separate equations, one for $\Psi_L$ and one for $\Psi_R$, with only two components. Letting

$$\Psi_L(x) = \begin{pmatrix} 0 \\ \Psi_L^{\mathrm{W}} \end{pmatrix}, \ \Psi_R(x) = \begin{pmatrix} \Psi_R^{\mathrm{W}} \\ 0 \end{pmatrix},$$

we then have

$$i\partial_t \Psi_L^{\mathrm{W}}(x) = ic\boldsymbol{\sigma}\boldsymbol{\nabla}\Psi_L^{\mathrm{W}}(x), \tag{3.9.4a}$$

$$i\partial_t \Psi_R^{\mathrm{W}}(x) = -ic\boldsymbol{\sigma}\boldsymbol{\nabla}\Psi_R^{\mathrm{W}}. \tag{3.9.4b}$$

Equations (3.9.4) are known as *Weyl equations*. They are covariant under Lorentz transformations, but *not* invariant under parity. Under this transformation $\Psi_L$ and $\Psi_R$ are exchanged:

$$\Psi_l \overset{\mathcal{P}}{\leftrightarrow} \Psi_R; \tag{3.9.5}$$

recall that $\mathcal{P}\Psi(\mathbf{r}, t) = \eta_P\gamma_0\Psi(\mathbf{r}, t)$ and, in the Weyl realization, $\gamma_0^{\mathrm{W}}$ is anti-diagonal. This is not, however, a problem: quite the opposite, in fact, since it so happens experimentally that the interactions that involve neutrinos (weak interactions) do *not* preserve parity.

The functions $\Psi_L$, $\Psi_R$ are eigenfunctions of the operator multiplication by $\frac{1}{2}\gamma_5$, called the *chirality operator*:

$$\frac{1}{2}\gamma_5\Psi_L = \frac{-1}{2}\Psi_L, \quad \frac{1}{2}\gamma_5\Psi_R = +\frac{1}{2}\Psi_R, \tag{3.9.6}$$

or, together, $\frac{1}{2}\gamma_5\Psi_X = \eta_X\Psi_X$, $X = L, R$; $\eta_L = -1/2$, $\eta_R = +1/2$. The physical meaning of the quantum numbers $\eta_L$, $\eta_R$ (and the reason for the names "left handed" and "right handed") is found by rewriting (3.9.4) for particles with well-defined momentum $\mathbf{p}$ as

$$|\mathbf{p}|\Psi_L^{\mathbf{WP}}(x) = -\mathbf{p}\boldsymbol{\sigma}\Psi_L^{\mathbf{WP}}(x), \quad |\mathbf{p}|\Psi_R^{\mathbf{WP}}(x) = \mathbf{p}\boldsymbol{\sigma}\Psi_R^{\mathbf{WP}}(x); \tag{3.9.7}$$

remembering that that, for massless particles, $E(\mathbf{p}) = c|\mathbf{p}|$. Now, the helicity operator (3.5.14b) is, in the Weyl representation,

$$S_p = \frac{\hbar}{2|\mathbf{p}|}\begin{pmatrix} \boldsymbol{\sigma}\mathbf{p} & 0 \\ 0 & \boldsymbol{\sigma}\mathbf{p} \end{pmatrix}.$$

Therefore, (3.9.7) imply that $\Psi_L$ (respectively, $\Psi_R$) corresponds to helicity $-\hbar/2$ (respectively, $+\hbar/2$), that is to say,

$$S_p\Psi_L = -\frac{\hbar}{2}\Psi_L, \quad S_p\Psi_R = +\frac{\hbar}{2}\Psi_R.$$

In fact, one can easily verify, using (3.9.7), that, for particles with well-defined momentum, we can replace $S_p$ by $\hbar\gamma_5/2$: the quantum numbers $\eta_L$, $\eta_R$ are simply the values of the helicity. "Left handed" or "right handed" refers to the corkscrew rule for rotations.

In general, and because the chirality (or helicity) is now Lorentz invariant, particles with different chirality may be *different*. We say that a particle is a *Majorana particle* if the states with helicity $\pm\hbar/2$ belong to the *same* particle. If the particle with helicity $+\hbar/2$ is different from that with helicity $-\hbar/2$, we talk about *Weyl particles*. The three kinds of neutrino that exist all appear to be Weyl particles; they all have helicity $-\hbar/2$. The particles with helicity $+\hbar/2$ are the corresponding *antineutrinos*, different from the neutrinos. A mathematical possibility, viz., that the particle with chirality opposite to a given one did not exist, seems not to occur in nature.


## Problems

**P.3.1.** The conjugate spinor is defined by

$$v(p, \lambda) = i\gamma_2 u^*(p, \lambda).$$

(A) Find its explicit expression. (B) Show that

$$\overline{v}(p', \lambda')\gamma_\mu v(p, \lambda) = \overline{u}(p, \lambda)\gamma_\mu u(p', \lambda').$$

**P.3.2.** Check that $D^{-1}(\Lambda)\gamma_\mu D(\Lambda) = \sum \Lambda_{\mu\nu}\gamma_\nu$.

*Solution.* For infinitesimal transformations,

$$\Lambda \simeq 1 - \sum \omega_{\alpha\beta} X^{(\alpha\beta)}, \quad D(\Lambda) \simeq 1 - \frac{i}{2}\sum g_{\mu\mu}g_{\nu\nu}\omega_{\mu\nu}\sigma_{\mu\nu},$$

with (cf. (1.5.6))

$$X^{(\alpha\beta)}_{\mu\nu} = -(\delta_{\mu\alpha}g_{\nu\beta} - \delta_{\mu\beta}g_{\nu\alpha});$$

then use (3.2.11b) and (3.2.12).

**P.3.3.** Verify (3.6.11); e.g., that

$$\frac{1}{|\mathbf{r}|}\mathbf{r}\sigma\, \mathcal{Y}_M^{l+}(\theta,\varphi) = -\mathcal{Y}_M^{l+1,-}(\theta,\varphi).$$

*Solution.* Using the explicit values of the Clebsch–Gordan coefficients (Appendix A.1), we obtain

$$\mathcal{Y}_M^{l+}(\Omega) = \sqrt{\frac{l+M+1/2}{2l+1}}Y^l_{M-1/2}(\Omega)\chi\left(\frac{1}{2}\right)$$
$$+ \sqrt{\frac{l-M+1/2}{2l+1}}Y^l_{M+1/2}(\Omega)\chi\left(-\frac{1}{2}\right),$$

$$\mathcal{Y}_M^{l+1,-}(\Omega) = -\sqrt{\frac{l-M+3/2}{2l+3}}Y^l_{M-1/2}(\Omega)\chi\left(\frac{1}{2}\right)$$
$$+ \sqrt{\frac{l+M+3/2}{2l+3}}Y^{l+1}_{M+1/2}(\Omega)\chi\left(-\frac{1}{2}\right).$$

Because both sides of (3.6.11) transform in the same way, we may choose whatever reference system we please. Then take $\mathbf{r}$ along $OZ$, so that $\mathbf{r}\sigma/r = \sigma_3$. Here also $Y^l_\lambda(\theta = \varphi = 0) = \delta_{\lambda 0}Y^l_0(\theta = \varphi = 0)$, so we can restrict ourselves to the values $\pm 1/2$ for $M$. Then,

$$Y_0^{l+1}(\theta = \varphi = 0) = \sqrt{\frac{2l+3}{2l+M}}Y_0^l(\theta = \varphi = 0),$$

$$\sigma_3\chi\left(\pm\frac{1}{2}\right) = \pm\chi\left(\pm\frac{1}{2}\right),$$

and the verification that $\sigma_3\mathcal{Y}_{\pm 1/2}^{l+} = -\mathcal{Y}_{\pm 1/2}^{l+1,-}$ follows.

**P.3.4.** Verify (3.8.1).

*Solution.* Consider, e.g., $\overline{\Psi}\gamma_\mu\Psi$ and parity. We have

$$\overline{\Psi}(x)\gamma_\mu\Psi(x) \to \overline{\Psi}_P\gamma_\mu\Psi_P, \quad \Psi_P \equiv \eta_P\gamma_0\Psi(I_s x).$$

Therefore,

$$\overline{\Psi}(x)\gamma_\mu\Psi(x) \to \eta_P^*\eta_P\overline{\Psi}(I_sx)\gamma_0\gamma_\mu\gamma_0\Psi(I_sx)$$

$$= \sum_\nu I_{s\mu\nu}\overline{\Psi}(I_sx)\gamma_\nu\Psi(I_sx).$$

**P.3.5.** Evaluate the extra term in (3.4.11), that is to say, find the explicit expression for

$$\frac{\hbar e\delta}{4mc}\sigma \cdot F,$$

and its contribution to the magnetic moment.

*Solution.*

$$\frac{\hbar e\delta}{4mc}\sigma \cdot F = \frac{\hbar e\delta}{2mc}(i\alpha\mathcal{E} - \Sigma\mathcal{B})$$

$$= -\frac{\hbar e\delta}{2mc}\begin{pmatrix} \sigma\mathcal{B} & -i\sigma\mathcal{E} \\ -i\sigma\mathcal{E} & \sigma\mathcal{B} \end{pmatrix}.$$

Because the "ordinary" contribution to the magnetic moment produces a term

$$-\frac{\hbar e}{2mc}\sigma\mathcal{B},$$

it follows that our extra expression modifies this to

$$-\frac{\hbar e}{2mc}(1 + \delta)\sigma\mathcal{B},$$

plus electric terms (not contributing to a magnetic moment).

**P.3.6.** (*Gordon decomposition*) Show that, for arbitrary $p, \lambda$, with $p^2 = m^2$, one has the identity

$$\overline{u}(p_2, \lambda_2)\gamma_\mu u(p_1, \lambda_1)$$

$$= \frac{1}{2mc}\{\overline{u}(p_2, \lambda_2)(p_1 + p_2)_\mu u(p_1, \lambda_1)$$

$$-i\overline{u}(p_2, \lambda_2)\sum_\nu g_{\nu\nu}(p_1 - p_2)_\nu\sigma_{\mu\nu}u(p_1, \lambda_1)\}.$$

*Solution.* One has, from the Dirac equation,

$$\overline{u}(p_2, \lambda_2)\gamma_\mu u(p_1, \lambda_1) = \frac{1}{2mc}\overline{u}(p_2, \lambda_2)(\gamma_\mu\not{p}_1 + \not{p}_2\gamma_\mu)u(p_1, \lambda_1);$$

then write

$$\gamma_\mu\not{p} = 2p_\mu - \sum g_{\nu\nu}p_\nu\gamma_\nu\gamma_\mu.$$

# 4. Dirac Particle in a Potential

## 4.1 Dirac Particle in a Spherical Well

The study of a Dirac particle in a spherical well is similar to that of the nonrelativistic case, up to a few complications. First, we must work with two coupled equations. Second, we have to distinguish between scalar potentials and "vector" potentials, for which the potential is the fourth component of a Minkowski four-vector. We will study the first case only; the second may be found in the treatise of Greiner, Müller and Rafelski (1985). We then consider a scalar well (more precisely, a barrier) with potential

$$V_S(r) = \begin{cases} v_0, r > R, \\ 0, r < R. \end{cases} \tag{4.1.1}$$

This situation is important because, in the limit $R \to \infty$, it gives the first approximation to the bag model for bound states of quarks[1]. A proper well is studied in the text of Akhiezer and Berestetskii (1963). Substituting $V_S$ into (3.6.12) we find that

$$i\hbar c \left( \partial_r g_\kappa + \frac{1}{r} g_\kappa + \frac{\kappa}{r} g_\kappa \right) + (mc^2 + v_0) f_\kappa = E f_\kappa,$$

$$i\hbar c \left( \partial_r f_\kappa + \frac{1}{r} f_\kappa - \frac{\kappa}{r} f_\kappa \right) - (mc^2 + v_0) g_\kappa = E g_\kappa, \quad r > R, \tag{4.1.2}$$

and the free equations for $r < R$. Let us consider bound states, the only case of practical importance, so we take $E < mc^2 + v_0$. We will also assume that $E > 0$. For $r < R$ the solution is the free one, (3.6.13), (3.6.14):

$$g_{l+}(r) = \frac{-i\hbar c k}{mc^2 + E} f_{l+1,+}(r), \quad f_{l+}(r) = j_l(kr);$$

$$g_{l-}(r) = \frac{i\hbar c k}{mc^2 + E} f_{l-1,-}(r), \quad f_{l-}(r) = j_{l+1}(kr); \tag{4.1.3}$$

$$k = \frac{1}{\hbar c} \sqrt{E^2 - m^2 c^4}, \quad l = j - 1/2, \quad \kappa = \pm(j + 1/2).$$

---

[1] The text of Álvarez-Estrada et al. (1986) contains references and further details on this.

For $r > R$ we define

$$\overline{k} = \frac{1}{\hbar c}\sqrt{(mc^2 + v_0)^2 - E^2}.$$

Equations (4.1.2) are formally equal to the free ones if we replace $mc^2$ by $mc^2 + v_0$, but we have to impose the condition of decrease at infinity. Thus,

$$g_{l+}(r) = \frac{-i\hbar c\overline{k}}{mc^2 + v_0 + E} f_{l+1,+}(r),$$

$$g_{l-}(r) = \frac{i\hbar c\overline{k}}{mc^2 + v_0 + E} f_{l-1,-}(r),$$

$$(4.1.4a)$$

with

$$f_{l+}(r) = C\sqrt{\frac{\pi}{2r\overline{k}}} K_{l+1/2}(\overline{k}r),$$

$$f_{l-}(r) = C'\sqrt{\frac{\pi}{2r\overline{k}}} K_{l+3/2}(\overline{k}r);$$

$$(4.1.4b)$$

$K$ is the Bessel function of the second kind (Appendix A.2), given by

$$K_{n+1/2}(z) = \left(\frac{\pi}{2z}\right)^{1/2} z^{n+1} \left(-\frac{1}{z}\frac{d}{dz}\right)^n \frac{e^{-z}}{z}.$$

Explicit formulas for the $g_{l\pm}$ of (4.1.4a) can be found by using the differentiation properties for $k_n(z) \equiv (\pi/2z)^{1/2} K_{n+1/2}(z)$,

$$k_n' = \frac{n}{z} k_n - k_{n+1}.$$

Matching $f(r)$, $f'(r)$ at $r = R$ one finds the constants $C$, $C'$ and the energy values. For the simple case of the $S$ wave, $l = 0$, the quantization condition is

$$\tan kR = -k/\overline{k},$$

quite analogous to the nonrelativistic one. In the limit $v_0 \to \infty$, and if we neglect the mass of the particles (a case of practical interest for bound states of light quarks, $u, d$), the quantization condition becomes $k_n R = n\pi$, $n = 1, 2, \ldots$, so we get the energies

$$E_n = n\frac{\pi\hbar c}{R}.$$

The relation between $E_n$ and $n$ is linear, in good agreement with the experimental results. Requiring $E_1 = \frac{1}{2}m_\rho c^2$, with $\rho$ a particle made of two quarks with a mass some 1500 times the electron mass, we obtain $R \simeq 1.7$ fm, which is not bad for such a crude model (*Bogoliubov model*).

## 4.2 Particle in a Coulomb Potential: Continuum States

In this and the next two sections we consider a particle with charge $e_1$ in the Coulomb potential[2] created by a pointlike charge $e_2$, so that the potential is $V(r) = e_1 e_2/r$. We then define $\alpha_0$ by $e_1 e_2 = -\hbar c \alpha_0$. If the particle is an electron, and the potential that created by a nucleus (assumed pointlike) with $Z$ protons, then $\alpha_0 = Z\alpha$, with $\alpha$ the usual fine-structure constant,

$$\alpha = |e|^2/\hbar c \simeq 1/137.036;$$

$|e|$ is the charge of the proton.

If we thus write $V(r) = -\hbar c \alpha_0/r$, (3.6.12) now become

$$i\hbar c\left(\partial_r + \frac{l+2}{r}\right) g_{l+}(r) + \left(mc^2 - E - \frac{\hbar c \alpha_0}{r}\right) f_{l+}(r) = 0,$$

$$i\hbar c\left(\partial_r - \frac{l}{r}\right) f_{l+}(r) + \left(-mc^2 - E - \frac{\hbar c \alpha_0}{r}\right) g_{l+}(r) = 0; \tag{4.2.1a}$$

$$i\hbar c\left(\partial_r - \frac{l}{r}\right) g_{l-}(r) + \left(mc^2 - E - \frac{\hbar c \alpha_0}{r}\right) f_{l-}(r) = 0,$$

$$i\hbar c\left(\partial_r + \frac{l+2}{r}\right) f_{l-}(r) + \left(-mc^2 - E - \frac{\hbar c \alpha_0}{r}\right) g_{l-}(r) = 0. \tag{4.2.1b}$$

We will mainly consider explicitly the case $(+)$; the case $(-)$ is totally analogous and, in general, we will only indicate the results for it. The method of solution is fairly similar to that of the nonrelativistic case. We start by considering small values of $r$. If we take the ansatz

$$f_{l+}(r) \underset{r\to 0}{\simeq} C_+ r^{\lambda_+},$$

$$f_{l-}(r) \underset{r\to 0}{\simeq} C_- r^{\lambda_-}, \tag{4.2.2a}$$

then (4.2.1) tell us that

$$g_{l+}(r) \underset{r\to 0}{\simeq} i\frac{\lambda_+ - l}{\alpha_0} C_+ r^{\lambda_+}, \quad g_{l-}(r) \underset{r\to 0}{\simeq} i\frac{\lambda_- + l + 2}{\alpha_0} C_- r^{\lambda_-},$$

that is to say, the $g_{l\pm}$ behave as the $f_{l\pm}$:

$$g_{l\pm}(r) \underset{r\to 0}{\simeq} C'_\pm r^{\lambda_\pm}. \tag{4.2.2b}$$

Substituting (4.2.2) into (4.2.1) we obtain the consistency conditions

$$i\hbar c\left(i\frac{\lambda_+ - l}{\alpha_0}\right)(\lambda_+ + l + 2) - \hbar c\alpha_0 = 0,$$

$$i\hbar c\left(i\frac{\lambda_- + l + 2}{\alpha_0}\right)(\lambda_- - l) - \hbar c\alpha_0 = 0.$$

---

[2] The solution of the Dirac equation in a Coulomb potential is due to Gordon (1928).

each of these equations has two solutions. Of these we select

$$\lambda_+ = \lambda_- = -1 + \gamma, \ \ \gamma = \sqrt{(l+1)^2 - \alpha_0^2}. \tag{4.2.2c}$$

The other solutions produce nonintegrable wave functions for $l \neq 0$ and, for $l = 0$, wave functions that do *not* become free wave functions as $\alpha_0 \to 0$; see Problem 4.6.

For $l = 0$, the wave function is infinite at the origin, although the infinity is integrable and of little practical importance (see Problem 4.5). When $\alpha_0^2 > 1$, $\lambda_\pm$ become imaginary, which makes no physical sense. Luckily enough, for electrons and nuclei, this requires $Z \geq 137$, something that does not occur in nature.

Let us next consider the limit $r \to \infty$. The equations are now simply

$$i\hbar c \partial_r g_{l+}(r) \underset{r \to \infty}{\simeq} (E - mc^2) f_{l+}(r), \ \ i\hbar c \partial_r f_{l+}(r) \underset{r \to \infty}{\simeq} (E + mc^2) g_{l+}(r),$$

or, if we substitute the second into the first,

$$-\hbar^2 c^2 \partial_r^2 f_{l+}(r) \underset{r \to \infty}{\simeq} (E^2 - m^2 c^2) f_{l+}(r). \tag{4.2.3}$$

*For continuum states, $E > mc^2$*; if we define the wave number $k$ by

$$E^2 - m^2 c^4 = c^2 \hbar^2 k^2, \ \ k = \frac{1}{c\hbar} \sqrt{E^2 - m^2 c^4}, \tag{4.2.4}$$

we have the solutions

$$f_{l\pm}(r) \simeq e^{\pm ikr}, \ \ g_{l\pm}(r) \simeq e^{\pm ikr}, \tag{4.2.5}$$

with the signs $\pm$ *uncorrelated* here.

The behaviours as $r \to 0$, $r \to \infty$ suggest that, as in the nonrelativistic case, we define functions $F_{kl}$, $G_{kl}$ by

$$f_{l+}(r) \equiv e^{-ikr}(2kr)^{\lambda_+} F_{kl}^+(r),$$

$$\tag{4.2.6a}$$

$$g_{l+}(r) \equiv e^{-ikr}(2kr)^{\lambda_+} G_{kl}^+(r).$$

Changing the variables as well,

$$z = 2ikr, \ \ \partial_z \equiv \partial/\partial z, \tag{4.2.6b}$$

we can rewrite (4.2.1a) as

$$i\left(\partial_z - \frac{1}{2} + \frac{\gamma + (l+1)}{z}\right) G_{kl}^+(z)$$

$$+ \left(\frac{mc^2 - E}{2i\hbar ck} - \frac{\alpha_0}{z}\right) F_{kl}^+(z) = 0, \tag{4.2.7a}$$

$$i\left(\partial_z - \frac{1}{2} + \frac{\gamma - (l+1)}{z}\right) F_{kl}^+(z)$$

$$+ \left(-\frac{mc^2 + E}{2i\hbar ck} - \frac{\alpha_0}{z}\right) G_{kl}^+(z) = 0. \tag{4.2.7b}$$

Because we know that, in the NR limit $G$ is of the order of

$$\sqrt{(E - mc^2)/(E + mc^2)}$$

with respect to $F$, we take this into account and define

$$F_{kl}^+ = (E + mc^2)^{1/2}\overline{F}_{kl}^+, \quad G_{kl}^+ = (E - mc^2)^{1/2}\overline{G}_{kl}^+, \tag{4.2.8}$$

for which we have the equations

$$\left(\partial_z + \frac{\gamma + (l+1)}{z} - \frac{1}{2}\right)\overline{G}_{kl}^+(z)$$
$$+ \left(\frac{1}{2} + \frac{i\alpha_0}{c\hbar k z}(E + mc^2)\right)\overline{F}_{kl}^+(z) = 0,$$

$$\tag{4.2.9a}$$

$$\left(\partial_z + \frac{\gamma - (l+1)}{z} - \frac{1}{2}\right)\overline{F}_{kl}^+(z)$$
$$+ \left(\frac{1}{2} + \frac{i\alpha_0}{c\hbar k z}(E - mc^2)\right)\overline{G}_{kl}^+(z) = 0.$$

The form of (4.2.9a) suggests simpler equations for the half sum, $\Phi_s = \frac{1}{2}(\overline{F} + \overline{G})$, and the half difference, $\Phi_d = \frac{1}{2}(\overline{F} - \overline{G})$: indeed one has

$$\left(\partial_z + \frac{\xi}{z}\right)\Phi_s(z) - \frac{\varrho}{z}\Phi_d(z) = 0,$$

$$\left(\partial_z + \frac{\xi^*}{z} - 1\right)\Phi_d(z) - \frac{\varrho^*}{z}\Phi_s(z) = 0, \tag{4.2.9b}$$

$$\xi \equiv \gamma + \frac{iE\alpha_0}{c\hbar k}, \quad \varrho \equiv l + 1 - \frac{imc^2\alpha_0}{c\hbar k}.$$

From the first equation in (4.2.9b),

$$\Phi_d(z) = \frac{1}{\varrho}(z\partial_z + \xi)\Phi_s(z); \tag{4.2.10a}$$

substituting this into the second we obtain a second-degree equation for $\Phi_s$:

$$\left\{z\partial_z^2 + (1 + \xi + \xi^* - z)\partial_z - \xi + \frac{\xi\xi^* - \varrho\varrho^*}{z}\right\}\Phi_s(z) = 0. \tag{4.2.10b}$$

Now, $\xi\xi^* - \varrho\varrho^* = 0$, owing to condition (4.2.2c). This was to be expected: choosing $\gamma$ given by (4.2.2c), we are eliminating the most singular part in the behaviour at the origin. The equation for $\Phi_s$ then becomes

$$\Phi_s''(z) + \left(\frac{1 + 2\gamma}{z} - 1\right)\Phi_s'(z) - \frac{1}{z}\left(\gamma + \frac{iE\alpha_0}{c\hbar k}\right)\Phi_s(z) = 0. \tag{4.2.11}$$

This is the equation that defines Kummer's function $M$ (see Appendix A.2) so that

$$\Phi_s(r) = N_+ \, M\!\left(\gamma + \frac{iE\alpha_0}{c\hbar k}, 1 + 2\gamma, 2ikr\right), \qquad (4.2.12a)$$

and, using (4.2.10a) and the properties of $M$

$$\partial_z M(a,b,z) = \frac{a}{b} M(a+1, b+1, z),$$

$$zM(a+1, b+1, z) = b\{M(a+1, b, z) - M(a, b, z)\},$$

we also obtain

$$\Phi_d(r) = N_+ \frac{\gamma + iE\alpha_0/c\hbar k}{l + 1 - imc\alpha_0/\hbar k}$$

$$\times M\left(\gamma + 1 + \frac{iE\alpha_0}{c\hbar k}, \; 1 + 2\gamma, 2ikr\right). \qquad (4.2.12b)$$

Finally, we can undo the changes of variables, and functions, to give the explicit solutions (which we present simultaneously for $\omega = \pm$),

$$\left(\begin{array}{c} f_{l\omega}(r) \\ g_{l\omega}(r) \end{array}\right) = N_\omega e^{-ikr}(2kr)^{\gamma-1}$$

$$\times \left(\begin{array}{c} (E + mc^2)^{1/2}\left\{ M\left(\gamma + \dfrac{iE\alpha_0}{c\hbar k}, \; 1 + 2\gamma, 2ikr\right) \right. \\[2ex] (E - mc^2)^{1/2}\left\{ M\left(\gamma + \dfrac{iE\alpha_0}{c\hbar k}, \; 1 + 2\gamma, 2ikr\right) \right. \end{array}\right.$$

$$\left. + \frac{\gamma + iE\alpha_0/c\hbar k}{\omega(l+1) - imc\alpha_0/\hbar k} M\left(\gamma + 1 + \frac{iE\alpha_0}{c\hbar k}, \; 1 + 2\gamma, 2ikr\right)\right\} $$

$$\left. - \frac{\gamma + iE\alpha_0/c\hbar k}{\omega(l+1) - imc\alpha_0/\hbar k} M\left(\gamma + 1 + \frac{iE\alpha_0}{c\hbar k}, \; 1 + 2\gamma, 2ikr\right)\right\} \right) ;$$

$$\qquad (4.2.13)$$

$$\gamma = ((l+1)^2 - \alpha_0^2)^{1/2}.$$

The NR limit can be found in Problem P.4.3. In comparison with other texts, note that some authors exchange the names of $f$, $g$, others extract an $i$ explicitly from the latter, and finally other forms of writing (4.2.13) are used, notably some obtainable taking into account the relation

$$\frac{\gamma + iE\alpha_0/c\hbar k}{\omega(l+1) - imc\alpha_0/\hbar k} = \frac{\omega(l+1) + imc\alpha_0/\hbar k}{\gamma - iE\alpha_0/c\hbar k}. \qquad (4.2.14)$$

## 4.3 Scattering States. Phase Shifts. Cross-sections. Wave Function at the Origin

### 4.3.1 Scattering States. Phase Shifts

The asymptotic behaviour of the radial wave functions follows from (4.2.13) using the formula,

$$M(a,b,z) \underset{z\to\infty}{\simeq} \Gamma(b)\left\{\frac{e^{i\pi a}z^{-a}}{\Gamma(b-a)} + \frac{e^{z}z^{a-b}}{\Gamma(a)}\right\}, \quad -\frac{\pi}{2} < \arg z < \frac{3\pi}{2}.$$

We find that

$$\begin{pmatrix} f_{l+} \\ g_{l+} \end{pmatrix} \underset{r\to\infty}{\simeq} N_+ \frac{\Gamma(1+2\gamma)}{2kr}(E + (\pm)mc^2)^{1/2}$$

$$\times \frac{e^{-\pi E\alpha_0/2c\hbar k}}{|\Gamma(\gamma+1+iE\alpha_0/c\hbar k)|}\left\{\exp i\left[\frac{\pi}{2}\gamma - \frac{E\alpha_0\log 2kr}{c\hbar k} - kr - \overline{\delta}\right]\right.$$

$$\left. +(\pm)\frac{\gamma + iE\alpha_0/c\hbar k}{l+1-imc\alpha_0/\hbar k}\exp i\left[-\frac{\pi}{2}\gamma + \frac{E\alpha_0\log 2kr}{c\hbar k} + kr + \overline{\delta}\right]\right\},$$

where the sign $(+)$ in $(\pm)$ here goes with $f_{l+}$, and the sign $(-)$ with $g_{l+}$. Moreover, $\overline{\delta}$ is given by

$$\Gamma(\gamma+1+iE\alpha_0/c\hbar k) = e^{-i\overline{\delta}}|\Gamma(\gamma+1+iE\alpha_0/c\hbar k)|.$$

Next define

$$\phi(r) = kr - \frac{\pi}{2}\gamma + \frac{E\alpha_0}{c\hbar k}\log 2kr + \overline{\delta},$$

$$-e^{2i\eta_\omega} = \frac{\gamma + iE\alpha_0/c\hbar k}{\omega(l+1) - imc\alpha_0/\hbar k} = \frac{\omega(l+1) + imc\alpha_0/\hbar k}{\gamma - iE\alpha_0/c\hbar k};$$

that the latter is a phase follows from (4.2.14). In terms of this, and writing now the solution for both values of $\omega = \pm$, we get

$$\begin{pmatrix} f_{l\omega}(r) \\ g_{l\omega}(r) \end{pmatrix} \underset{r\to\infty}{\simeq} N_\omega\sqrt{E + mc^2}\,\frac{\Gamma(1+2\gamma)e^{-\pi E\alpha_0/2c\hbar k}e^{i\eta_\omega}}{i|\Gamma(\gamma+1+iE\alpha_0/c\hbar k)|}$$

$$\times \begin{pmatrix} \dfrac{\sin(\phi(r) + \eta_\omega)}{kr} \\[2ex] \dfrac{ihck}{E + mc^2}\dfrac{\cos(\phi(r) + \eta_\omega)}{kr} \end{pmatrix}. \tag{4.3.1}$$

On comparing this with the free solutions, (3.6.14)–(3.6.23), we can fix the constants $N_\omega$. For the normalization

$$\int_0^\infty dr\, r^2(f_{l\omega}^* f_{l\omega}' + g_{l\omega}^* g_{l\omega}') = 2E\frac{\pi}{2k^2}\delta(k - k'), \tag{4.3.2a}$$

we have

$$N_\omega = \frac{|\Gamma(\gamma + 1 + iE\alpha_0/c\hbar k)|e^{\pi E\alpha_0/2c\hbar k}}{|\Gamma(1 + 2\gamma)|}e^{i\beta}, \qquad (4.3.2b)$$

$\beta$ being an arbitrary phase. Moreover, if we again compare this with the free solutions, it follows that we can define the *relativistic Coulomb phase shifts* $\delta_{l\omega}(k)$ by

$$\phi(r) + \eta_+ = kr - \frac{\pi l}{2} + \frac{E\alpha_0}{c\hbar k}\log 2kr + \delta_{l+}(k), \qquad (4.3.3a)$$

$$\phi(r) + \eta_- = kr - \frac{\pi(l+1)}{2} + \frac{E\alpha_0}{c\hbar k}\log 2kr + \delta_{l-}(k), \qquad (4.3.3b)$$

with

$$e^{2i\delta_{l\omega}} = \omega e^{i\pi(l+1-\gamma)}\frac{\gamma + iE\alpha_0/c\hbar k}{\omega(l+1) - imc\alpha_0/\hbar k}\frac{\Gamma(\gamma + 1 - iE\alpha_0/c\hbar k)}{\Gamma(\gamma + 1 + iE\alpha_0/c\hbar k)}, \qquad (4.3.3c)$$

$$\gamma = ((l+1)^2 - \alpha_0^2)^{1/2}.$$

### 4.3.2 Cross-sections

To obtain the cross-sections for scattering by a Coulomb field we have to generalize the ordinary nonrelativistic description of scattering of spinless particles to relativistic spin 1/2–spin 0 (the source of the Coulomb field) scattering. This we will do in Sect. 7.7.2. Borrowing formulas from there, we use the Coulomb relativistic phase shifts $\delta_{l\omega}$ just found, (4.3.3), to define the amplitudes

$$F(\mathbf{k}_i, \lambda \rightarrow \mathbf{k}_f, \lambda)$$

$$= k^{-1}\sum_{l=0}^{\infty}\{(l+1)\sin\delta_{l+}e^{i\delta_{l+}} + l\sin\delta_{l-}e^{i\delta_{l-}}\}P_l(\cos\theta), \qquad (4.3.4a)$$

$$F(\mathbf{k}_i, \lambda \rightarrow \mathbf{k}_f, -\lambda)$$

$$= k^{-1}\sum_{l=0}^{\infty}\{\sin\delta_{l+}e^{i\delta_{l+}} - \sin\delta_{l-}e^{i\delta_{l-}}\}P_l^1(\cos\theta), \qquad (4.3.4b)$$

$$P_l^1(\cos\theta) = -\sin\theta\, dP_l(\cos\theta)/d\cos\theta,$$

for scattering of an electron with third component of spin $\lambda$, and wave vector (in the centre of mass system) $\mathbf{k}_i$ into a state with spin component $\pm\lambda$, and wave vector $\mathbf{k}_f$. The corresponding cross-sections are then

$$\frac{d\sigma(\lambda \rightarrow \lambda)}{d\Omega} = |F(\mathbf{k}_i, \lambda \rightarrow \mathbf{k}_f, \lambda|^2,$$

$$\frac{d\sigma(\lambda \rightarrow -\lambda)}{d\Omega} = |F(\mathbf{k}_i, \lambda \rightarrow \mathbf{k}_f, -\lambda)|^2. \qquad (4.3.4c)$$

More details on the spin complications for Coulomb scattering may be found in Akhiezer and Berestetskii (1963) or Rose (1961).

### 4.3.3 Wave Function at the Origin

A quantity that appears in many applications (for example, in evaluations of Coulomb corrections to $\beta$ decay) is the ratio (*Fermi factor*)

$$F = |\psi(\mathbf{r})|^2/|\psi^{(0)}(\mathbf{r})|^2, \ r \simeq 0,$$

where $\psi$ is the wave function of, for example, an electron in the Coulomb field of a nucleus, and $\psi^{(0)}$ is the wave function without Coulomb field (free particle). We need only consider $S$ waves, since all the others vanish relative to them for $r \to 0$. Then, using (4.2.13) with $N_\omega$ given by (4.3.2b), and (3.6.14)–(3.6.17) for the free waves, with $N_\omega$ given by (3.6.20), and for $\omega = +$, for example,

$$\begin{pmatrix} f_{0+}^{(0)}(r) \\ g_{0+}^{(0)}(r) \end{pmatrix} \underset{r \to 0}{\simeq} (E + mc^2)^{1/2} \begin{pmatrix} 1 \\ 0 \end{pmatrix},$$

$$\begin{pmatrix} f_{0+}(r) \\ g_{0+}(r) \end{pmatrix} \underset{r \to 0}{\simeq} \frac{|\Gamma(\gamma + 1 + iE\alpha_0/c\hbar k)|e^{\pi E\alpha_0/2c\hbar k}}{|\Gamma(1 + 2\gamma)|} (2kr)^{\gamma - 1}$$

$$\times (E + mc^2)^{1/2}$$

$$\times \begin{pmatrix} 1 + (\gamma + iE\alpha_0/c\hbar k)/(1 - imc\alpha_0/\hbar k) \\ \left(\dfrac{E - mc^2}{E + mc^2}\right)^{1/2} \left\{1 - \left(\gamma + \dfrac{iE\alpha_0}{c\hbar k}\right) \Big/ \left(1 - \dfrac{imc\alpha_0}{\hbar k}\right)\right\} \end{pmatrix},$$

with $\gamma = \gamma(l = 0) = \sqrt{1 - \alpha_0^2}$. With this, and after some work, we find the explicit expression

$$F = (|f_{0+}|^2 + |g_{0+}|^2)/(|f_{0+}^{(0)}|^2 + |g_{0+}^{(0)}|^2)$$

$$= 4\frac{E + mc^2\sqrt{1 - \alpha_0^2}}{E + mc^2}(2kr)^{2(\gamma - 1)}e^{\pi E\alpha_0/c\hbar k} \tag{4.3.5}$$

$$\times \frac{|\Gamma(\sqrt{1 - \alpha_0^2} + iE\alpha_0/c\hbar k)|^2}{|\Gamma(1 + 2\sqrt{1 - \alpha_0^2})|^2}.$$

In applications it is customary to simplify this formula by approximating

$$4(E + mc^2\sqrt{1 - \alpha_0^2})/(E + mc^2) \simeq 2(1 + \sqrt{1 - \alpha_0^2}).$$

In the NR limit we may use the relation

$$|\Gamma(1 + iy)| = \frac{2\pi y}{e^{\pi y} - e^{-\pi y}}$$

to find

$$F \underset{NR}{\simeq} \frac{2\pi\sigma/a_0 k}{1 - e^{-2\pi\sigma/a_0 k}}, \ \ \sigma \equiv \alpha_0/|\alpha_0|, \ a_0^{-1} \equiv mc|\alpha_0|/\hbar,$$

which of course agrees with the standard NR formula[3].

The expression (4.3.5) possesses the property of diverging for $r$ strictly zero, owing to the factor

$$(2kr)^{2(\sqrt{1-\alpha_0}-1)}.$$

For practical applications, one should remember that the source of the Coulomb field is not exactly pointlike, but in (for example) the case of $\beta$ decay, is of the size of a nucleus. If $\overline{R}$ is the mean nuclear radius, we should then make the replacement

$$(2kr)^{2(\sqrt{1-\alpha_0^2}-1)} \simeq (2k\overline{R})^{2(\sqrt{1-\alpha_0^2}-1)}.$$

The text of Blatt and Weisskopf (1952) may be consulted for detailed applications to $\beta$ decay.


## 4.4 Bound States in a Coulomb Potential

We now consider the bound states of a Dirac particle in a Coulomb potential. The analysis of the previous sections can then be repeated, with appropriate modifications. At large $r$,

$$f_{l\pm}(r) \underset{r\to\infty}{\simeq} C_1 e^{ikr} + C_2 e^{-ikr}, \ g_{l\pm}(r) \underset{r\to\infty}{\simeq} D_1 e^{ikr} + D_2 e^{-ikr},$$

$$(4.4.1)$$

$$k = \frac{1}{c\hbar}\sqrt{E^2 - m^2c^4}.$$

Since now $E < mc^2$, $k$ will be pure imaginary. We choose the determination

$$k = -i\overline{k}, \ \overline{k} \equiv -\frac{\sqrt{m^2c^4 - E^2}}{c\hbar},$$

$$(4.4.2)$$

and then the requirement of decrease of the wave functions at infinity tells us that one must have $C_2 = D_2 = 0$ in (4.4.1), so that

$$f_{l\pm} \underset{r\to\infty}{\simeq} C_1 e^{-\overline{k}r}, \ g_{l\pm}(r) = D_1 e^{-\overline{k}r}.$$

$$(4.4.3)$$

We may still use (4.2.6b), which we rewrite as $z = 2|k|r$: $z$ is now *real*. The analysis then proceeds as in Sect. 4.2, up to the point where one finds the function $\Phi_s$; now

$$\Phi_s(r) = \overline{N}_+ M(\gamma + \frac{iE\alpha_0}{c\hbar k}, \ 1 + 2\gamma, 2ikr)$$

$$(4.4.4)$$

$$= \overline{N}_+ M(\gamma - E\alpha_0/c\hbar\overline{k}, \ 1 + 2\gamma, 2\overline{k}r).$$

---

[3] Blatt and Weisskopf (1952); Ynduráin (1988).

However, here $M$ must be a polynomial; otherwise the behaviour (4.4.3) would be disrupted. The condition that $M$ in (4.4.4) be a polynomial[4] is that $\gamma - E\alpha_0/c\hbar\bar{k}$ = negative integer. This is then the quantization condition

$$\gamma - E\alpha_0/c\hbar\bar{k} = -n', \quad n' = 0, 1, 2, \ldots, \tag{4.4.5a}$$

or, if we solve for $E$, and select the *positive* square root,

$$E = E_{n'j} = mc^2 \left( 1 + \left( \frac{\alpha_0}{n' + \sqrt{(j+1/2)^2 - \alpha_0^2}} \right)^2 \right)^{1/2}. \tag{4.4.5b}$$

Defining the principal quantum number $n = n' + (j+1/2)$, and expanding in powers of $1/c$ (recall that, for a hydrogenlike atom, $\alpha_0 = Ze^2/\hbar c = 0(1/c)$), we obtain

$$E_{nj} \simeq mc^2 - \frac{\alpha_0^2}{2n^2} \left[ 1 + \frac{\alpha_0^2}{n} \left( \frac{1}{j+1/2} - \frac{3}{4n} \right) \right] mc^2 + 0(\alpha_0^6). \tag{4.4.6}$$

This agrees, to order $\alpha_0^2$, with the standard NR energy levels (up to the additive term $mc^2$), and the order $\alpha_0^4$ terms are the relativistic corrections[5], also called *fine structure*. They are in perfect agreement with experimental data, unlike what would have happened with those based upon the Klein–Gordon equation (cf. Sect. 2.5).

When $a = -n'$, the expression for $M(a, b, z)$ becomes

$$M(-n', b, z) = \Gamma(b) \sum_{n=0}^{n'} \frac{\Gamma(n'+1)(-z)^n}{n! \Gamma(n'-n-1)\Gamma(b+n)}$$
$$= \frac{\Gamma(b)\Gamma(n'+1)}{\Gamma(n'+b)} L_{n'}^{b-1}(z), \tag{4.4.7}$$

where the $L_N^\nu$ are the Laguerre polynomials. Thus we obtain the explicit expression for the bound-state wave functions,

$$\begin{pmatrix} f_{n'l\omega}(r) \\ g_{n'l\omega}(r) \end{pmatrix} = C_\omega e^{-\bar{k}r}(2\bar{k}r)^{\gamma-1}$$

$$\times \begin{pmatrix} L_{n'}^{2\gamma}(2\bar{k}r) - \dfrac{n'+2\gamma}{\omega(l+1) + mc\alpha_0/\hbar\bar{k}} L_{n'-1}^{2\gamma}(2\bar{k}r) \\[2ex] \dfrac{-ic\hbar\bar{k}}{mc^2+E} \left\{ L_{n'}^{2\gamma}(2\bar{k}r) + \dfrac{n'+2\gamma}{\omega(l+1) + mc\alpha_0/\hbar\bar{k}} L_{n'-1}^{2\gamma}(2\bar{k}r) \right\} \end{pmatrix}, \tag{4.4.8a}$$

and the constant $C_\omega$ will be found from the normalization condition,

---

[4] See Appendix A.2 for details on properties of the Kummer function.

[5] These corrections had in fact been obtained semi-phenomenologically before the exact solution to the Dirac equation was known. The history of these corrections may be found in the book of Rose (1961).

$$\int_0^\infty dr\, r^2 (f^*_{n'l\omega} f_{\overline{n}'l\omega} + g^*_{n'l\omega} g_{\overline{n}'l\omega}) = \delta_{n'\overline{n}'};$$
(4.4.8b)

this gives, if we take $C_\omega$ to be positive,

$$C_\omega = \sqrt{\frac{\Gamma(n'+1)}{\Gamma(n'+2\gamma+1)}}(2\overline{k})^{3/2}\sqrt{\frac{mc^2+E}{2mc^2}}\sqrt{\frac{mc\alpha_0+\hbar\overline{k}\omega(l+1)}{2mc\alpha_0}}.$$
(4.4.8c)

Before we verify (4.4.8c), a few words have to be said about the meaning of (4.4.8a) for $n' = 0$. In this case, in the combination of Kummer functions (4.2.13),

$$M(\gamma + iE\alpha_0/c\hbar k, 1+2\gamma, 2ikr)$$
$$+ \frac{\gamma + iE\alpha_0/c\hbar k}{\omega(l+1) - i\alpha_0 mc/\hbar k} M\left(\gamma + 1 + \frac{iE\alpha_0}{c\hbar k}, 1+2\gamma, 2ikr\right)$$
$$= M(-n', 1+2\gamma, 2\overline{k}r) + \frac{-n'}{\omega(l+1) + \alpha_0 mc/\hbar \overline{k}} M(1 - n', 1 + \gamma, 2\overline{k}r),$$

only the first $M$ can be replaced by a Laguerre polynomial; but this does not matter, because the coefficient of the second vanishes for $n' = 0$. We may therefore use (4.4.8a) for $n' = 0$ as well, provided we adopt the convention that

$$L^{2\gamma}_{-1}(z) \equiv 0.$$
(4.4.8d)

Let us now evaluate $C_\omega$. From (4.4.8b), and using (4.4.8a), we find that

$$1 = |C|^2 \frac{1}{(2\overline{k})^3} \int_0^\infty dx\, x^2 x^{2\gamma-2} e^{-x}$$

$$\times \left\{ \left[ L^{2\gamma}_{n'}(x) - \frac{n'+2\gamma}{\kappa + mc\alpha/\hbar\overline{k}} L^{2\gamma}_{n'-1}(x) \right]^2 \right.$$

$$\left. + \frac{\hbar^2 c^2 \overline{k}^2}{(mc^2+E)^2} \left[ L^{2\gamma}_{n'}(x) + \frac{n'+2\gamma}{\kappa + mc\alpha_0/\hbar\overline{k}} L^{2\gamma}_{n'-1}(x) \right]^2 \right\};$$

we have now used the definition $\omega(l+1) \equiv \kappa$. If we take into account the quantization condition, and the orthogonality properties of the Laguerre polynomials,

$$\int_0^\infty dx\, x^\nu e^{-x} L^\nu_N(x) L^\nu_{N'}(x) = \frac{\Gamma(N+\nu+1)}{\Gamma(N+1)} \delta_{NN'},$$

the normalization condition becomes

$$1 = |C|^2 \frac{2mc^2}{(mc^2+E)(2\overline{k})^3} \frac{\Gamma(n'+2\gamma+1)}{\Gamma(n'+1)} \frac{\nu}{(\kappa + mc\alpha_0/\hbar\overline{k})^2},$$

$$\nu = (\kappa + mc\alpha_0/\hbar\overline{k})^2 + n'(n'+2\gamma) =$$

$$= \frac{2mc\alpha_0}{\hbar\overline{k}}\left(\kappa + \frac{mc\alpha_0}{\hbar\overline{k}}\right),$$

so that, finally,

$$1 = |C|^2 \frac{2mc^2}{(mc^2 + E)(2\overline{k})^3} \cdot \frac{\Gamma(n' + 2\gamma + 1)}{\Gamma(n' + 1)} \cdot \frac{1}{\kappa + mc\alpha_0/\hbar\overline{k}} \cdot \frac{2\alpha_0 mc}{\hbar\overline{k}},$$

from which (4.4.8c) follows directly.

It is instructive to consider in some detail the NR limit of (4.4.8). Taking into account that, in this limit,

$$\gamma \simeq l + 1, \ \overline{k} \simeq 1/na_B, \ n \simeq n' + l + 1, \ mc\alpha_0/\hbar\overline{k} \simeq n,$$

with $a_B$ the Bohr radius, we find that

$$\begin{pmatrix} f_{n'l\omega}(r) \\ g_{n'l\omega}(r) \end{pmatrix} \underset{NR}{\simeq} C_\omega e^{-r/na_B}(2\overline{k}r)^l$$

$$\times \begin{pmatrix} L^{2l+2}_{n-l-1}(2r/na_B) - \frac{n+l+1}{n+\omega(l+1)} L^{2l+2}_{n-l-2}(2r/na_B) \\ 0 \end{pmatrix}.$$

For $\omega = +$ we use

$$L^\nu_N(x) = L^{\nu+1}_N(x) - L^{\nu+1}_{N-1}(x), \tag{4.4.9}$$

and the limit

$$C \simeq C_{NR} \equiv \frac{2}{n^2 a_B^{3/2}} \sqrt{\frac{(n - l - 1)!}{(n + l)!}},$$

to obtain

$$\begin{pmatrix} f_{n'l+}(r) \\ g_{n'l+}(r) \end{pmatrix} \underset{NR}{\simeq} \begin{pmatrix} R_{nl}(r) \\ 0 \end{pmatrix}; \tag{4.4.10a}$$

likewise, for $\omega = -$, again using (4.4.9) and

$$(N + \nu)L^{\nu-1}_N(x) = (N + 1)L^\nu_{N+1}(x) - (N + 1 - x)L^\nu_N(x),$$

we get

$$\begin{pmatrix} f_{n'l-}(r) \\ g_{n'l-}(r) \end{pmatrix} \underset{NR}{\simeq} \begin{pmatrix} R_{n,l+1}(r) \\ 0 \end{pmatrix}; \tag{4.4.10b}$$

here $R_{nl}$ are the standard radial $NR$ wave functions,

$$R_{nl}(r) = C_{NR}(2r/na_B)^l e^{-r/na_B} L^{2l+1}_{n-l-1}(2r/na_B),$$

$$C_{NR} = \frac{2}{n^2 a^{3/2}} \sqrt{\frac{(n - l - 1)!}{(n + l)!}} \tag{4.4.11}$$

# 4.5 Semirelativistic Approximation: the Foldy–Wouthuysen Transformation

A spin 1/2 particle has only two polarization states; therefore, of the four solutions of the Dirac equation (with or without a potential) only two are physical. The other two correspond to negative energies. It should in principle be possible to transform the Hamiltonian for a Dirac particle so that this separation into positive and negative states is diagonal, i.e., so that

$$H \to \overline{H} = \begin{pmatrix} H_1 & 0 \\ 0 & -H_2 \end{pmatrix},$$

with $H_i$ positive definite. This is the *Foldy–Wouthuysen* transformation. It is possible to give the exact form of the transformation only in a few cases: those in which the Dirac equation can be solved exactly. In general, we give it iteratively, diagonalizing $H$ to any desired order in $1/c$. This means that the method will provide a systematic way to evaluate the relativistic corrections to the NR Schrödinger equation.

## 4.5.1 General Method

We consider the Dirac Hamiltonian for a particle in a potential, the sum of a Minkowski vector $(\mathbf{A}, A_0)$, a Minkowski fourth component, $V_0$, and a scalar one, $\beta V_S$:

$$H = mc^2\beta + V + V_S\beta + c\boldsymbol{\alpha}\left(\mathbf{P} - \frac{e}{c}\mathbf{A}\right); \tag{4.5.1}$$

we have here defined $V \equiv A_0 + V_0$. This Hamiltonian is of the form

$$H = mc^2\left\{\beta + \frac{1}{c^2}D + \frac{1}{c}\Omega\right\}, \tag{4.5.2}$$

where the operator $D$ is "even", or "diagonal", i.e., it can be written as

$$D = \begin{pmatrix} d_1 & 0 \\ 0 & d_2 \end{pmatrix},$$

and $\Omega$ is "odd" or "antidiagonal",

$$\Omega = \begin{pmatrix} 0 & \omega_1 \\ \omega_2 & 0 \end{pmatrix},$$

where the $d_i$, $\omega_j$ are $2 \times 2$ matrices. Of course, $\beta$ and the unit operator are "diagonal".

Given an operator like (4.5.2), the sum of a diagonal and an antidiagonal one, one can find a unitary transformation, $e^T$ (with $T^+ = -T$), such that

$$H' = e^T H e^{-T}$$

is also of the form (4.5.2), but with the odd term of higher order in $1/c$; in our specific case (4.5.2) we will get

$$H' = mc^2 \left\{ \beta + \frac{1}{c^2}D' + \frac{1}{c^3}\Omega' \right\}.$$

Iterating these transformations, we can get the odd term to be of order $1/c^n$, with $n$ as large as we wish. Therefore, to a given order, the resulting Hamiltonian will be equivalent to one where positive and negative energies are separated. This is the Foldy–Wouthuysen method.

To implement this programme, we note that, for any operators $F$, $T$ we have the relation

$$e^T F e^{-T} = F + [T, F] + \frac{1}{2!}[T, [T, F]]$$

$$+ \frac{1}{3!}[T, [T, [T, F]]] + \ldots .$$

(4.5.3a)

This can be proved easily by induction by considering $F(\lambda) = e^{\lambda T} F e^{-\lambda T}$ and evaluating $\partial^n F(\lambda)/\partial \lambda^n$. On the other hand, for any odd operator $O$,

$$[\beta O, \beta] = -2O : \tag{4.5.3b}$$

we can apply a transformation like (4.5.3a) with $H$ replacing $F$, and profit from (4.5.3b) to choose $T$ proportional to $\beta O$ so that the dominating odd term of $H$ gets cancelled by a piece of $[T, H]$.

To be precise, let $h = (1/mc^2)H$,

$$h = \beta + \frac{1}{c^2}D + \frac{1}{c}\Omega,$$

$$D = \frac{1}{m}(V + V_S\beta), \quad \Omega = \frac{1}{m}\alpha(\mathbf{P} - \frac{e}{c}\mathbf{A});$$

(4.5.4)

we then apply to it the transformation characterized by[6]

$$T = \frac{1}{2c}\beta\Omega.$$

*All the transformations will have this structure.* Using (4.5.3a) we find the transformed operator,

$$h' \equiv e^T h e^{-T} = \beta + \frac{1}{c^2}D + \frac{1}{c}\Omega + \frac{1}{2c}[\beta\Omega, h]$$

$$+ \frac{1}{2!}\frac{1}{(2c)^2}[\beta\Omega, [\beta\Omega, h]] + \frac{1}{3!}\frac{1}{(2c)^3}[\beta\Omega, [\beta\Omega, [\beta\Omega, h]]]$$

(4.5.5)

$$+ \frac{1}{4!}\frac{1}{(2c)^4}[\beta\Omega, [\beta\Omega, [\beta\Omega, [\beta\Omega, h]]]] + O(c^{-5}).$$

We want to evaluate the first relativistic corrections. This means that we have to go to order $1/c^2$ in $H$, i.e., we have to be exact to order $1/c^4$ in $h$,

---

[6] It is easy to check that $(\beta O)^+ = -(O\beta)$ for any Hermitean, odd operator $O$. From this it follows that $\exp \beta\Omega/2C$ is unitary.

$h'$, $h''$ .... Now the commutator $[\beta\Omega, h]$ equals $[\beta\Omega, \beta]$ up to higher orders in $1/c$; because of (4.5.3), this term,

$$\frac{1}{2c}[\beta\Omega, \beta] = -\frac{1}{c}\Omega,$$

will cancel the term $(1/c)\Omega$ in $h$. Therefore, (4.5.5) gives us

$$h' = \beta + \frac{1}{c^2}D + \frac{1}{2c^2}[\beta\Omega, \Omega] + \frac{1}{2c^3}[\beta\Omega, D]$$

$$+\frac{1}{2!}\frac{1}{(2c)^2}[\beta\Omega, [\beta\Omega, h]] + \frac{1}{3!}\frac{1}{(2c)^3}[\beta\Omega, [\beta\Omega, [\beta\Omega, h]]]$$

$$+\frac{1}{4!}\frac{1}{(2c)^4}[\beta\Omega, [\beta\Omega, [\beta\Omega, [\beta\Omega, h]]]]$$

$$+O(c^{-5}).$$

(4.5.6)

We have even$\times$even = even = odd$\times$odd, odd$\times$even = odd, so, as announced, the odd term in (4.5.6) is $O(c^{-3})$. For the explicit calculation we can use the fact that, in our case,

$$\{\Omega, \beta\} = 0, \ [D, \beta] = 0,$$

to obtain, after straightforward manipulations, the relations

$$[\beta\Omega, \beta] = -2\Omega, [\beta\Omega, [\beta\Omega, \beta]] = -4\beta\Omega^2,$$

$$[\beta\Omega, [\beta\Omega, [\beta\Omega, \beta]]] = 8\Omega^3,$$

$$[\beta\Omega, [\beta\Omega[\beta\Omega[\beta\Omega, \beta]]]] = 16\beta\Omega^4,$$

$$[\beta\Omega, \Omega] = 2\beta\Omega^2, [\beta\Omega, [\beta\Omega, \Omega]] = -4\Omega^3,$$

$$[\beta\Omega, [\beta\Omega, [\beta\Omega, \Omega]]] = -8\beta\Omega^4,$$

$$[\beta\Omega, [\beta\Omega, D]] = -[\Omega, [\Omega, D]].$$

With these we get

$$h' = \beta + \frac{1}{2c^2}D + \frac{1}{2c^2}\beta\Omega^2 - \frac{1}{8c^4}\beta\Omega^4 - \frac{1}{4c^4}[\Omega, [\Omega, D]]$$

$$-\frac{1}{3c^3}\Omega^3 + \frac{1}{2c^3}\beta[\Omega, D] + O(c^{-5}),$$

that we write as

$$h' = \beta + \frac{1}{c^2}D' + \frac{1}{c^3}\Omega' + O(c^{-5}),$$

$$D' = D + \frac{1}{2}\beta\Omega^2 - \frac{1}{8c^2}\beta\Omega^4 - \frac{1}{8c^4}[\Omega, [\Omega, D]],$$

(4.5.7)

$$\Omega' = -\frac{1}{3}\Omega^3 + \frac{1}{2}\beta[\Omega, D].$$

Performing a new transformation,

$$h'' = e^{T'} h' e^{-T'},$$

$$T' = \frac{1}{2c^3}\beta\Omega',$$

we eliminate the term $(1/c^3)\Omega'$ in $h'$ and obtain

$$h'' = h' + \frac{1}{2c^3}[\beta\Omega, h'] + \frac{1}{(2c^3)^2}[\beta\Omega', [\beta\Omega', h']] + \dots$$

$$= \beta + \frac{1}{c^2}D' + 0(c^{-5}) :$$

we have found that, up to terms of order $c^{-3}$, the Hamiltonian $H$ in (4.5.1) is unitarily related to $H''$,

$$H'' = mc^2 h'' = mc^2\beta + mD' + 0(c^{-3}),$$

$$D' = D + \frac{1}{2}\beta\Omega^2 - \frac{1}{8c^2}\beta\Omega^4 - \frac{1}{8c^2}[\Omega, [\Omega, D]].$$

$$(4.5.8)$$

With the explicit expressions for $D$, $\Omega$, we have the values

$$\Omega^4 = \frac{1}{m^4}(\mathbf{P}^2)^2 + 0(1/c),$$

$$\Omega^2 = \frac{1}{m^2}\left[\alpha\left(\mathbf{P} - \frac{e}{c}\mathbf{A}\right)\right]^2$$

$$= \frac{1}{m^2}\left(\mathbf{P} - \frac{e}{c}\mathbf{A}\right)^2 - \frac{e}{m^2c^2}\sigma\mathcal{B},$$

and therefore we see that $H'' = H_{\mathrm{FW}} + 0(c^{-3})$, where the Foldy–Wouthuysen Hamiltonian is defined as

$$H_{\mathrm{FW}} = mc^2\beta + V + V_S\beta + \frac{1}{2m}\beta\left(\mathbf{P} - \frac{e}{c}\mathbf{A}\right)^2 - \frac{e}{2mc^2}\beta\sigma\mathcal{B}$$

$$- \frac{1}{8m^3c^2}\beta\mathbf{P}^4$$

$$(4.5.9)$$

$$- \frac{1}{8m^2c^2}[\alpha\mathbf{P}, [\alpha\mathbf{P}, V]] - \frac{1}{8m^2c^2}\beta\{\alpha\mathbf{P}, \{\alpha\mathbf{P}, V_S\}\}.$$

The positive energy solutions of the Dirac equation will be represented by wave functions $\Psi_{\mathrm{FW}}$,

$$\Psi_{\mathrm{FW}} = e^{T'} e^T \Psi + 0(c^{-3}),$$

with $\beta\Psi_{\mathrm{FW}} = \Psi_{\mathrm{FW}}$. In this physical subspace we can replace $\beta$ by unity and use two-component wave functions,

$$\Psi_{\mathrm{FW}} = \begin{pmatrix} \Psi_{\mathrm{s.r.}} \\ 0 \end{pmatrix},$$

where $\Psi_{\mathrm{s.r.}}$ is the semi-relativistic wave function, correct to order $c^{-2}$.

### 4.5.2 Electromagnetic Interactions

For the electromagnetic case, $V = e\phi$, $\mathcal{E} = -\nabla\phi$, $V_S = 0$. Then

$$[\boldsymbol{\alpha}\mathbf{P}, [\boldsymbol{\alpha}\mathbf{P}, e\phi]] = 2\hbar e\boldsymbol{\sigma}(\mathcal{E} \times \mathbf{P}) + \hbar^2 e\nabla\mathcal{E}$$

$$+ i\hbar e\boldsymbol{\sigma}(\nabla \times \mathcal{E}). \tag{4.5.10}$$

The corresponding expression for $H_{\mathrm{FW}}$ is particularly interesting in the case of a Coulomb potential,

$$\mathbf{A} = 0, V = -\alpha_0\hbar c/r.$$

We have

$$\nabla \times \mathcal{E} = 0, e\mathcal{E} = -\alpha_0\hbar c\frac{1}{r^2}\mathbf{r},$$

$$e\nabla\mathcal{E} = -\Delta V = +\alpha_0\hbar c\Delta\frac{1}{r} = -4\pi\alpha_0\hbar c\delta(\mathbf{r}),$$

so that

$$-\frac{1}{8m^2c^2}[\boldsymbol{\alpha}\mathbf{P}, [\boldsymbol{\alpha}\mathbf{P}, e\phi]] = \frac{-\hbar}{4m^2c^2}\cdot\frac{\partial V(r)/\partial r}{r}\boldsymbol{\sigma}\mathbf{L} + \frac{\hbar^2}{8m^2c^2}\Delta V(\mathbf{r})$$

$$= \frac{\hbar^2\alpha_0}{4m^2cr^3}\boldsymbol{\sigma}\mathbf{L} + \frac{\pi\hbar^3\alpha_0}{2m^2c}\delta(\mathbf{r}).$$

The resulting Hamiltonian is thus

$$H_{\mathrm{FW}} = H_{NR} - \frac{1}{8m^3c^2}\mathbf{P}^4 + \frac{\hbar^2\alpha_0}{4m^2cr^3}\boldsymbol{\sigma}\mathbf{L} + \frac{\pi\hbar^3\alpha_0}{2m^2c}\delta(\mathbf{r}),$$

$$H_{NR} = mc^2 + \frac{1}{2m}\mathbf{P}^2 - \frac{\hbar c\alpha_0}{r}; \tag{4.5.11}$$

$H_{NR}$ is, up to the constant $mc^2$, the standard nonrelativistic Coulombic Hamiltonian. The correction terms are the relativistic correction to the kinetic energy, and the so-called Thomas term and Darwin term[7]. Their contributions to the enegy levels are easily evaluated, since we can treat them as first-order perturbations. In this way we obtain

$$E_{nj} \simeq mc^2 - \frac{\alpha_0^2}{2n^2}\left[1 + \frac{\alpha_0^2}{n}\left(\frac{1}{j+1/2} - \frac{3}{4n}\right)\right]mc^2,$$

in expected agreement with (4.4.6).

---

[7] These are discussed in some detail in Galindo and Pascual (1978) or Gottfried (1966).

For *constant* electric and magnetic fields, $\mathcal{E}$, $\mathcal{B}$, it is convenient to choose the gauge such that

$$A_0(\mathbf{r}) = -\mathbf{r}\mathcal{E}, \quad \mathbf{A}(\mathbf{r}) = \frac{1}{2}\mathbf{r} \times \mathcal{B}. \tag{4.5.12a}$$

The *FW* Hamiltonian is then evaluated straightforwardly. In units such that $\hbar = c = 1$,

$$
\begin{aligned}
H_{\mathrm{FW}} &= m + \frac{1}{2m}(\mathbf{P} - e\mathbf{A})^2 - \frac{1}{8m^3}\mathbf{P}^4 \\[2mm]
&\quad -e\mathbf{r}\mathcal{E} - \frac{e}{2m}\boldsymbol{\sigma}\mathcal{B} + \frac{e}{4m^2}(\boldsymbol{\sigma} \times \mathbf{P})\mathcal{E}.
\end{aligned}
\tag{4.5.12b}
$$

### 4.5.3 Free Particle

For free particles the Foldy–Wouthuysen transformation can be carried over exactly. We have

$$H_{0\mathrm{FW}} = S_0 H_0 S_0^{-1}, \tag{4.5.13a}$$

with

$$S_0 = \frac{1}{\sqrt{2E(E + mc^2)}}\begin{pmatrix} mc^2 + E & c\mathbf{p}\boldsymbol{\sigma} \\ -c\mathbf{p}\boldsymbol{\sigma} & mc^2 + E \end{pmatrix}, \tag{4.5.13b}$$

and the not surprising expression for $H_{0\mathrm{FW}}$,

$$H_{0\mathrm{FW}} = (m^2 c^4 + c^2\mathbf{P}^2)^{1/2}\beta, \tag{4.5.13c}$$

which exhibits quite explicitly the positive and negative energy pieces of the Dirac Hamiltonian.

**Exercise.** Check (4.5.13). Check that $S_0$ is unitary ∎

## Problems

**P.4.1.** Solve the Dirac equation for a particle in a homogeneous magnetic field.

*Solution.* Choosing the $OZ$ axis along the field $\mathcal{B}$, we can write the vector potential as $A_z = A_x = 0$, $A_y = x\mathcal{B}$. The Dirac equation is thus, in units with $c = 1$,

$$(\boldsymbol{\alpha}(\mathbf{P} - e\mathbf{A}) + \beta m)\psi = E\psi.$$

A solution with well-defined $p_z$, $p_y$ components of the momentum has the form

$$\psi = \begin{pmatrix} \psi_b \\ \psi_s \end{pmatrix}, \quad \psi_b \sim Ce^{-\xi^{3/2}} H_n(\xi) e^{i(p_y y + p_z z)},$$

$$\xi = \sqrt{eB}(x - p_y/eB).$$

For the details, see Akhiezer and Berestetskii (1963).

**P.4.2.** Solve the Dirac equation for a particle in a plane electromagnetic wave.

*Solution.* This problem was solved first by Volkov; the details may be found in the text of Berestetskii, Lifshitz and Pitayevskii (1979).

**P.4.3.** Study the nonrelativistic limit of the solutions of the Dirac equation for $E > mc^2$, (4.2.13).

*Solution.* (i) $\omega = +1$. Clearly, $g_{l+} \to 0$ and

$$f_{l+} \simeq N_+ e^{-ikr} (2kr)^l (2mc^2)^{1/2} (l + 1 - i\sigma/a_0 k)^{-1}$$

$$\times \left\{ \left( l + 1 - \frac{i\sigma}{a_0 k} \right) M \left( l + 1 + \frac{i\sigma}{a_0 k}, \; 2l + 3, 2ikr \right) \right.$$

$$\left. + \left( l + 1 + \frac{i\sigma}{a_0 k} \right) M \left( l + 2 + \frac{i\sigma}{a_0 k}, \; 2l + 3, 2ikr \right) \right\},$$

where $\sigma = \text{sign } \alpha_0$ and $a_0$ is the Bohr radius. Using recursion formulas for the $M$ (Appendix A.2), and the value of $N_+$ (4.3.2b) in the NR limit,

$$f_{l+} \simeq N_+ (2mc^2)^{1/2} \frac{2l + 2}{l + 1 - i\sigma/a_0 k} e^{-ikr} (2kr)^l$$

$$\times M \left( l + 1 + \frac{i\sigma}{a_0 k}, \; 2l + 2, 2ikr \right),$$

$$N_+ \simeq \frac{|\Gamma(l + 2 - i\sigma/a_0 k)|}{\Gamma(2l + 3)} e^{\pi\sigma/2a_0 k} e^{i\beta},$$

we see that $f_{l+}$ coincides with $\sqrt{2mc^2}$ times the nonrelativistic function,

$$f_{El}^{NR} = \frac{|\Gamma(l + 2 - i\sigma/a_0 k)|}{\Gamma(2l + 2)} e^{i\delta_l} e^{\pi\sigma/2a_0 k} e^{-ikr} (2kr)^l$$

$$\times M \left( l + 1 + \frac{i\sigma}{a_0 k}, \; 2l + 2, 2ikr \right),$$

up to an (arbitrary) phase.

(ii) $\omega = -1$. Here also $g_{l-} \to 0$. Now,

$$f_{l-} \sim (\text{const}) e^{-ikr} (2kr)^{l+1} \frac{1}{2l + 3} M \left( l + 1 + \frac{i\sigma}{a_0 k}, \; 2l + 4, 2ikr \right),$$

and we end up with $f_{l-} \sim (2mc^2)^{1/2} f_{E,l+1}^{NR}$.

**P.4.4.** (*The Klein paradox*) Solve the Dirac equation in a step potential along $OZ$,

$$V(z) = \begin{cases} 0, & z < 0, \\ v_0, & z > 0. \end{cases}$$

Evaluate the current reflected and transmitted.

*Solution.* We consider $0 < E < v_0$. For $z < 0$, $\psi = \psi_{\text{in}} + \psi_{\text{r}}$,

$$\psi_{\text{in}} = a e^{ikz} \begin{pmatrix} 1 \\ 0 \\ \lambda \\ 0 \end{pmatrix},$$

$$\psi_{\text{r}} = b e^{-ikz} \begin{pmatrix} 1 \\ 0 \\ -\lambda \\ 0 \end{pmatrix} + b' e^{-ikz} \begin{pmatrix} 0 \\ 1 \\ 0 \\ \lambda \end{pmatrix};$$

$$\lambda = \frac{c\hbar k}{E + mc^2}, \quad k = \frac{\sqrt{E^2 - m^2 c^4}}{\hbar};$$

for $z > 0$, $\psi = \psi_{\text{trans}}$,

$$\psi_{\text{trans}} = d\, e^{ik_v z} \begin{pmatrix} 1 \\ 0 \\ \lambda_v \\ 0 \end{pmatrix} + d'\, e^{ik_v z} \begin{pmatrix} 0 \\ 1 \\ 0 \\ -\lambda_v \end{pmatrix},$$

$$\lambda_v = \frac{c\hbar k_v}{E - v_0 + mc^2}, \quad k_v = \frac{\sqrt{(E - v_0)^2 - m^2 c^4}}{\hbar c}.$$

The matching conditions give $a + b = d$, $b' = d' = 0$,

$$a - b = \frac{k_v}{k} \frac{E + mc^2}{E - v_0 + mc^2} d \equiv \rho d.$$

From this we have,

$$\frac{j_{\text{trans}}}{j_{\text{in}}} = \frac{4\rho}{(1 + \rho)^2}, \quad \frac{j_{\text{r}}}{j_{\text{in}}} = 1 - \frac{4\rho}{(1 + \rho)^2}.$$

When $v_0 > E + mc^2$, $\rho$ becomes *negative* and we find the absurd result (*the Klein paradox*) that $j_{\text{trans}} < 0$, $j_{\text{r}}/j_{\text{in}} > 1$.

**P.4.5.** Take into account the finite size of the nucleus when solving the relativistic hydrogenlike atom. (See Akhiezer and Berestetskii (1963); Blatt and Weisskopf (1952).)

**P.4.6.** Check that the continuum Coulombic wave functions $f_{l\omega}$, $g_{l\omega}$ given in (4.2.13) and (4.3.2b) do indeed tend to the free ones $f_{l\omega}^{(0)}$, $g_{l\omega}^{(0)}$ given in (3.6.14)–(3.6.20), when $\alpha_0 \to 0$.

*Solution.* The problem is less trivial than may appear at first sight. One has to use recursion formulas for the $M(a, b, z)$, the relation between these and Bessel functions,

$$M(n+1, 2n+2, 2iz) = \Gamma(n+3/2)e^{iz}\left(\frac{z}{2}\right)^{-n-1/2} J_{n+1/2}(z),$$

and the duplication formula for the $\Gamma$ function,

$$\Gamma(2z) = (2\pi)^{-1/2} 2^{2z-1/2} \Gamma(z)\Gamma(z+1/2).$$

**P.4.7.** Use *Kummer's transformation,*

$$M(a, b, z) = e^z M(b-a, b, -z),$$

to verify that, up to a common, $r$-independent phase, $f_{l\omega}$ is real and $g_{l\omega}$ is pure imaginary ($f$, $g$ being the continuum Coulombic functions, (4.2.13)).

# 5. Massive Particles with Spin 1. Massless Spin 1 Particle: Photon Wave Functions. Particles with Higher Spins (3/2, 2, ... )

## 5.1 Particle with Spin 1 and Mass $m \neq 0$

A nonrelativistic particle with spin 1 can be described by a three-component wave function, $\mathbf{V}(\mathbf{r}, t)$ or, in momentum space, $\mathbf{V}(\mathbf{p}, t)$. Under Lorentz transformations a three-vector will develop a fourth component; therefore, to describe a relativistic particle with spin 1 (and mass $m \neq 0$) we will need a four-vector, $V_\mu(x)$. This wave function has one component too many, so we will have to subject it to a supplementary condition. As we shall see in a moment, the one leading to a correct interpretation is that of (four-) transversality, $\partial \cdot V(x) = 0$. $V(x)$ will also have to verify the Klein–Gordon equation, so that we have, in natural units $\hbar = c = 1$,

$$(\partial \cdot \partial + m^2)V_\mu(x) = 0, \ \partial \cdot V_\mu = 0. \tag{5.1.1}$$

These equations are sometimes called the *Proca equations*[1]. If we are only interested in positive energies, we can convert the Klein–Gordon equation into a KGS equation, just as for scalar particles:

$$i\partial_\mu V_\mu(\mathbf{r}, t) = \sqrt{m^2 - \triangle} \ V_\mu(\mathbf{r}, t). \tag{5.1.2}$$

For the transformation properties under the Lorentz group we take

$$\Lambda \ : \ V_\mu(x) \rightarrow \sum_\nu \Lambda_{\mu\nu} V_\nu(\Lambda^{-1}x), \tag{5.1.3}$$

and (5.1.1) are manifestly invariant. For parity and time reversal,

$$\mathcal{P}V_\mu(x) = \eta_P \eta_\mu V_\mu(I_s x), \ \mathcal{T}V_\mu(x) = -\eta_T \eta_\mu V_\mu^*(I_s x), \tag{5.1.4}$$

with $\eta_0 = 1$, $\eta_i = -1$, and $\eta_P$, $\eta_T$ phases. Equation (5.1.4) is consistent with the nonrelativistic limit (Problem P.5.1). States with well-defined momentum,

$$\mathbf{P}V_\mu^{(\mathbf{P})}(x) = \mathbf{p}V_\mu^{(\mathbf{P})}(x),$$

are given by

$$V_\mu^{(\mathbf{P})}(x) = \frac{1}{(2\pi)^{3/2}}\epsilon_\mu(p)e^{-ip\cdot x}, \tag{5.1.5a}$$

---

[1] The name is also reserved for equations (5.1.11) below; they are, however, fully equivalent to (5.1.1).

and, because of (5.1.1), (5.1.2) we must have

$$p_0 = \sqrt{m^2 + \mathbf{p}^2},$$

$$(5.1.5b)$$

$$p \cdot \epsilon(p) = 0.$$

In the reference system in which the particle is at rest (or, likewise, in the NR limit $|\mathbf{p}| \gg p_0$) the transversality condition (5.1.5b) tells us that

$$\epsilon_0(\mathbf{p} \simeq 0) \simeq 0,$$

so that, in the NR limit, the wave function in fact becomes a three-vector since the zeroth component vanishes.

To fix the $\epsilon$ we require definite values for quantum numbers other than momentum; we then consider wave functions

$$V_\mu^{(p,\lambda,\mathbf{n})}(x)$$

with momentum $\mathbf{p}$ and spin component along $\mathbf{n}$ equal to $\lambda$. We have

$$V_\mu^{(p,\lambda,\mathbf{n})}(x) = \frac{1}{(2\pi)^{3/2}} \epsilon_\mu(p, \lambda, \mathbf{n}) e^{-ip \cdot x}, \quad \lambda = 0, \pm 1. \qquad (5.1.6a)$$

If $\mathbf{n} = \mathbf{z}$ (so that $\lambda$ is the three-component of spin), we have the explicit expressions

$$\epsilon_\mu(p, 0, \mathbf{z}) = L_{\mu 3}(p),$$

$$(5.1.6b)$$

$$\epsilon_\mu(p, \pm 1, \mathbf{z}) = \frac{1}{\sqrt{2}}(L_{\mu 1}(p) \pm i L_{\mu 2}(p)).$$

$L(p)$ is the Lorentz boost that accelerates from rest to momentum $p$. The verification of (5.1.6b) may be found in Problem P.5.2.

At times it is useful to use, rather than polarization vectors $\epsilon_\mu(p, \lambda, \mathbf{n})$ corresponding to given spin projection along $\mathbf{n}$, the *transverse* polarization vectors, $\epsilon_\mu(\mathbf{p}, T\alpha)$, $\alpha = 1, 2$, and the *longitudinal* polarization vector $\epsilon_\mu(\mathbf{p}, L)$. These correspond to spin directed along two axes perpendicular to $\mathbf{p}$ (the transverse ones) and to spin along $\mathbf{p}$ (the longitudinal). We have

$$\epsilon_0(\mathbf{p}, L) = |\mathbf{p}|/m, \quad \epsilon(\mathbf{p}, L) = p_0 \mathbf{p}/m|\mathbf{p}|, \qquad (5.1.7a)$$

and a possible choice for the transverse vectors is

$$\epsilon_0(\mathbf{p}, T\alpha) = 0,$$

$$(5.1.7b)$$

$$\epsilon(\mathbf{p}, T1) = \frac{\mathbf{p} \times \mathbf{u}}{|\mathbf{p}| \sin\theta}, \quad \epsilon(\mathbf{p}, T2) = \frac{\mathbf{p} \times (\mathbf{p} \times \mathbf{u})}{\mathbf{p}^2 \sin\theta}.$$

Here $\mathbf{u}$ is any unit vector *not* parallel to $\mathbf{p}$, and $\theta$ is the angle between $\mathbf{p}$ and $\mathbf{u}$.

The $\epsilon$ constructed in (5.1.6) and (5.1.7) are orthonormal in the sense that, if we denote the relevant quantum numbers by $a$, so that $a = \lambda$ for case (5.1.6), and $a = L, \alpha$ for case (5.1.7), we have

$$\epsilon^*(p, a) \cdot \epsilon(p, a') = -\delta_{aa'}. \tag{5.1.8a}$$

Moreover, in the case where $a$ is $\lambda$,

$$\sum_\lambda \epsilon_\mu(p, \lambda)\epsilon_\nu(p, \lambda)^* = -g_{\mu\nu} + p_\mu p_\nu/m^2. \tag{5.1.8b}$$

Equation (5.1.8a) is immediate. To prove (5.1.8b), we use (5.1.6b) to write

$$\sum_\nu \epsilon_\mu(p, \lambda)\epsilon_\nu^*(p, \lambda) = \sum_{j=1}^{3} L_{\mu j}(p)L_{\nu j}(p)$$

$$= -g_{\mu\nu} + L_{\mu 0}(p)L_{\nu 0}(p),$$

where we have used the fact that $LGL^T = G$. Defining $n_t, n_{t\mu} = \delta_{\mu 0}$, we then have

$$L_{\mu 0}(p) = (L(p)n_t)_\mu = (p/m)_\mu,$$

and (5.1.8b) follows.

The wave functions that we are using do not possess a smooth limit as $m \to 0$; this is obvious for (5.1.8b), which blows up in that limit. One can get smoother functions by replacing the $V_\mu(x)$ by other wave functions, $V_{\mu\nu}(x)$. These can be defined in terms of the $V_\mu$ by

$$V_{\mu\nu}(x) \equiv \partial_\mu V_\nu(x) - \partial_\nu V_\mu(x). \tag{5.1.9}$$

For particles with well-defined momentum and spin,

$$V_{\mu\nu}^{(p,\lambda,\mathbf{n})}(x) = \frac{1}{(2\pi)^{3/2}} e^{-ip\cdot x} f_{\mu\nu}(p, \lambda, \mathbf{n}), \tag{5.1.10a}$$

where, in terms of the $\epsilon$,

$$f_{\mu\nu}(p, \lambda, n) = -i\{p_\mu \epsilon_\nu(p, \lambda, \mathbf{n}) - p_\nu \epsilon_\mu(p, \lambda, \mathbf{n})\}. \tag{5.1.10b}$$

For $m \neq 0$ the $V_\mu$ and $V_{\mu\nu}$ are totally equivalent. For example, the inverse of (5.1.10b) is

$$\epsilon_\mu(p, \lambda, \mathbf{n}) = \frac{-i}{m^2} \sum g_{\nu\nu} p_\nu f_{\mu\nu}(p, \lambda, \mathbf{n}). \tag{5.1.10c}$$

The limit as $m \to 0$ is less singular for the $V_{\mu\nu}$, although it is certainly not well defined, since, as $m \to 0$, $V_\mu$ and $V_{\mu\nu}$ cease to be equivalent. The case $m = 0$ has to be studied on its own, something we will do in the two coming sections.

**Exercise.** Show that the equations satisfied by the $V_{\mu\nu}$ are

$$\sum \partial_\mu V_{\mu\nu}(x) = 0, \quad (\partial^2 + m^2)V_{\mu\nu}(x) = 0. \tag{5.1.11}$$

Verify that they are equivalent to (5.1.1) ∎

There exist in nature three elementary particles with spin 1, the $W^+$, $W^-$ and $Z$, with masses about one hundred times the proton mass. They are highly unstable, so the wave function formalism will be useful for them only in the extreme relativistic region. There, the concept of "particle in a potential" is not of much use, which is the reason we will not consider the interaction of these particles with a potential.

## 5.2 Particle with Spin 1 and Zero Mass: The Photon. Plane Waves. Photon Spin

It is at first sight straightforward to interpret the Maxwell equations as being the equivalent of the Schrödinger equation for photons. Nevertheless, this presents problems, one of which we will encounter now, and others that will be discussed in later chapters. We will, in spite of these problems, follow this path, since it gives indications for a fully satisfactory quantum treatment of photons.

The first problem appears when we try to decide whether the photon wave function should be identified with the electric and magnetic fields (in the covariant version, the tensor $F_{\mu\nu}$) or with the potentials $A_\mu$. *Directly observable* effects only depend on the $F_{\mu\nu}$; thus, the probability of finding a photon at position $\mathbf{r}$ at time $t$ should be proportional to $|F_{\mu\nu}(x)|^2$. The choice of $F_{\mu\nu}$, however, is untenable. Not only is there no bilinear in $F_{\mu\nu}$, $F_{\mu\nu}^*$ with appropriate relativistic transformation properties to be interpreted as a probability density[2], but the Akharonov–Bohm effect or the photon spectrum in bremsstrahlung radiation (Sect. 8.4) implies that the basic quantum mechanical objects are the $A_\mu$.

### 5.2.1 Photon Wave Function. Gauge Fixing. Transformation Properties

We will then consider the $A_\mu$ to represent the wave function of the photon, which immediately poses the question of gauge indeterminacy: physics should not change under the replacement

$$A_\mu(x) \to A_\mu(x) - \partial_\mu f(x). \tag{5.2.1}$$

We now have two alternatives: either to assume that a photon is not described by a single wave function, but by a class of functions given by (5.2.1), as we will do in Sect. 9.6, or to make $A_\mu$ unique by requiring sufficiently many subsidiary conditions (*fixing the gauge*). This will be the procedure we will follow now. It should be noted, however, that the quantity $|\mathbf{A}(\mathbf{r}, t)|^2$ can only in a loose sense be considered a probability density for photons; it does not

---

[2] More details about these questions may be found in the text of Akhiezer and Berestetskii (1963).

even have the right relativistic properties. As we will see in coming sections, the deep reason why no function has all the desired properties for a photon wave function is that it makes no sense to speak about isolated, localized photons. In spite of these problems we will continue with the discussion.

We will work in the so-called *Coulomb* or *radiation gauge*. Here, for the electromagnetic field in the absence of charges we require

$$A_0(x) \equiv 0. \tag{5.2.2a}$$

From the Maxwell equations it then follows that div $\mathbf{A}$ is independent of time, and we complete the gauge fixing with the condition

$$\boldsymbol{\nabla}\mathbf{A} = 0. \tag{5.2.2b}$$

Given an arbitrary $A_\mu(x)$ we can obtain a gauge-equivalent $A_\mu^C(x)$ satisfying (5.2.2) by a gauge transformation,

$$A_\mu^C(x) = A_\mu(x) - \partial_\mu f_C(x), \tag{5.2.3a}$$

where $f_C$ is constructed as follows. First define

$$f_C'(\mathbf{r},t) = \int_{t_0}^{t} dt'\, A_0(\mathbf{r},t'), \tag{5.2.3b}$$

$$A_\mu'(\mathbf{r},t) = A_\mu(\mathbf{r},t) - \partial_\mu f_C'(\mathbf{r},t);$$

$t_0$ is any fixed time (for example one can take $t_0 = 0$, or $t_0 = -\infty$). Equation (5.2.3b) mean that $A_0' \equiv 0$, and, by virtue of the Maxwell equations, div $\mathbf{A}(\mathbf{r},t)$ will be time independent. Then, we set

$$f_C''(\mathbf{r}) = \int \frac{d^3 r'}{4\pi|\mathbf{r}' - \mathbf{r}|}\mathrm{div}\ \mathbf{A}'(\mathbf{r}',t_0); \tag{5.2.3c}$$

with the equation

$$\triangle\frac{1}{r} = -4\pi\delta(\mathbf{r})$$

it follows that

$$\mathrm{div}\ f'' = -\mathrm{div}\ \mathbf{A}'.$$

Moreover, $f''$ is time independent. We set

$$f_C = f_C' + f_C'', \quad A_\mu^C = A_\mu' - \partial_\mu f_C'' \tag{5.2.3d}$$

and obtain $A_0^C = \mathrm{div}\ \mathbf{A}^C = 0$ .

The Maxwell equations, in this gauge, reduce to

$$\partial_0^2 - \triangle\mathbf{A} = 0, \quad \boldsymbol{\nabla}\mathbf{A} = 0 \tag{5.2.4a}$$

or, if we take the positive square root of the first, to

$$i\partial_t\mathbf{A}(x) = H_0\mathbf{A}(x), \quad H_0 = +\sqrt{-\triangle}, \tag{5.2.4b}$$

and, of course, $\boldsymbol{\nabla}\mathbf{A} = 0$.

In momentum space, by defining

$$\boldsymbol{\Psi}(\mathbf{p},t) \equiv \frac{1}{(2\pi)^{3/2}} \int d^3r \; e^{-i\mathbf{pr}} \mathbf{A}(\mathbf{r},t),$$

we have the equations

$$i\hbar\partial_t\boldsymbol{\Psi}(\mathbf{p},t) = c|\mathbf{p}|\boldsymbol{\Psi}(\mathbf{p},t), \; \mathbf{p}\boldsymbol{\Psi}(\mathbf{p},t) = 0. \tag{5.2.5}$$

We recognize $c|\mathbf{p}|$ as the energy of a massless particle with momentum $\mathbf{p}$.

Let us consider the transformation properties of the wave function. Under rotations, and working in momentum space, we note that $\boldsymbol{\Psi}$ should transform as a vector. Therefore, we write

$$R \; : \Psi_j(\mathbf{p},t) \to U(R)\Psi_j(\mathbf{p},t) = \sum_{j'} R_{jj'}\Psi_{j'}(\mathbf{p},t). \tag{5.2.6}$$

If $R = R(\boldsymbol{\alpha})$, where $\boldsymbol{\alpha}$ is infinitesimal, we use (1.1.1b) to find

$$U(R)\Psi_j(\mathbf{p},t) = \Psi_j(\mathbf{p},t) - \sum_l \alpha_l \sum_{j'} \epsilon_{jlj'}\Psi_{j'}(\mathbf{p},t)$$

$$- \frac{i}{\hbar} \sum_l \alpha_l L_l\Psi_j(\mathbf{p},t) + 0(\alpha^2),$$

where $\mathbf{L}$ is the momentum space orbital angular momentum operator,

$$\mathbf{L} = (i\hbar\boldsymbol{\nabla}_{\mathbf{p}}) \times \mathbf{p}.$$

Writing

$$U(R) \simeq 1 - \frac{i}{\hbar}\boldsymbol{\alpha}\mathbf{J}, \; \mathbf{J} = \mathbf{L} + \mathbf{S},$$

we find the spin operators for photons $S_a$, $a = 1,2,3$,

$$\mathbf{S}\Psi_j = \sum_{j'} \mathbf{S}_{jj'}\Psi_{j'},$$
$$\tag{5.2.7}$$
$$(S_a)_{jj'} = -i\hbar\epsilon_{ajj'}.$$

For general Lorentz transformations we encounter the difficulty that boosts mix a three-vector with a fourth component, so we have

$$(A) = \begin{pmatrix} 0 \\ \mathbf{A}(x) \end{pmatrix} \overset{L(\boldsymbol{\xi})}{\to} \begin{pmatrix} \dfrac{\sinh\xi}{\xi}\boldsymbol{\xi}\mathbf{A}(L^{-1}(\boldsymbol{\xi})x) \\ \mathbf{A}(L^{-1}(\boldsymbol{\xi})x) + \dfrac{\cosh\xi - 1}{\xi^2}(\boldsymbol{\xi}\mathbf{A}(L^{-1}(\boldsymbol{\xi})x)\boldsymbol{\xi} \end{pmatrix}$$

$$= \begin{pmatrix} A_0^\xi \\ \mathbf{A}^\xi \end{pmatrix}. \tag{5.2.8a}$$

Unlike for the case $m \neq 0$ of the previous section, this is now acceptable since we can use gauge transformations to redress the situation. Consistent transformation properties are obtained by postulating that a Lorentz boost *must include a gauge transformation.* So we *define*

$$A_\mu \xrightarrow{L} A_\mu^\xi - \partial_\mu f_C^\xi \equiv A_\mu^L, \tag{5.2.8b}$$

with $A_\mu^\xi$ that of (5.2.8a) and $f_C^\xi$ as constructed in (5.2.3), so that, after the boost, we again have

$$A_0^L = A_0^\xi - \partial_0 f_C^\xi = 0, \ \ \nabla \mathbf{A}^L = \mathrm{div}\ (\mathbf{A}^\xi + \nabla f_C^\xi) = 0, \tag{5.2.8c}$$

i.e., the gauge conditions are restored.

### 5.2.2 Plane Waves. Helicity States

Photon states with well-defined momentum are (with units $\hbar = c = 1$)

$$\boldsymbol{\Psi}^{(p)}(\mathbf{r}, t) = \frac{\boldsymbol{\epsilon}(p)}{(2\pi)^{3/2}} e^{-ip\cdot x}, \ \ \mathbf{p}\boldsymbol{\epsilon}(p) = 0, \tag{5.2.9}$$

where (5.2.4) tells us that $p_0 = |\mathbf{p}|$. The operator for spin along $\mathbf{n}$, $|\mathbf{n}| = 1$ is

$$S_\mathbf{n} = \mathbf{n}\mathbf{S}; \ \ S_\mathbf{n}\Psi_j = i \sum_{a,k=1}^{3} n_a \epsilon_{ajk}\Psi_k. \tag{5.2.10}$$

In particular, the helicity operator is $S_\mathbf{p} \equiv S_\mathbf{n}$ for $\mathbf{n} = \mathbf{p}/|\mathbf{p}|$. Plane waves with well-defined helicity are

$$\Psi^{(p,\eta)}(\mathbf{r}, t) = \boldsymbol{\epsilon}(p, \eta)e^{-ip\cdot x}/(2\pi)^{3/2}.$$

The *polarization vectors* $\boldsymbol{\epsilon}$ are found by requiring

$$S_\mathbf{p}\boldsymbol{\epsilon}(p, \eta) = \eta\boldsymbol{\epsilon}(p, \eta).$$

To find the explicit form of these $\boldsymbol{\epsilon}$ we start by working in the reference system in which $\mathbf{p}$ is directed along $OZ$. Then $S_\mathbf{p}$ coincides with $S_3$ and we have the equation

$$S_3\boldsymbol{\epsilon}^{(0)}(\eta) = \eta\boldsymbol{\epsilon}^{(0)}(\eta), \tag{5.2.11a}$$

$\boldsymbol{\epsilon}^{(0)}(\eta)$ being $\boldsymbol{\epsilon}(p, \eta)$ for $\mathbf{p}$ along $OZ$. Moreover, the condition $\mathbf{p}\boldsymbol{\epsilon} = 0$ implies that

$$\epsilon_3^{(0)}(\eta) = 0.$$

Equation (5.2.11a) gives the conditions

$$\epsilon_2^{(0)}(\eta) = i\eta\epsilon_1^{(0)}, \ \ \epsilon_1^{(0)} = -i\eta\epsilon_2^{(0)}, \tag{5.2.11b}$$

from which it follows that $\eta = \pm 1$, and the solution, unique up to a constant, is

$$\epsilon_1^{(0)}(\eta) = \frac{1}{\sqrt{2}}, \ \ \epsilon_2^{(0)}(\eta) = \frac{i\eta}{\sqrt{2}}, \ \ \epsilon_3^{(0)}(\eta) = 0. \tag{5.2.12a}$$

Note that $\eta = 0$ is impossible; (5.2.11b) would imply that $\epsilon_2^{(0)} = \epsilon_1^{(0)} = 0$, and hence $\epsilon \equiv 0$. This is a peculiarity of *massless* spin 1 particles, and one of the reasons why the limit $m \to 0$ cannot be totally smooth: we lose a degree of freedom on the way.

For $\mathbf{p}$ in any direction, let $R(\mathbf{z} \to \mathbf{p})$ be the rotation around the $\mathbf{z} \times \mathbf{p}$ axis which brings $\mathbf{z}$ over $\mathbf{p}$ ($\mathbf{z}$ is a unit vector along $OZ$). The rotation may be characterized as

$$R(\mathbf{z} \to \mathbf{p}) = R(\boldsymbol{\theta}), \ \boldsymbol{\theta} = (\mathbf{z} \times \mathbf{p})\frac{\theta}{p\sin\theta},$$

$$\cos\theta = p_3/|\mathbf{p}|.$$

Then, $\epsilon(p,\eta) = R(\mathbf{z} \to \mathbf{p})\epsilon^{(0)}$, so that, using (1.1.1), we get the explicit expression

$$\epsilon(p,\eta) = \frac{p_3}{|\mathbf{p}|}\epsilon^{(0)}(\eta) + \frac{i\eta p_1 - p_2}{\sqrt{2}|\mathbf{p}|(|\mathbf{p}| + p_3)}\mathbf{z} \times \mathbf{p} - \frac{p_1 + i\eta p_2}{\sqrt{2}|\mathbf{p}|}\mathbf{z}. \qquad (5.2.12b)$$

The phase conventions for $\epsilon^{(0)}$, $\epsilon(p,\eta)$ differ from those of Condon and Shortley (1951): one has the relation

$$\epsilon^{CS}(p,\eta) = -\eta\epsilon(p,\eta).$$

We have chosen the $\epsilon$ normalized so that

$$\epsilon^*(p,\eta)\epsilon(p,\eta') = \delta_{\eta\eta'}, \qquad (5.2.13a)$$

and, as is easily verified, by defining $\epsilon_0(p,\eta) \equiv 0$, we have

$$\sum_\eta \epsilon_\mu(p,\eta)\epsilon_\nu(p,\eta)^* = -g_{\mu\nu}$$

$$+\frac{1}{n_t \cdot p}(n_{t\mu}p_\nu + n_{t\nu}p_\mu) - \frac{p_\mu p_\nu}{(n_t \cdot p)^2}, \ n_{t\mu} = \delta_{\mu 0}. \qquad (5.2.13b)$$

The plane waves,

$$\boldsymbol{\Psi}^{(p,\eta)}(\mathbf{r},t) = \frac{1}{(2\pi)^{3/2}}\epsilon(p,\eta)e^{-ip\cdot x}, \qquad (5.2.14a)$$

are then normalized to

$$\int d^3r \, \boldsymbol{\Psi}^{(p,\eta)}(x)^*\boldsymbol{\Psi}^{(p',\eta')}(x) = \delta_{\eta\eta'}\delta(\mathbf{p} - \mathbf{p}'), \qquad (5.2.14b)$$

a nonrelativistic normalization.

### 5.2.3 Field Variables as Wave Functions for the Photon. The Schwinger Gauge

Because the observable quantities for the electromagnetic field are the field intensities, there is justification in using $F_{\mu\nu}$ as the photon wave functions, and likewise employing the $f_{\mu\nu}(k) = k_\mu\epsilon_\nu - k_\nu\epsilon_\mu$ in lieu of the polarization vectors. The corresponding formalism is developed in, for example, Akhiezer and Berestetskii (1963) or Weinberg (1964). Of course, one goes from the $A_\mu$ to the $F_{\mu\nu}$ by $F_{\mu\nu} = \partial_\mu A_\nu - \partial_\nu A_\mu$. The inverse may be constructed as follows. Choose the gauge-fixing condition to be

$$\sum g_{\mu\mu} x_\mu A_\mu^S(x) = 0 \tag{5.2.15}$$

*(Schwinger's gauge)*. From any $A$ one goes to an equivalent $A^S$ in this gauge by

$$A_\mu^S(x) = A_\mu(x) - \partial_\mu \left[ \int_0^1 ds\, x \cdot A(sx) \right]. \tag{5.2.16}$$

One then has the relation

$$A_\mu^S(x) = \int_0^1 ds\, s \sum g_{\nu\nu} x_\nu F_{\nu\mu}(sx). \tag{5.2.17}$$

The (straightforward) verification of these formulas is left to the reader. Note that, as opposed to the massive case, the relation between $F$ and $A$, (5.2.17) is now nonlocal: $A_\mu^S(x)$ depends on $F_{\nu\mu}(x')$ with $x' \neq x$.


## 5.3 Angular Momentum Eigenstates for the Photon. Vector Spherical Harmonics. Multipoles

### 5.3.1 General Useful Formulas

In this section we describe the construction of total angular momentum eigenstates for the photon, i.e., of vector functions that will be denoted by[3] $\mathbf{Y}_\mu^{(I)l}$, with $I$ an index to be specified later, such that, in units with $\hbar = 1$,

$$\mathbf{J}^2 \mathbf{Y}_\mu^{(I)l} = l(l+1)\mathbf{Y}_\mu^{(I)l}, \quad J_3 \mathbf{Y}_\mu^{(I)l} = \mu Y_\mu^{(I)l}; \tag{5.3.1}$$

here $\mathbf{J}$ is the total angular momentum operator, $\mathbf{J} = \mathbf{L} + \mathbf{S}$, with $\mathbf{L}$ the orbital angular momentum and $\mathbf{S}$ the spin operator. We recall that, in $x$ space,

$$\mathbf{L} = -i\hbar \mathbf{r} \times \boldsymbol{\nabla}_\mathbf{r}, \tag{5.3.2a}$$

and in $p$ space,

$$\mathbf{L} = -i\hbar \mathbf{p} \times \boldsymbol{\nabla}_\mathbf{p}. \tag{5.3.2b}$$

Moreover, for an arbitrary vector $\mathbf{v}$, the component $S_a$ of $\mathbf{S}$ acts by

$$(S_a \mathbf{v})_j = -i \sum \epsilon_{ajk} v_k. \tag{5.3.2c}$$

The construction of the $\mathbf{Y}$ is complicated by the gauge condition, which we write both in $x$ and $p$ space:

---

[3] In all of this section the index $\mu$ will denote the eigenvalues of $J_3$; it should not be mistaken for a Minkowski index.

$$\boldsymbol{\nabla}_\mathbf{r}\mathbf{Y}_\mu^{(I)l}(\mathbf{r}) = \mathbf{p}\mathbf{Y}_\mu^{(I)l}(\mathbf{p}) = 0. \tag{5.3.3}$$

For an arbitrary vector $\mathbf{v}$ we define its *spherical components* $v(\lambda)$, $\lambda = 0, \pm 1$ in terms of the Cartesian components $v_j$ by

$$v(0) = v_3, \;\; v(\pm 1) = \mp\frac{1}{\sqrt{2}}(v_1 \pm iv_2). \tag{5.3.4}$$

If we introduce the standard vectors $\boldsymbol{\chi}(\lambda)$ (with Condon–Shortley phase conventions)

$$\boldsymbol{\chi}(0) = \mathbf{n}_3, \;\; \boldsymbol{\chi}(\pm 1) = \mp\frac{1}{\sqrt{2}}(\mathbf{n}_1 \pm i\mathbf{n}_2), \tag{5.3.5}$$

where $\mathbf{n}_j$ is a unit vector along the axis $Oj$, then we have

$$\mathbf{v} = \sum_{\lambda=0,\pm 1} \boldsymbol{\chi}(\lambda)^* v(\lambda); \;\; \boldsymbol{\chi}(\lambda)^* = (-1)^\lambda \boldsymbol{\chi}(-\lambda). \tag{5.3.6}$$

Moreover the polarization vectors of Sect. 5.2 are simply related to the $\boldsymbol{\chi}$:

$$\boldsymbol{\chi}(\pm 1) = \mp\boldsymbol{\epsilon}^{(0)}(\pm 1). \tag{5.3.7}$$

Thus the $\boldsymbol{\chi}$ are eigenstates of $S_3$:

$$S_3\boldsymbol{\chi}(\lambda) = \lambda\boldsymbol{\chi}(\lambda). \tag{5.3.8}$$

Let us systematically denote, in this section, by $\hat{\mathbf{v}}$ the vector $\mathbf{v}/|\mathbf{v}|$. Then, for any arbitrary function $f(\mathbf{p})$ we have

$$J_a\hat{\mathbf{p}}f(\mathbf{p}) = \hat{\mathbf{p}}L_a f(\mathbf{p}), \tag{5.3.9a}$$

$$J_a\hat{\boldsymbol{\nabla}}_p f(\mathbf{p}) = \hat{\boldsymbol{\nabla}}_p L_a f(\mathbf{p}), \tag{5.3.9b}$$

$$J_a\mathbf{L}f(\mathbf{p}) = \mathbf{L}L_a f(\mathbf{p}), \tag{5.3.9c}$$

where

$$\hat{\boldsymbol{\nabla}}_p \equiv |\mathbf{p}|\boldsymbol{\nabla}_\mathbf{p}. \tag{5.3.9d}$$

**Exercise.** Show that, for any vector operator $\mathbf{F}$,

$$[L_a, F_k] = i\sum \epsilon_{akj}F_j = -(S_a\mathbf{F})_k \;\blacksquare$$

**Exercise.** Verify (5.3.9a–c).

*Hint.* Work in $\mathbf{p}$-space and use the relation just proved $\blacksquare$

Defining the spherical components $L(\lambda)$ of the orbital angular momentum operator as in (5.3.4), we have, for any (ordinary) spherical harmonic $Y_M^l$, and using (5.3.6),

$$\mathbf{L}Y_M^l = \sum_\lambda (-1)^\lambda \boldsymbol{\chi}(-\lambda) L(\lambda) Y_M^l$$

$$= M Y_M^l \boldsymbol{\chi}(0) - \sqrt{\frac{l(l+1) - M(M-1)}{2}} Y_{M-1}^l \boldsymbol{\chi}(+1)$$

$$+ \sqrt{\frac{l(l+1) - M(M+1)}{2}} Y_{M+1}^l \boldsymbol{\chi}(-1),$$

where we have used the standard formulas for the action[4] of the $L(\lambda)$ on the $Y_M^l$. We can thus write, identifying the Clebsch–Gordan coefficients,

$$\mathbf{L}Y_M^l = \sqrt{l(l+1)} \sum_\lambda (l, M - \lambda; 1, \lambda | l) Y_{M-\lambda}^l \boldsymbol{\chi}(\lambda). \tag{5.3.10}$$

With a little more effort we can also verify the relations

$$\hat{\mathbf{r}} Y_M^l(\mathbf{r}) = \frac{1}{\sqrt{2l+1}} \sum_\lambda \Big\{ -\sqrt{l+1}(l+1, M - \lambda; 1, \lambda | l) Y_{M-\lambda}^{l+1}(\mathbf{r})$$

$$+ \sqrt{l}(l-1, M - \lambda; 1, \lambda | l) Y_{M-\lambda}^{l-1}(\mathbf{r}) \Big\} \boldsymbol{\chi}(\lambda), \tag{5.3.11a}$$

$$\hat{\mathbf{p}} Y_M^l(\mathbf{p}) = \frac{1}{\sqrt{2l+1}} \sum_\lambda \Big\{ -\sqrt{l+1}(l+1, M - \lambda; 1, \lambda | l) Y_{M-\lambda}^{l+1}(\mathbf{p})$$

$$+ \sqrt{l}(l-1, M - \lambda; 1, \lambda | l) Y_{M-\lambda}^{l-1}(\mathbf{p}) \Big\} \boldsymbol{\chi}(\lambda), \tag{5.3.11b}$$

$$\hat{\boldsymbol{\nabla}}_r Y_M^l(\mathbf{r}) = \sqrt{\frac{l(l+1)}{2l+1}} \sum_\lambda \Big\{ \sqrt{l}(l+1, M - \lambda; 1, \lambda | l) Y_{M-\lambda}^{l+1}(\mathbf{r})$$

$$+ \sqrt{l+1}(l-1, M - \lambda; 1, \lambda | l) Y_{M-\lambda}^{l-1}(\mathbf{r}) \Big\} \boldsymbol{\chi}(\lambda), \tag{5.3.12a}$$

$$\hat{\boldsymbol{\nabla}}_p Y_M^l(\mathbf{p}) = \sqrt{\frac{l(l+1)}{2l+1}} \sum_\lambda \Big\{ \sqrt{l}(l+1, M - \lambda; 1, \lambda | l) Y_{M-\lambda}^{l+1}(\mathbf{p})$$

$$+ \sqrt{l+1}(l-1, M - \lambda; 1, \lambda | l) Y_{M-\lambda}^{l-1}(\mathbf{p}) \Big\} \boldsymbol{\chi}(\lambda). \tag{5.3.12b}$$

**Exercise.** Verify (5.3.11), (5.3.12).

*Hint.* For the first, for example, use the fact that

$$r(\lambda) = r\sqrt{4\pi/3}\, Y_\lambda^1(\mathbf{r}) \tag{5.3.13}$$

and the product formula for spherical harmonics, Appendix A.1 ∎

---

[4] See Appendix A.1 for the $Y_M^l$, and the Clebsch–Gordan coefficients. For the first, note in particular that owing to the symmetry between (5.3.2a) and (5.3.2b) we can choose them perfectly symmetric under $\mathbf{r} \leftrightarrow \mathbf{p}$. Also, we write $Y_M^l(\mathbf{r})$, $Y_M^l(\hat{\mathbf{r}})$ or $Y_M^l(\Omega_\mathbf{r})$ interchangeably.

Finally, we also leave as a simple exercise to verify that, if $f$ depends only on $\hat{\mathbf{p}} = \mathbf{p}/|\mathbf{p}|$, then $\hat{\boldsymbol{\nabla}}_p f$ is transverse:

$$\mathbf{p}\,\hat{\boldsymbol{\nabla}}_p f(\hat{\mathbf{p}}) = 0.$$

### 5.3.2 Multipoles

We are now in a position to find the desired total angular momentum eigenfunctions, called *multipoles*. From general considerations, it is obvious that the general solution of (5.3.1) is an arbitrary combination of three basic solutions:

$$\mathbf{Y}_\mu^{(I)l} = a\mathbf{V}_\mu^{(l)l} + b\mathbf{V}_\mu^{(l+1)l} + c\mathbf{V}_\mu^{(l-1)l},$$

where the *vector spherical harmonic* $\mathbf{V}^{(l_0)l}$ is obtained by combining orbital angular momentum $l_0 = l,\ l\pm 1$ with spin 1 to get total angular momentum $l$:

$$\mathbf{V}_\mu^{(l_0)l} = \sum (l_0, M - \lambda; 1, \lambda | l) Y_{M-\lambda}^{l_0} \boldsymbol{\chi}(\lambda). \tag{5.3.14}$$

We can further pin down the solutions by requiring *definite* parity, so we have either a state with $\mathbf{V}^{(l)l}$, or two orthogonal combinations of $\mathbf{V}^{(l-1)l}$ and $\mathbf{V}^{(l+1)l}$. Finally, of these we can pick a *transverse* and a *longitudinal* one. This can be done by solving the transversality conditions by brute force; but we shall profit from the work done in Subsection 5.3.1 to present the solutions directly. In $p$ space, we then define

$$\mathbf{Y}_\mu^{(e)l}(\mathbf{p}) \equiv \frac{1}{\sqrt{l(l+1)}}\hat{\boldsymbol{\nabla}}_p Y_\mu^l(\mathbf{p}), \tag{5.3.15a}$$

$$\mathbf{Y}_\mu^{(m)l}(\mathbf{p}) \equiv \frac{i}{\sqrt{l(l+1)}}\mathbf{L}Y_\mu^l(\mathbf{p}); \tag{5.3.15b}$$

these two are transverse,

$$\mathbf{p}\,\mathbf{Y}_\mu^{(e,m)l}(\mathbf{p}) = 0.$$

The multipoles $\mathbf{Y}_\mu^{(e,m)l}$ are called respectively *electric* and *magnetic* because it so happens that the corresponding photons couple like an electric, or respectively magnetic, field. The third combination is longitudinal, and hence *cannot* represent a photon state. Because it is nevertheless useful, we also present it here:

$$\mathbf{Y}_\mu^{(L)l}(\mathbf{p}) \equiv \hat{\mathbf{p}}Y_\mu^l(\mathbf{p}); \tag{5.3.16}$$

it is obviously parallel to $\mathbf{p}$. It is also elementary to verify that, letting $\mathcal{P}$ be the parity operator, and defining the intrinsic parity to be $\eta$ (for photons experiment gives $\eta = -1$),

$$\mathcal{P}\mathbf{Y}_\mu^{(e,L)l} = (-1)^{l+1}\eta\mathbf{Y}_\mu^{(e,L)l};$$

$$\mathcal{P}\mathbf{Y}_\mu^{(m)l} = (-1)^l\eta\mathbf{Y}_\mu^{(m)l}. \tag{5.3.17}$$

**Exercise.** Verify the alternative expression for (5.3.15b):

$$\mathbf{Y}_\mu^{(m)l}(\mathbf{p}) = \frac{1}{\sqrt{l(l+1)}}\hat{\mathbf{p}}\times\hat{\boldsymbol{\nabla}}_p Y_\mu^l(\mathbf{p})\ \blacksquare$$

The proof that the $\mathbf{Y}_\mu^{(I)l}$, $I = e, m, L$ do verify (5.3.1) is simple recalling (5.3.9)–(5.3.12). Indeed, take, for example, $\mathbf{Y}_\mu^{(e)l}$. We obtain, by applying (5.3.9b) repeatedly,

$$\mathbf{J}^2\mathbf{Y}_\mu^{(e)l} = \sum_a J_a J_a \frac{1}{\sqrt{l(l+1)}}\hat{\boldsymbol{\nabla}}_p Y_\mu^l(\mathbf{p})$$

$$= \frac{1}{\sqrt{l(l+1)}}\sum_a J_a\hat{\boldsymbol{\nabla}}_p L_a Y_\mu^l(\mathbf{p}) = \frac{1}{\sqrt{l(l+1)}}\sum_a \hat{\boldsymbol{\nabla}}_p L_a^2 Y_\mu^l$$

$$= \frac{1}{\sqrt{l(l+1)}}\hat{\boldsymbol{\nabla}}_p l(l+1)Y_\mu^l = l(l+1)\mathbf{Y}_\mu^{(e)l},$$

as was to be shown. To prove (5.3.1) for $I = m, L$, use (5.3.9a,c).

The explicit expression of the multipoles in terms of the vector spherical harmonics is found using the definitions and relations (5.3.10)–(5.3.12). From (5.3.10)

$$\mathbf{Y}_\mu^{(m)l} = i\sum_\lambda (l, M-\lambda; 1, \lambda|l)Y_{M-\lambda}^l\boldsymbol{\chi}(\lambda); \tag{5.3.18a}$$

likewise, (5.3.11), (5.3.12) give

$$\begin{aligned}
\mathbf{Y}_\mu^{(e)l} &= \frac{1}{\sqrt{2l+1}}\sum_\lambda \Big\{\sqrt{l}(l+1, \mu-\lambda; 1, \lambda|l)Y_{\mu-\lambda}^{l+1}\\
&\quad + \sqrt{l+1}(l-1, \mu-\lambda; 1, \lambda|l)Y_{\mu-\lambda}^{l-1}\Big\}\boldsymbol{\chi}(\lambda),
\end{aligned} \tag{5.3.18b}$$

$$\begin{aligned}
\mathbf{Y}_\mu^{(L)l} &= \frac{1}{\sqrt{2l+1}}\sum_\lambda \Big\{-\sqrt{l+1}(l+1, \mu-\lambda; 1, \lambda|l)Y_{\mu-\lambda}^{l+1}\\
&\quad + \sqrt{l}(l-1, \mu-\lambda; 1, \lambda|l)Y_{\mu-\lambda}^{l-1}\Big\}\boldsymbol{\chi}(\lambda).
\end{aligned} \tag{5.3.18c}$$

Using these formulas it is also straightforward to check the orthonormality conditions:

$$\int d\Omega\, \mathbf{Y}_\mu^{(I)l*}\mathbf{Y}_{\mu'}^{(I')l'} = \delta_{II'}\delta_{ll'}\delta_{\mu\mu'}, I = e, m, L. \tag{5.3.19}$$

**Exercise.** Check the normalization directly. Use the fact that $\hat{\nabla}_r = 0$, $\hat{\nabla}_\theta = \partial_\theta$, $\hat{\nabla}_\varphi = (1/\sin\theta)\partial_\varphi$, and $\hat{\triangle} = (1/\sin^2\theta)\partial_\varphi^2 + (1/\sin\theta)\partial_\theta(\sin\theta\partial_\theta)$ is the angular part of the Laplacian, so that

$$\hat{\triangle}Y_M^l = -l(l+1)Y_M^l.$$

*Solution.* For example, with $n = \sqrt{ll'(l+1)(l'+1)}$

$$\int d\Omega_p \mathbf{Y}_\mu^{(e)l}(\mathbf{p})^* \mathbf{Y}_{\mu'}^{(e)l'}(\mathbf{p}) = \frac{1}{\sqrt{ll'(l+1)(l'+1)}}$$

$$\times \int d\Omega_p \left(\hat{\boldsymbol{\nabla}}_p Y_\mu^l(\mathbf{p})\right)^* \left(\hat{\boldsymbol{\nabla}}_p Y_{\mu'}^{l'}(\mathbf{p})\right)$$

$$= \frac{1}{n} \int d\Omega \left\{ \left(\partial_\theta Y_\mu^{l*}\right)\left(\partial_\theta Y_{\mu'}^{l'}\right) + \frac{1}{\sin^2\theta}\left(\partial_\varphi Y_\mu^{l*}\right)\left(\partial_\varphi Y_{\mu'}^{l'}\right) \right\}$$

$$= \frac{-1}{n} \int d\Omega Y_\mu^{l*}\hat{\triangle}Y_{\mu'}^{l'} = \frac{+l'(l'+1)}{n} \int d\Omega Y_\mu^{l*}Y_{\mu'}^{l'}$$

$$= \delta_{ll'}\delta_{\mu\mu'} \quad \blacksquare$$

### 5.3.3 Photon Wave Functions with Well-Defined Angular Momentum

We can determine the wave functions for photons with well-defined energy, parity, total angular momentum, and third component thereof. We thus require ($\hbar$ is still taken to be unity, but $c$ is made explicit)

$$i\partial_t \mathbf{X}_\mu^{(I,\omega,l)} = H_0 \mathbf{X}^{(I,\omega,l)} = c\omega \mathbf{X}^{(I,\omega,l)}, \tag{5.3.20a}$$

$$\mathbf{J}^2 \mathbf{X}_\mu^{(I,\omega,l)} = l(l+1)\mathbf{X}_\mu^{(I,\omega,l)}; \quad J_3 \mathbf{X}_\mu^{(I,\omega,l)} = \mu X_\mu^{(I,\omega,l)}, \tag{5.3.20b}$$

$I = e, m$. Moreover, the $\mathbf{X}$ are required to be transverse, and

$$\mathcal{P}\mathbf{X}_\mu^{(I,\omega,l)} = \eta(-1)^{l+\delta_I}\mathbf{X}_\mu^{(I,\omega,l)},$$

$$\tag{5.3.20c}$$

$$\delta_e = 1, \ \delta_m = 0.$$

The time dependence is trivial, and will not be written explicitly; one has simply

$$\mathbf{X}_{\mu(\text{time})}^{(I,\omega,l)} = e^{-i\omega t}\mathbf{X}_{\mu(\text{without time})}^{(I,\omega,l)}. \tag{5.3.21}$$

Because, in $p$ space, the free photon Hamiltonian is just $H_0 = c|\mathbf{p}|$, (5.3.20a) tells us that the $\mathbf{X}$ are proportional to $\delta(|\mathbf{p}| - \omega)$; while (5.3.20b, c) imply that the proportionality factors are the multipoles. Thus, we have the $p$ space wave functions

$$\mathbf{X}_\mu^{(e,\omega,l)}(\mathbf{p}) = \mathbf{Y}_\mu^{(e)l}(\mathbf{p})\delta(|\mathbf{p}| - \omega),$$

$$\mathbf{X}_\mu^{(m,\omega,l)}(\mathbf{p}) = \mathbf{Y}_\mu^{(m)l}(\mathbf{p})\delta(|\mathbf{p}| - \omega); \tag{5.3.22a}$$

$$\mathbf{X}_\mu^{(L,\omega,l)}(\mathbf{p}) = \mathbf{Y}_\mu^{(L)l}(\mathbf{p})\delta(|\mathbf{p}| - \omega).$$

$$\int d^3p\, \mathbf{X}^{(I,\omega,l)}(\mathbf{p})^* \mathbf{X}^{(I',\omega',l')}(\mathbf{p})$$

$$= \omega^2 \delta(\omega - \omega')\delta_{II'}\delta_{ll'}\delta_{\mu\mu'}. \tag{5.3.22b}$$

In $x$ space,

$$\mathbf{X}_\mu^{(I,\omega,l)}(\mathbf{r}) = \frac{1}{(2\pi)^{3/2}} \int d^3p\, e^{i\mathbf{pr}} \mathbf{X}_\mu^{(I,\omega,l)}(\mathbf{p})$$

$$= \frac{4\pi}{(2\pi)^{3/2}} \sum_{l_0,M} i^{l_0}\omega^2 j_{l_0}(\omega r)Y_M^{l_0}(\mathbf{r})$$

$$\times \int d\Omega_\mathbf{p} Y_M^{l_0}(\mathbf{p})^* \mathbf{Y}_M^{(I)}(\mathbf{p});\, |\mathbf{p}| = \omega,$$

with $j_l(z)$ the standard spherical Bessel functions. Substituting here expressions (5.3.18) for the multipoles, we get the explicit formulas

$$\mathbf{X}_\mu^{(e,\omega,l)}(\mathbf{r}) = \sqrt{\frac{2}{\pi}}i^{l+1}\omega^2 \left\{ \sqrt{\frac{l}{2l+1}}j_{l+1}(\omega r)\mathbf{V}_\mu^{(l+1)l}(\mathbf{r}) \right.$$

$$\left. - \sqrt{\frac{l+1}{2l+1}}j_{l-1}(\omega r)\mathbf{V}_\mu^{(l-1)l}(\mathbf{r}) \right\}, \tag{5.3.23a}$$

$$\mathbf{X}_\mu^{(m,\omega,l)}(\mathbf{r}) = \sqrt{\frac{2}{\pi}}i^{l+1}\omega^2 j_l(\omega r)\mathbf{V}_\mu^{(l)l}(\mathbf{r}), \tag{5.3.23b}$$

and, giving for completeness also the longitudinal case,

$$\mathbf{X}_\mu^{(L,\omega,l)}(\mathbf{r}) = -\sqrt{\frac{2}{\pi}}i^{l+1}\omega^2 \left\{ \sqrt{\frac{l+1}{2l+1}}j_{l+1}(\omega r)\mathbf{V}_\mu^{(l+1)l}(\mathbf{r}) \right.$$

$$\left. + \sqrt{\frac{l}{2l+1}}j_{l-1}(\omega r)\mathbf{V}_\mu^{(l-1)l}(\mathbf{r}) \right\}. \tag{5.3.23c}$$

We recall that the $\mathbf{V}^{(l_0)l}$ are defined in (5.3.14), where $\chi$ is the photon spin wave function (cf. (5.3.5)–(5.3.8)).

To finish this section we evaluate the matrix elements for the change from plane waves to radial photon wave functions. The plane wave function was defined in (5.2.9a) to be (omitting the time dependence, the same in both cases)

$$\boldsymbol{\Psi}^{(\mathbf{p},\eta)}(\mathbf{r}) = \frac{1}{(2\pi)^{3/2}}\boldsymbol{\epsilon}(\mathbf{p},\eta)e^{i\mathbf{p}\mathbf{r}},$$

so that

$$\langle\mathbf{p},\eta|I,\omega,l,\mu\rangle = \frac{1}{(2\pi)^{3/2}}\int d^3r\, e^{-i\mathbf{p}\mathbf{r}}$$

$$\times\boldsymbol{\epsilon}^*(\mathbf{p},\eta)\mathbf{X}_\mu^{(I,\omega,l)}(\mathbf{r})$$

$$= \boldsymbol{\epsilon}^*(\mathbf{p},\eta)\mathbf{X}_\mu^{(I,\omega,l)}(\mathbf{p})$$

$$= \delta(|\mathbf{p}| - \omega)\boldsymbol{\epsilon}\,^*(\mathbf{p},\eta)\mathbf{Y}_\mu^{(I)l}(\mathbf{p}),$$

from which an explicit solution follows by using (5.3.18).

## 5.4 Particles with Higher Spins. Rarita–Schwinger and Bargmann–Wigner Equations. The Graviton

Elementary particles with spins 3/2 and higher have been not found in nature. The graviton, with spin 2 and being massless, is supposed to be the quantum of the gravitational field, but only indirect evidence exists for it. Some theoretical speculations have involved spin 3/2 particles. Because of these facts, we will only present a brief description of relativistic wave functions for higher spin particles. Actually we will present two different, but equivalent, formalisms for particles with mass (the Rarita–Schwinger and Bargmann–Wigner ones) and then say a few words about the graviton.

### 5.4.1 Rarita–Schwinger Equations

We consider here[5] a wave function $\Psi_{\mu a}(x)$ with Minkowski index $\mu$ and Dirac index $a$. The evolution equation will be

$$i\sum_b \gamma_{ab}\cdot\partial\Psi_{\mu b}(x) - m\Psi_{\mu a}(x) = 0, \tag{5.4.1a}$$

and we require either of the subsidiary conditions

$$\sum_\mu g_{\mu\mu}\partial_\mu\Psi_{\mu a}(x) = 0, \quad \sum_\mu g_{\mu\mu}\gamma_\mu\Psi_\mu(x) = 0; \tag{5.4.1b}$$

these are equivalent for functions that satisfy (5.4.1a).

The spin operator along $\mathbf{n}$, $S_{\mathbf{n}}$ is such that, upon acting on (say) the space components, it gives

---

[5] Rarita and Schwinger (1941).

$$S_{\mathbf{n}} \;:\; \Psi_{ja}(x) \to \frac{-i}{2} \sum_{bj'k} (\mathbf{n}\Sigma)_{ab} n_k \epsilon_{kjj'} \Psi_{j'b'}(x).$$

With this it is not difficult to verify that the particles described by $\Psi_{\mu a}$ are spin 3/2 particles.

One can generalize this to describe particles of spin $n + 1/2$. Denoting by the symbol $\{\mu_1, \ldots, \mu_n\}$ the symmetrization of the indices $\mu_1, \ldots, \mu_n$, we see that the function

$$\Psi_{\{\mu_1,\ldots,\mu_n\}a}(x),$$

will describe a spin $n + 1/2$ particle if it satisfies the evolution equation and subsidiary conditions

$$(i\not{\partial} - m)\Psi_{\{\mu_1,\ldots,\mu_n\}a} = 0,$$

$$\sum g_{\mu_j\mu_j} \partial_{\mu_j} \Psi_{\{\mu_1,\ldots,\mu_n\}a} = 0, \tag{5.4.2}$$

$$\sum g_{\mu_j\mu_j} \gamma_{\mu_j} \Psi_{\{\mu_1,\ldots,\mu_n\}a} = 0, \quad j = 1, \ldots, n.$$

The limit $m = 0$ is not trivial for the particles of spin $s > 1$; a phenomenon like that described for photons (loss of degrees of freedom) also occurs.

### 5.4.2 Bargmann–Wigner Equations

We will describe particles of spin $n/2$. For this we now consider the wave function

$$\Psi_{\{a_1,\ldots,a_n\}}(x),$$

the $a_i$ being Dirac indices.

We then require the equations[6]

$$i \sum (\gamma_{a_j b_j} \cdot \partial)\Psi_{\{a_1,\ldots,b_j,\ldots,a_n\}}(x)$$

$$-m\Psi_{\{a_1,\ldots a_j,\ldots a_n\}}(x) = 0, \quad j = 1, \ldots, n. \tag{5.4.3}$$

The verification that the spin is actually $n/2$ is left to the reader.

The Rarita–Schwinger and the Bargmann–Wigner wave functions are equivalent, for the same value of the spin. In fact, for $m \neq 0$, any pair of wave functions $\tilde{\Psi}_\alpha(x)$ and $\Psi_a(x)$ corresponding to the same spin are equivalent in the following sense: there exist polynomials $P(\partial)$, $Q(\partial)$ in the derivatives such that

$$\tilde{\Psi}_\alpha(x) = \sum_a P_{\alpha a}(\partial)\Psi_a(x),$$

$$\Psi_a(x) = \sum_a Q_{a\alpha}(\partial)\tilde{\Psi}_\alpha(x).$$

---

[6] Bargmann and Wigner (1948)

The superficially surprising fact that $P$ and $Q$, each the inverse of the other, are both polynomials is due to the Klein–Gordon equation: since $\Psi$, $\tilde{\Psi}$ must both satisfy it, it follows that we can replace, *for nonzero mass*, $(\partial,\partial)^{-1}$ by $-m^{-2}$. The general proof of this equivalence theorem may be found in Ynduráin (1971); one case, the $V_\mu$ and $V_{\mu\nu}$ for spin 1, was worked out explicitly in the present text in Sect. 5.1.

### 5.4.3 The Graviton

A reasonably detailed treatment of spin 2 massless particles (*gravitons*) may be found in Weinberg (1964). Here only a brief description is presented.

The wave function for gravitons will be given by a symmetric tensor $g_{\mu\nu}(x) \equiv g_{\{\mu\nu\}}(x)$. Of the ten components only two will be independent, corresponding to the only two possible values of the helicity, $\pm 2$ (see Sect. 6.4). We can choose a gauge, called at times the *Newton gauge* defined by

$$g_{\mu\nu}(x) = 0, \ \mu \neq \nu;$$

$$g_{00}(x) = 0, \sum_i \frac{\partial}{\partial r_i} g_{ij}(x) = 0, \ j = 1, 2, 3.$$

(5.4.4)

The evolution equation is to be obtained from the Klein–Gordon equation,

$$\partial \cdot \partial g_{\mu\nu}(x) = 0,$$

i.e.,

$$i\partial_t g_{\mu\nu}(x) = \sqrt{-\triangle} \, g_{\mu\nu}(x).$$

A Lorentz transformation will entail a gauge shift to restore (5.4.4) in the new reference system.

A helicity operator may be constructed as for photons; with it one checks the assertion that $\eta = \pm 2$. The values $\pm 1, 0$ are not possible. This situation is quite general. If, for a massless particle, we define the spin $s$ as $s = \max|\eta|$, where $\eta$ is the helicity, then only the values $\eta = \pm s$ are allowed (sometimes only one of these). The proof will be presented in Sect. 6.4.

## Problems

**P.5.1.** Evaluate the action of $\mathcal{P}, \mathcal{T}$ for a nonrelativistic spin 1 particle. Compare with (5.1.4).

**P.5.2.** Check (5.1.6b). Prove that for $\mathbf{n} = \mathbf{p}/|\mathbf{p}| \equiv \hat{\mathbf{p}}$ one has

$$\epsilon_\mu(p, 0, \hat{\mathbf{p}}) = H_{\mu 3}(\mathbf{p}),$$

$$\epsilon_\mu(p, \pm 1, \hat{\mathbf{p}}) = \frac{1}{\sqrt{2}} \left( (H_{\mu 1}(\mathbf{p}) \pm i H_{\mu 2}(\mathbf{p})) \right).$$

Here $H(\mathbf{p}) = R(\mathbf{z} \to \mathbf{p})L(p^z)$ with $p_0^z = p_0 = p_3$, $p_1^z = p_2^z = 0$, and the rotation $R(\mathbf{z} \to \mathbf{p})$ around the axis $\mathbf{z} \times \mathbf{p}$ carries $OZ$ over $\mathbf{p}$. Check that both in (5.1.6b) and now one has

$$p \cdot \epsilon(p, \lambda, n) = 0.$$

*Hint.* Remember that $p = L(p)\overline{p}$.

**P.5.3.** Verify that the pure boost $L(\overline{k} \to k^z)$ such that $L(\overline{k} \to k^z)\overline{k} = k^z$, where $\overline{k}_0 = \overline{k}_3 = 1$, $\overline{k}_1 = \overline{k}_2 = 0$; $k_0^z = k_3^z = k_0$, $k_1^z = k_2^z = 0$, corresponds to values of the parameters $\boldsymbol{\xi}$, $L(\overline{k} \to k^z) = L(\boldsymbol{\xi})$, such that $\xi_1 = \xi_2 = 0$, $\xi_z = \xi$; $k_0 = e^\xi$. If $R(\mathbf{z} \to \mathbf{k})$ is the rotation that carries $OZ$ over $\mathbf{k}$, and $H(k) = R(\mathbf{z} \to \mathbf{k})L(\overline{k} \to k^z)$, where $k$ is an arbitrary lightlike vector, $k \sim (k_0, \mathbf{k})$, check that

$$H(k)\overline{k} = k,$$

and, if we define

$$H(k)n_t \equiv \epsilon(k, 0); \;\; H(k)n_z \equiv \epsilon(k, 3),$$

with $n_{t\mu} = \delta_{\mu 0}$, $n_{z\mu} = \delta_{\mu 3}$, then they verify the relations

$$\epsilon_0(k, 0) = \frac{k_0^2 + 1}{2k_0}, \;\; \boldsymbol{\epsilon}(k, 0) = \frac{k_0^2 - 1}{2k_0}\mathbf{k};$$

$$\epsilon_0(k, 3) = \frac{k_0^2 - 1}{2k_0}, \;\; \boldsymbol{\epsilon}(k, 3) = \frac{k_0^2 + 1}{2k_0^2}\mathbf{k}.$$

# 6. General Description of Relativistic States

## 6.1 Preliminaries

In spite of the successes of the Dirac equation and of its usefulness in the construction of relativistic quantum fields (to be discussed later, in Sect. 8), there is little doubt that the wave function formalism for relativistic particles is not quite satisfactory. First of all, the meaning of the variables $\mathbf{r}$ and $t$ in a wave function $\Psi(\mathbf{r}, t)$ is unclear; as we will show, $\mathbf{r}$ does not represent the position for a Dirac particle, and in fact a position operator does not even exist, *strictu senso*, for a photon. As for $t$, the interpretation of it as the time becomes less clear when we have several particles: which time? The proper time of each of the particles? Time as measured in the centre of mass reference system?

For these, and other reasons, it is convenient to introduce an abstract characterization of relativistic states, freeing it from the problems encountered in explicit realizations. We will thus describe the states by "safe" observables: momentum $\mathbf{p}$ and another one that we label $\zeta$ and that will be related to a spin component: our task will then be to construct the states, $|\mathbf{p}, \zeta\rangle$, and study their transformation properties under relativistic transformations. This we will do from the next section onwards; in what remains of the present section we will introduce some standard theorems on group representations, without proofs, and, at the end, describe the group of relativistic transformations, the Poincaré group.

**Definition.** *Given a group $G$ with elements $g_1, g_2, \ldots$ we say that the set of operators $U(g)$ acting on a Hilbert space form a* representation *of $G$ if we have*

$$U(g_1 g_2) = U(g_1)U(g_2), \quad U(g^{-1}) = U^{-1}(g).$$

*The representation is said to be* unitary *if the operators $U$ are unitary.*

**Definition.** *We say that the (unitary) representation is* reducible *if all the $U(g)$ may be written as*

$$U(g) = \begin{pmatrix} U_1(g) & 0 \\ 0 & U_2(g) \end{pmatrix}.$$

*If such a decomposition is not possible we say that the representation is* irreducible.

**Lemma (Schur's Lemma).** *If $U(g)$ is irreducible, and the operator $A$ commutes with all the $U(g)$, then $A$ is a multiple of the unit operator, $A = \alpha \cdot 1$.*

The invariance group of relativity is the Poincaré group, also called the inhomogeneous Lorentz group. Its elements are pairs $(a, \Lambda)$ with $a$ a four-translation consisting of a spatial translation by $\mathbf{a}$, and a time translation by $a_0/c$; and a (proper, orthochronous) Lorentz transformation, $\Lambda$. We recall the multiplication and inversion formulas:

$$(a, \Lambda)(a', \Lambda') = (a + \Lambda a', \Lambda\Lambda'),$$

$$(a, \Lambda)^{-1} = (-\Lambda^{-1}a, \Lambda^{-1}). \tag{6.1.1}$$

The generators of the Poincaré group may be described as generators of rotations, boosts and translations. Let us consider any representation, $U(a, \Lambda)$ of the Poincaré group; then, for infinitesimal transformations we write

$$U(0, R(\boldsymbol{\theta})) \simeq 1 - \frac{i}{\hbar}\boldsymbol{\theta}\mathbf{L},$$

$$U(0, L(\boldsymbol{\xi})) \simeq 1 - \frac{i}{\hbar}\boldsymbol{\xi}\mathbf{N}, \tag{6.1.2}$$

$$U(a, 1) \simeq 1 + \frac{i}{\hbar}a \cdot P.$$

The commutation relations may be evaluated in any (faithful) representation; indeed, since these respect product and inverse rules, commutators will also be respected. We may then choose the representation of the $U$ as acting on scalar functions (cf. Sect. 2.2) so that we can take

$$L_j = i\hbar \sum \epsilon_{jkl}x_k\partial_l,$$

$$N_j = i\hbar(x_0\partial_j - x_j\partial_0),$$

$$P_j = i\hbar\partial_j, \quad P_0 = i\hbar\partial_0$$

and evaluate the commutators with these explicit expressions. That way we find the relations, valid in any representation,

$$[L_k, L_j] = i\hbar \sum \epsilon_{kjl} L_l,$$

$$[L_k, N_j] = i\hbar \sum \epsilon_{kjl} N_l,$$

$$[L_k, P_j] = i\hbar \sum \epsilon_{kjl} P_l;$$

$$[L_k, P_0] = 0, \quad [P_\mu, P_\nu] = 0; \tag{6.1.3}$$

$$[N_k, N_j] = -i\hbar \sum \epsilon_{kjl} L_l,$$

$$[N_k, P_j] = -i\hbar \delta_{kj} P_0,$$

$$[N_k, P_0] = -i\hbar P_k.$$

We may also write them in covariant form. If, as in (2.2.8), we let

$$U(\Lambda) \simeq 1 - \frac{i}{\hbar} \sum_{\mu\mu} g_{\mu\mu} g_{\nu\nu} \omega_{\mu\nu} M_{\mu\nu}, \tag{6.1.4}$$

then a simple calculation, making use of the fact that

$$[\partial_\mu, x_\nu] = g_{\mu\nu}$$

allows us to write the commutation relations in the form

$$[M_{\mu\nu}, P_\alpha] = \quad i\hbar(g_{\nu\alpha} P_\mu - g_{\mu\alpha} P_\nu),$$

$$[M_{\mu\nu}, M_{\alpha\beta}] = \ i\hbar(g_{\mu\alpha} M_{\beta\nu} + g_{\mu\beta} M_{\nu\alpha}$$

$$+ g_{\nu\alpha} M_{\mu\beta} + g_{\nu\beta} M_{\alpha\mu}), \tag{6.1.5}$$

$$[P_\mu, P_\nu] = 0.$$

Consider now a quantum system represented by the state $|\Psi\rangle$. A Poincaré transformation $g$ will carry it over a new state, $|\Psi_g\rangle$. According to the rules of quantum mechanics, we expect that this will be implemented by a linear unitary operator,

$$U(g) = U(a, \Lambda):$$

$$|\Psi_g\rangle = U(a, \Lambda)|\Psi\rangle. \tag{6.1.6}$$

We will require that this be a representation of the Poincaré group. Actually, this is asking for too much; in principle, one could have, more generally, a representation up to a phase:

$$U(a, \Lambda) U(a', \Lambda') = e^{i\varphi} U(a + \Lambda a', \Lambda\Lambda').$$

In the following sections we will give an *explicit* construction with $\varphi = 0$; the proof that the result is general is fairly complicated and will not be given here.

We will then consider unitary representations of the Poincaré group. Since a *reducible* representation can be decomposed into orthogonal *irreducible* ones, we need only consider the latter, which may be identified as those describing elementary systems that we will call *particles*. Note that here "elementarity" is not used in a dynamical sense; it only means that the corresponding *isolated* system cannot be described as two or more systems, *also isolated*[1].

## 6.2 Relativistic One-Particle States: General Description

Let us denote by $\mathcal{H}$ the Hilbert space for *free* one-particle states. We will construct a basis of $\mathcal{H}$, working in the Heisenberg picture, the simplest one to use for our analysis.

Consider the operators that represent translations, $U(a, 1) \equiv U(a)$. If we write them in exponential form,

$$U(a) = \exp ia \cdot P, \tag{6.2.1}$$

then unitarity of $U$ implies Hermiticity of the $P_\mu$. We will identify $P_0$ with the energy[2] operator (the Hamiltonian), and $\mathbf{P}$ the ordinary momentum operator; the four $P_\mu$ form the four-momentum operator.

From the commutation relations, (6.1.3) or (6.1.5), it follows that the operator $P^2 = P \cdot P$ commutes with all the generators of the Poincaré group, and hence also with all the $U(a, \Lambda)$. Schur's lemma then implies that it is a constant, which we identify with the square of the mass (which can be zero):

$$m^2 = P \cdot P. \tag{6.2.2}$$

Because of this, it follows that, for free particles, the operator $P_0$ is actually a function of the $\mathbf{P}$:

$$P_0 = +(m^2 + \mathbf{P}^2)^{1/2}, . \tag{6.2.3}$$

where we have chosen the positive square root to get positive energies. If $\mathbf{p}$ are the eigenvalues of the $\mathbf{P}$, and $p_0$ those of $P_0$, we thus have

$$p_0 = +\sqrt{m^2 + \mathbf{p}^2}, \tag{6.2.4}$$

as was to be expected for a relativistic particle.

As we know, the $P_\mu$ commute among themselves. We can then diagonalize them simultaneously, and consider the corresponding eigenvectors as the desired base of $\mathcal{H}$, which we denote by $|p, \zeta\rangle$, with $\zeta$ being whatever extra quantum numbers necessary to specify the states; as we will see, the $\zeta$ will

---

[1] Our treatment will not be mathematically rigorous. Mathematical rigour can be provided by consulting the treatises of Bogoliubov, Logunov and Todorov (1975) or Wightman (1960). The problem of giving the general description of relativistically invariant systems was first fully solved by Wigner (1939), whose paper we will essentially follow.

[2] In the rest of this chapter we will use, unless otherwise explicitly stated, natural units with $\hbar = c = 1$.

be essentially a spin component. Note that the notation $|p, \zeta\rangle$, although convenient, is redundant; we could also write $|p, \zeta\rangle = |\mathbf{p}, \zeta\rangle$, since $p_0$ is fixed by (6.2.4) once $\mathbf{p}$ is given.

Because $|p, \zeta\rangle$ are eigensates of the $P_\mu$, we have

$$P_\mu |p, \zeta\rangle = p_\mu |p, \zeta\rangle, \tag{6.2.5}$$

and, exponentiating, and writing $U(a)$ for $U(a, 1)$,

$$U(a)|p, \zeta\rangle = e^{ia \cdot P}|p, \zeta\rangle = e^{ia \cdot p}|p, \zeta\rangle. \tag{6.2.6}$$

Let us select a fixed momentum, $\bar{p}$, with $\bar{p} \cdot \bar{p} = m^2$, $\bar{p}_0 > 0$. This means that we are choosing a fixed reference system. Any admissible four-vector for the particle, $p$, may be written as

$$p = \Lambda(p)\bar{p},$$

where $\Lambda(p)$ is a (not unique) Lorentz transformation. That this is so may be seen by looking at Problem P.6.6. We then *choose* a family of such Lorentz transformations, $\Lambda(p)$, one for each $p$. The basis we will find will depend on the family of $\Lambda(p)$ we choose; but the choice will be left unspecified for the moment. Then, we *define* the basis $|\Lambda(p), \zeta\rangle$ by

$$|\Lambda(p), \zeta\rangle \equiv U(\Lambda(p))|\bar{p}, \zeta\rangle, \tag{6.2.7}$$

i.e., by accelerating via $\Lambda(p)$ to momentum $p$. In (6.2.7), and to simplify the notation, we write $U(\Lambda)$ for $U(0, \Lambda)$, just as in (6.2.6) we wrote $U(a)$ for $U(a, 1)$.

Let us first prove that the state $|\Lambda(p), \zeta\rangle$ corresponds to four-momentum $p$. To see this, we evaluate

$$U(a)|\Lambda(p), \zeta\rangle = U(a)U(\Lambda(p))|\bar{p}, \zeta\rangle.$$

Using the identity

$$U(a)U(\Lambda(p)) = U(a, \Lambda(p)) = U(\Lambda(p))U(\Lambda(p)^{-1}a),$$

we obtain

$$U(a)|\Lambda(p), \zeta\rangle = U(\Lambda(p))U(\Lambda(p)^{-1}a)|\bar{p}, \zeta\rangle.$$

From (6.2.6) for $p = \bar{p}$, and taking into account that

$$(\Lambda(p)^{-1}a) \cdot \bar{p} = a \cdot \Lambda(p)\bar{p} = a \cdot p,$$

we get

$$U(\Lambda(p))U(\Lambda(p)^{-1}a)|\bar{p}, \zeta\rangle$$

$$= U(\Lambda(p))e^{i(\Lambda(p)^{-1}a) \cdot \bar{p}}|\bar{p}, \zeta\rangle$$

$$= e^{ip \cdot a}U(\Lambda(p))|\bar{p}, \zeta\rangle$$

$$= e^{ip \cdot a}|\Lambda(p), \zeta\rangle.$$

We have thus shown that

$$U(a)|A(p), \zeta\rangle = e^{ia \cdot p}|A(p), \zeta\rangle, \tag{6.2.8a}$$

and (for example, by differentiating with respect to $a_\mu$ at $a = 0$) that $|A(p), \zeta\rangle$ is a state with momentum $p$, as claimed above:

$$P_\mu|A(p), \zeta\rangle = p_\mu|A(p), \zeta\rangle. \tag{6.2.8b}$$

Equation (6.2.8a) tells us how the translations act upon our basis of state vectors, $|A(p), \zeta\rangle$. We will now deduce corresponding formulas for Lorentz transformations. To do so, we start by considering transformations, which we will denote by $\Gamma, \Gamma', \ldots$, contained in the little group of $\bar{p}$, $\mathcal{W}(\bar{p})$, which was defined in Sect. 1.6 as the set of transformations that leave $\bar{p}$ invariant; and we will let these transformations act on $|\bar{p}, \zeta\rangle \equiv |A(\bar{p}), \zeta\rangle$ itself. Because the $\Gamma$ leave $\bar{p}$ invariant, it follows that the state vector $U(\Gamma)|\bar{p}, \zeta\rangle$ still corresponds to momentum $\bar{p}$. Therefore, it will have to be a linear combination of vectors $|\bar{p}, \zeta'\rangle$:

$$U(\Gamma)|\bar{p}, \zeta\rangle = \sum_{\zeta'} D_{\zeta'\zeta}(\Gamma)|\bar{p}, \zeta'\rangle, \tag{6.2.9}$$

where the $D_{\zeta'\zeta}$ are certain coefficients[3] that we may take to be the components of a matrix $D$:

$$D(\Gamma)^T = (D_{\zeta'\zeta}(\Gamma)), \quad \text{i.e.,} \quad D(\Gamma) = (D_{\zeta\zeta'}(\Gamma)).$$

It is easy to verify that the conditions

$$U(\Gamma)U(\Gamma') = U(\Gamma\Gamma'), \quad U(\Gamma^{-1}) = U^{-1}(\Gamma), \quad U^+(\Gamma) = U^{-1}(\Gamma)$$

imply that

$$D(\Gamma)D(\Gamma') = D(\Gamma\Gamma'),$$

$$D(\Gamma^{-1}) = D(\Gamma)^{-1}, \tag{6.2.10}$$

$$D^+(\Gamma) = D(\Gamma)^{-1};$$

it follows that the matrices $D$ build up a *unitary representation of the little group, $\mathcal{W}(\bar{p})$*. From the "elementarity" of the system, that is to say, from the fact that $U(a, A)$ is irreducible, we can deduce that the representation $D$ *must also be irreducible*.

---

[3] The parameter $\zeta$ will, for example, represent the third component of spin. Thus, for a particle with total spin $1/2$, we can have $\zeta = \pm 1/2$. At times it may be convenient to label the matrix elements not with the indices $\pm 1/2$, but with indices 1, 2. We thus identify

$$\begin{pmatrix} D_{1/2,1/2} & D_{1/2,-1/2} \\ D_{1/2,1/2} & D_{-1/2,-1/2} \end{pmatrix} \equiv \begin{pmatrix} D_{11} & D_{12} \\ D_{21} & D_{22} \end{pmatrix}.$$

The specific form of the $D$ will be given in the next two sections. For the moment we will assume that we have such a representation, so that we know the values of the coefficients $D_{\zeta'\zeta}(\Gamma)$; with their help we will be able to solve in full generality the problem of finding how arbitrary Lorentz transformations act. In fact, we have,

$$U(\Lambda)|\Lambda(p),\zeta\rangle = U(\Lambda)U(\Lambda(p))|\bar{p},\zeta\rangle$$

$$= U(\Lambda(\Lambda p))U(\Lambda(\Lambda p))^{-1}U(\Lambda\Lambda(p))|\bar{p},\zeta\rangle \tag{6.2.11a}$$

$$= U(\Lambda(\Lambda p))U((\Lambda(\Lambda p))^{-1}\Lambda\Lambda(p))|\bar{p},\zeta\rangle,$$

where $\Lambda(\Lambda p)\bar{p} = \Lambda p$, and we have introduced a term $U(\Lambda(\Lambda p))U(\Lambda(\Lambda p))^{-1} = 1$ and used the group properties of the $U$. Now,

$$(\Lambda(\Lambda p))^{-1}\Lambda\Lambda(p)\bar{p} = (\Lambda(\Lambda p))^{-1}\Lambda p = \bar{p},$$

so that the transformation $(\Lambda(\Lambda p))^{-1}\Lambda\Lambda(p)$, which we will write as $\Gamma(p,\Lambda)$, is in $\mathcal{W}(\bar{p})$, since it leaves $\bar{p}$ invariant. We can thus apply (6.2.9) to it and get

$$U(\Gamma(p,\Lambda))|\bar{p},\zeta\rangle = \sum_{\zeta'} D_{\zeta'\zeta}(\Gamma(p,\Lambda))|\bar{p},\zeta'\rangle; \tag{6.2.11b}$$

substituting this into (6.2.11a) we get the explicit formula

$$U(\Lambda)|\Lambda(p),\zeta\rangle = \sum_{\zeta'} D_{\zeta'\zeta}(\Gamma(p,\Lambda))|\Lambda(\Lambda p),\zeta'\rangle,$$

$$\Gamma(p,\Lambda) \equiv (\Lambda(\Lambda p))^{-1}\Lambda\Lambda(p). \tag{6.2.12}$$

Besides choosing the family of $\Lambda(p)$, and finding the explicit values of the $D_{\zeta'\zeta}$, the only thing that we need to have the problem totally solved is to find the normalization of the states $|\Lambda(p),\zeta\rangle$ such that relativistic transformations leave it invariant, i.e., such that the $U(a,\Lambda)$ are unitary. The $U(a)$ are unitary by construction. If we assume the $\zeta$ to be eigenvalues of an observable, we will have

$$\langle \Lambda(p),\zeta|\Lambda(p'),\zeta'\rangle = N(p)\delta(\mathbf{p}-\mathbf{p}')\delta_{\zeta\zeta'}, \tag{6.2.13a}$$

where $N$ is a factor to be determined by the requirement that, for any $\Lambda$,

$$\langle U(\Lambda)(\Lambda(p),\zeta)|U(\Lambda)(\Lambda(p'),\zeta')\rangle$$

$$= \langle \Lambda(p),\zeta|\Lambda(p'),\zeta'\rangle \tag{6.2.13b}$$

(*unitarity*). Substituting (6.2.12) into (6.2.13) and recalling that the matrix $D = (D_{\zeta'\zeta})$ is unitary, we find the condition

$$N(\Lambda p)\delta(\Lambda\mathbf{p}-\Lambda\mathbf{p}') = N(p)\delta(\mathbf{p}-\mathbf{p}'). \tag{6.2.14}$$

If $\Lambda$ is a rotation $R$, and since $\delta(R\mathbf{p}) = \delta(\mathbf{p})$, it follows that $N$ can only depend on $|\mathbf{p}|$, or, equivalently, on $p_0$, $N = N(p_0)$. Considering next a boost along $OZ$, $L_z$, with parameter $\xi$,

$$L_z: \quad p_0 \to (\cosh\xi)p_0 + (\sinh\xi)p_3,$$

$$p_3 \to (\cosh\xi)p_3 + (\sinh\xi)p_0,$$

$$p_1 \to p_1, \quad p_2 \to p_2,$$

(6.2.14) gives

$$N((\cosh\xi)p_0)\frac{1}{(\cosh\xi)p_0}\delta(\mathbf{p} - \mathbf{p}') = N(p_0)\delta(\mathbf{p} - \mathbf{p}'),$$

for any $\xi$, so that we get $N(p_0) = $ constant $\times$ $p_0$. We will follow custom in choosing this constant equal to 2, so the invariant form of the scalar product is finally

$$\langle \Lambda(p), \zeta | \Lambda(p'), \zeta' \rangle = 2p_0\delta(\mathbf{p} - \mathbf{p}')\delta_{\zeta\zeta'},$$

$$p_0 = +\sqrt{m^2 + \mathbf{p}^2},$$

$$(6.2.15)$$

to be compared with (2.3.14) and (3.5.15).

Before moving on to the detailed analysis of the various different cases, a few more words on general matters are in order. First of all we again remark that the analysis of this section is valid for massive as well as massless particles; for the latter it is sufficient to set $m = 0$ in the appropriate formulas. Secondly, it may appear that our analysis is dependent on the fixed vector (or reference system) $\bar{p}$, from which we build the basis. This is not so; because the little groups of two $\bar{p}, \bar{p}'$ are isomorphic, it follows that substituting $\bar{p}'$ for $\bar{p}$ merely result in a change of basis in $\mathcal{H}$. The same is true if we replace the family $\Lambda(p)$ by another family, $\Lambda'(p)$.

**Exercise.** Find the operators that implement the changes of basis (A) when replacing $\bar{p}$ by $\bar{p}'$, and (B) when replacing $\Lambda(p)$ by $\Lambda'(p)$ ∎

Finally, the analysis of this section may appear excessively abstract to the reader. This could be overcome by returning to it *after* having gone over the next two sections.


## 6.3 Relativistic States of Massive ($m \neq 0$) Particles

The idea behind Wigner's method is actually very simple, at least for particles with mass. In this case, one chooses a reference system with $\bar{p}_0 = m$, $\bar{p}_i = 0$, that is to say, the reference system in which the particle is at rest. Here, nonrelativistic quantum mechanics is manifestly valid, which suggests to us that we take the quantum numbers $\zeta$ to be the values of the third component of spin. In this case, we will use the label $\lambda$ instead of $\zeta$. We thus start by considering the states at rest,

$$|\bar{p}, \lambda\rangle.$$

The little group of $\bar{p}$ consists of ordinary three-dimensional rotations, which we denote by $R$ rather than $\Gamma$. The matrices $D(R)$ are just the standard $D^{(s)}(R(\boldsymbol{\theta}))$, for a particle with total spin $s$. They are

$$D^{(s)}(R(\boldsymbol{\theta})) = \exp \frac{-i}{\hbar} \boldsymbol{\theta} \mathbf{S},$$

where $\mathbf{S}$ are the familiar spin operators. For $s = 1/2$,

$$D^{(1/2)}(R(\boldsymbol{\theta})) = e^{-i\boldsymbol{\sigma}\boldsymbol{\theta}/2}.$$

For arbitrary $s$, the values of the matrix elements $D^{(s)}_{\lambda\lambda'}(R)$ of $D^{(s)}$ can be found in Appendix A.1. We then have

$$U(R)|\bar{p}, \lambda\rangle = \sum_{\lambda'} D^{(s)}_{\lambda'\lambda}(R)|\bar{p}, \lambda'\rangle. \tag{6.3.1}$$

For states in an arbitrary reference system, with momentum $p$, we may boost by $L(p)$ as given by, for example,(3.5.9) (there we used the notation $L_{\mathbf{p}}$ for $L(p)$) such that $L(p)\bar{p} = p$.

Then the states $|L(p), \lambda\rangle$ are *defined* as

$$|L(p), \lambda\rangle \equiv U(L(p))|\bar{p}, \lambda\rangle, \tag{6.3.2}$$

and we normalize them to

$$\langle L(p), \lambda | L(p'), \lambda'\rangle = 2p_0 \delta(\mathbf{p} - \mathbf{p}')\delta_{\lambda\lambda'}. \tag{6.3.3}$$

To find the transformation properties of the $|L(p), \lambda\rangle$ under an arbitrary Lorentz transformation $\Lambda$, we proceed as follows: $\Lambda$ will carry $p$ over $\Lambda p$. Therefore we (a) go to the reference system where the particle is at rest decelerating by $L^{-1}(p)$, (b) see how the state transforms there and (c) boost now by $L(\Lambda p)$. In formulas,

$$U(\Lambda)|L(p), \lambda\rangle = U(\Lambda)U(L(p))|\bar{p}, \lambda\rangle$$

$$= U(L(\Lambda p))U(L(\Lambda p)^{-1})U(\Lambda)U(L(p))|\bar{p}, \lambda\rangle$$

$$= U(L(\Lambda p))U(R(p, \Lambda))|\bar{p}, \lambda\rangle,$$

where

$$R(p, \Lambda) = L(\Lambda p)^{-1}\Lambda L(p)$$

is called a *Wigner rotation*; it *is* a rotation since $R(p, \Lambda)\bar{p} = \bar{p}$. Using (6.3.1), (6.3.2) now, we obtain

$$U(\Lambda)|L(p), \lambda\rangle = U(L(\Lambda p))U(R(p, \Lambda))|\bar{p}, \lambda\rangle$$

$$= U(L(\Lambda p)) \sum_{\lambda'} D^{(s)}_{\lambda'\lambda}(R(p, \Lambda))|\bar{p}, \lambda'\rangle$$

$$= \sum_{\lambda'} D^{(s)}_{\lambda'\lambda}(R(p, \Lambda))|L(\Lambda p), \lambda'\rangle,$$

so that

$$U(\Lambda)|\Lambda(p), \lambda\rangle = \sum_{\lambda'} D^{(s)}_{\lambda'\lambda}(R(p,\Lambda))|L(\Lambda p), \lambda'\rangle,$$

$$R(p, \Lambda) = L(\Lambda p)^{-1}\Lambda L(p). \tag{6.3.4}$$

Of course, we have already seen this in the previous section. The basis $|L(p), \lambda\rangle$ is sometimes called the *covariant spin* basis. Another useful basis is the *helicity* basis. To build it, we choose, instead of pure boosts $L(p)$, the transformations $H(p)$ defined as follows: first, take a pure boost $L(p^z)$ that carries $\bar{p}$ over $p^z$ with $p^z_0 = p_0$, $p^z_1 = p^z_2 = 0$, $p^z_3 = p_3$. Then, let $R(\mathbf{z} \to \mathbf{p})$ be a rotation around the axis $\mathbf{z} \times \mathbf{p}$ that carries the $OZ$ axis over $\mathbf{p}$. We define

$$H(p) \equiv R(\mathbf{z} \to \mathbf{p})L(p^z), \quad |H(p), \ \eta = \zeta\rangle = U(H(p))|\bar{p}, \zeta\rangle.$$

The corresponding states $|H(p), \ \eta = \zeta\rangle$ are the helicity states, since $\eta$ is the projection of the spin on $\mathbf{p}$.

The analysis is fairly straightforward for massive particles. The reason why we gave the general discussion of the previous section is its usefulness in studying the case of massless particles.

The nonrelativistic limit is obtained when $|\mathbf{p}| \ll m$, so that $p_0 \simeq m$. The normalization becomes (taking the covariant spin case for definiteness)

$$\langle L(p), \lambda | L(p'), \lambda'\rangle \underset{NR}{\simeq} 2m\delta_{\lambda\lambda'}\delta(\mathbf{p} - \mathbf{p}'),$$

so that

$$|L(p), \lambda\rangle = \sqrt{2p_0}|\mathbf{p}, \lambda\rangle_{NR} \underset{NR}{\simeq} \sqrt{2m}|\mathbf{p}, \lambda\rangle_{NR}, \quad {}_{NR}\langle \mathbf{p}, \lambda | \mathbf{p}', \lambda'\rangle_{NR}$$

$$= \delta_{\lambda\lambda'}\delta(\mathbf{p} - \mathbf{p}'). \tag{6.3.5}$$

Because of this some authors define

$$\text{(I)} \qquad |L(p), \lambda\rangle_{\mathrm{I}} = \frac{1}{\sqrt{2m}}|L(p), \lambda\rangle,$$

or

$$\text{(II)} \qquad |L(p), \lambda\rangle_{\mathrm{II}} = \frac{1}{\sqrt{2p_0}}|L(p), \lambda\rangle.$$

Here we will stick to our conventions. Choice I presents the problem of collapsing for massless particles; choice II is not relativistically invariant. The choice of (6.3.3) is valid for massless as well as massive particles, and is relativistically invariant; the price to pay is a factor $\sqrt{2p_0}$ between relativistic and NR normalization, a price that is quite justified.

Next we turn to the discrete symmetries $\mathcal{C}, \mathcal{P}, \mathcal{T}$. $\mathcal{C}$ is defined trivially by setting

$$\mathcal{C}|p, \lambda\rangle \equiv \eta_C|\overline{p, \lambda}\rangle,$$

where $|\overline{p, \lambda}\rangle$ denotes the state of an antiparticle with the same momentum $p$ and spin $\lambda$ as the particle $|p, \lambda\rangle$. $\mathcal{P}$ and $\mathcal{T}$ are not given by the previous analysis; but we can use the same method, with slight modifications. Beginning with parity, we define the operator $\mathcal{P}$ by considering that it is the representative of space reversal, $I_s$, $(I_s x)_\mu = g_{\mu\mu} x_\mu$: $\mathcal{P} = U(I_s)$. We then write

$$\mathcal{P}|L(p), \lambda\rangle = U(I_s) U(L(p))|\bar{p}, \lambda\rangle$$

$$= U(L(I_s p)) U(L(I_s p)^{-1} I_s L(p))|\bar{p}, \lambda\rangle.$$

Now, $L(I_s p)^{-1} I_s L(p)$ leaves $\bar{p}$ invariant. It is *not* a rotation, because its determinant is $(-1)$; but then

$$R(p, I_s) \equiv L(I_s p)^{-1} I_s L(p) I_s \tag{6.3.6a}$$

*is* a rotation. In the nonrelativistic case (Sect. 1.1),

$$\mathcal{P}|\bar{p}, \lambda\rangle = \eta_P |\bar{p}, \lambda\rangle,$$

so that, finally,

$$\mathcal{P}|L(p), \lambda\rangle = \eta_P \sum_{\lambda'} D^{(s)}_{\lambda'\lambda}(R(p, I_s))|L(I_s p), \lambda\rangle. \tag{6.3.6b}$$

For time reversal we can repeat the analysis with the modifications due to the *antiunitary* character of $\mathcal{T}$. Using (1.1.14), and

$$\mathcal{T} P_\mu \mathcal{T}^{-1} = (I_s P)_\mu,$$

we find that

$$\mathcal{T}|L(p), \lambda\rangle = \eta_T \sum_{\lambda'} D^{(s)}_{\lambda', -\lambda}(R(p, I_s))(-i)^{2\lambda}|L(I_s p), \lambda'\rangle. \tag{6.3.7}$$

**Exercise.** Evaluate $\mathcal{P}|H(p), \zeta\rangle, \mathcal{T}|H(p), \zeta\rangle$ ∎

## 6.4 Massless Particles

This case is *essentially* different from the previous one, not merely the limit as $m \to 0$, something that could already have been imagined from what we found for massless particles with the wave function formalism. To begin with, since a particle without mass cannot be at rest, the choice of $\bar{p}$ is less helpful than before. What we do is merely define our spatial axes so that $\bar{\mathbf{p}}$ points in a convenient direction, say, along $OZ$: we thus take

$$\bar{p}_1 = \bar{p}_2 = 0, \; \bar{p}_3 = \bar{p}_0.$$

The particular value of $\bar{p}_0$ is (for systems with a single particle) irrelevant; we may get $\bar{p}_0 = 1$ by a boost, or by just taking $\bar{p}_0$ as the unit of energy.

Let us now consider the little group of this $\bar{p}$, $\mathcal{W}(\bar{p})$. If $\Gamma$ is in $\mathcal{W}(\bar{p})$, we can represent it as in (1.6.1). We then decompose $\Gamma$ as

$$\Gamma = \Lambda_t R_z(\theta), \tag{6.4.1a}$$

where $R_z(\theta)$ is a rotation around $OZ$ by an angle $\theta$, so that the corresponding matrix $(\gamma)$ is

$$(\gamma) = \begin{pmatrix} 1 & 0 & 0 \\ 0 & 1 & 0 \\ \xi & \eta & 1 \end{pmatrix} \begin{pmatrix} \cos\theta & \sin\theta & 0 \\ -\sin\theta & \cos\theta & 0 \\ 0 & 0 & 1 \end{pmatrix},$$

$$\gamma_{31} = \xi\cos\theta - \eta\sin\theta, \ \ \gamma_{32} = \xi\sin\theta + \eta\cos\theta.$$

The first term in the expression for $(\gamma)$, viz.,

$$\begin{pmatrix} 1 & 0 & 0 \\ 0 & 1 & 0 \\ \xi & \eta & 1 \end{pmatrix},$$

corresponds to $\Lambda_t$; the second one to $R_z(\theta)$. Because the product of two transformations $\Gamma_1$, $\Gamma_2$ in $\mathcal{W}(\bar{p})$ lies in $\mathcal{W}(\bar{p})$, it follows that we can write

$$\Gamma_i = \Lambda_{it}R_z(\theta_i), \ \ i = 1, 2, \tag{6.4.2a}$$

and

$$\Gamma_1\Gamma_2 = \Lambda_{12t}R_z(\theta_{12}), \tag{6.4.2b}$$

where the angle $\theta_{12}$ will depend on $\Gamma_1$, $\Gamma_2$:

$$\theta_{12} = \theta_{12}(\Gamma_1, \Gamma_2).$$

To get a representation of the Poincaré group we require a representation of this little group, $\mathcal{W}(\bar{p})$. This little group itself has a structure very similar to that of $\mathcal{JL}$ ($\mathcal{W}(\bar{p})$ is actually isomorphic to the Euclidean group in two dimensions), and its representations can be studied by the same methods we are using to find the representations of the Poincaré group. The details may be found in Wigner (1939)[4]; we will take from there, and without proof, the following result. If we want to have particles with *discrete* spin values, then the representation must be of the form

$$D(\Gamma) = D(R_z(\theta)), \tag{6.4.3a}$$

i.e., we must have

$$D(\Lambda_t) \equiv 1. \tag{6.4.3b}$$

Moreover, the representation $D(R_z(\theta))$ must be at most double-valued, so that

$$D(R_z(2\pi)) = \pm 1. \tag{6.4.3c}$$

There is no *a priori* reason for excluding particles with continuous spins (which have been studied by Wigner, 1963); but it is a fact that all particles found in nature have discrete spin values. We will therefore *require* (6.4.3).

With the help of (6.4.3) the analysis is easily completed. The *irreducible* representations of the $R_z(\theta)$, rotations around a fixed *(OZ)* axis, are trivial.

---

[4] Or in Wightman (1960), Bogoliubov, Logunov and Todorov (1975). The proof of (6.4.3a,b) is easy; that of (6.4.3c) is nontrivial. It is based on the fact that, because the Lorentz group is doubly connected, its representations are, at most, double-valued.

Since the group is Abelian, Schur's lemma implies that these representations must be one-dimensional. From this it follows that the index $\lambda$ in the classification of the states,

$$|\bar{p}, \lambda\rangle,$$

can only take on *one* value. The matrices $D_{\lambda'\lambda}(\Gamma)$ are therefore just numbers, equal to $\delta_{\lambda\lambda'}d_\lambda(\theta)$. Because the representation has to be unitary, these numbers are of modulus unity and we can write

$$d_\lambda(\theta) = e^{-i\lambda\theta}.$$

The fact that the representation is at most two-valued, (6.4.3c), implies that the number $\lambda$ is integer or half integer. Its interpretation is readily accomplished by comparing the expression for $d(\theta)$ with that for a rotation around the $OZ$ axis in terms of the $S_z$ component of the spin operator,

$$U(R_z(\theta)) = e^{-i\theta S_z/\hbar} :$$

$\lambda$ is the spin component along $OZ$ (or along $\bar{\mathbf{p}}$, since it coincides with the $OZ$ axis). This is the *helicity*. Because there is only one possible value of $\lambda$, it follows that, *for massless particles, the helicity is relativistically invariant*, something we had found in specific cases with the wave function formalism.

Once the transformation properties of the states $|\bar{p}, \lambda\rangle$ under the little group $\mathcal{W}(\bar{p})$,

$$U(\Gamma)|\bar{p}, \lambda\rangle = e^{-i\lambda\theta(\Gamma)}|\bar{p}, \lambda\rangle, \tag{6.4.4}$$

are known, we have to specify the family of transformations $\Lambda(p)$ with $\Lambda(p)\bar{p} = p$ to extend the analysis to arbitary transformations. Choose $\bar{p}_0 = 1$; for an arbitrary $p$ we set

$$\Lambda(p) = H(p),$$

$$H(p) = R(\mathbf{z} \to \mathbf{p})L(p^z).$$

$L(p^z)$ is the pure boost along $OZ$ such that

$$L(p^z)\bar{p} = p^z,$$

$$p_0^z = p_0, \ p_1^z = p_2^z = 0, \ p_3^z = p_0;$$

$R(\mathbf{z} \to \mathbf{p})$ is the rotation around the axis $\mathbf{z} \times \mathbf{p}$ that carries $OZ$ over $\mathbf{p}$. According to (6.2.7) we then define

$$|p, \lambda\rangle \equiv U(H(p))|\bar{p}, \lambda\rangle, \tag{6.4.5}$$

so that, using (6.2.12), we find that

$$U(\Lambda)|p, \lambda\rangle = e^{-i\lambda\theta(p,\Lambda)}|\Lambda p, \lambda\rangle; \tag{6.4.6}$$

the angle $\theta(p, \Lambda)$ is the angle of the $OZ$ rotation contained in

$$\Gamma(p, \Lambda) = H(\Lambda p)^{-1}\Lambda H(p),$$

when we decompose it as in (6.4.1),

$$\Gamma(p, \Lambda) = \Lambda_t R_z(\theta(p, \Lambda)).$$

The normalization is

$$\langle p, \lambda | p, \lambda \rangle = 2p_0 \delta(\mathbf{p}, \mathbf{p}').$$

Next we consider the discrete symmetries $\mathcal{P}$, $\mathcal{T}$. Starting with parity, the corresponding operator should satisfy

$$\mathcal{P} P_0 \mathcal{P}^{-1} = P_0, \quad \mathcal{P} \mathbf{P} \mathcal{P}^{-1} = -\mathbf{P},$$

$$\mathcal{P} \mathbf{L} \mathcal{P}^{-1} = \mathbf{L}, \quad \mathcal{P} \mathbf{S} \mathcal{P}^{-1} = \mathbf{S};$$

from this, and for the helicity operator

$$S_{\mathbf{p}} = (\mathbf{p}/|\mathbf{p}|)\mathbf{S},$$

we obtain

$$\mathcal{P} S_{\mathbf{p}} \mathcal{P}^{-1} = -S_{\mathbf{p}}.$$

Therefore we would have to postulate that

$$\mathcal{P}|p, \lambda\rangle = \eta_P |I_s p, -\lambda\rangle. \tag{6.4.7}$$

In general this will be impossible: because the value of $\lambda$ is *invariant*, (6.4.7) requires that there exist *two independent* states, a state with helicity $\lambda$ and another with $-\lambda$. In nature we find two kinds of particle. In one class we have particles like the photon, gluons or, presumably, the graviton, which can exist in the two helicity states: $\pm 1$ for the first two, $\pm 2$ for the last. In the second class we have particles, like the neutrinos, which exist only with helicity $-1/2$; or the *antineutrinos* which always carry helicity $+1/2$. *For these particles parity is not defined* and indeed the interactions that involve them violate parity.

For neutrinos and antineutrinos we can define a combined operation, $\mathcal{CP}$, the product of parity and particle–antiparticle conjugation that carries neutrinos (with helicity $-1/2$) into antineutrinos (with helicity $+1/2$), and vice versa[5]. There is a third class, that of particles with helicity $\lambda$ for which neither particles or antiparticles with helicity $-\lambda$ existed, which is mathematically possible but of which no representative has been found in nature.

For time reversal,

$$\mathcal{T} \mathbf{S} \mathcal{T}^{-1} = -\mathbf{S}, \ \mathcal{T} \mathbf{P} \mathcal{T}^{-1} = -\mathbf{P},$$

so that

$$\mathcal{T} S_{\mathbf{p}} \mathcal{T}^{-1} = S_{\mathbf{p}},$$

---

[5] One can prove quite generally that the product $\mathcal{CPT}$ is always a symmetry for any relativistic theory of local fields. For the proof see, for example, the text of Bogoliubov, Logunov and Todorov (1975).

and we can define the antiunitary operator $\mathcal{T}$ with

$$\mathcal{T}|p,\lambda\rangle = \eta_T(-i)^{2\lambda}|I_s p,\lambda\rangle; \tag{6.4.8}$$

the phase $(-i)^{2\lambda}$ is introduced for aesthetic reasons, to make the massless case similar to the massive one (6.3.7).

Let us return to parity. If the state $|I_s p, -\lambda\rangle$ exists, we will have to double our Hilbert space of states to make room for it. We define total spin as $s = \max|\lambda|$, and chirality $\delta$ as $\delta = \lambda/s = \pm 1$. We may label the states as

$$|p,s,\delta\rangle,$$

and (6.4.6), (6.4.7) can then be written as

$$U(\Lambda)|p,s,\delta\rangle = e^{-i\delta s\theta(p,\Lambda)}|\Lambda p,s,\delta\rangle,$$

$$\mathcal{P}|p,s,\delta\rangle = \eta_P|I_s p,s,-\delta\rangle. \tag{6.4.9}$$

The representation is *reducible* as a representation of the Poincaré group because the subspaces with $\delta = 1$ and $\delta = -1$ are separately invariant; it is *irreducible* as a representation of the orthochronous (but *not* proper) group obtained adjoining space reversal, $I_s$, with $U(I_s) \equiv \mathcal{P}$, to the orthochronous, proper Poincaré group.


# 6.5 Many-Particle States. Creation–Annihilation Operators. Fock Space

In a relativistic theory we have to take into account that we can create and annihilate particles with appropriate energy transfers. It is then useful to consider assemblies of several particles, and describe them in terms of creation–annihilation operators.

Let us denote the state of a particle with four-momentum $p$ and spin component $\lambda$ by $|p,\lambda\rangle$; we omit the specification of the Lorentz transformation $\Lambda(p)$, implicit in all that follows, and easily deduced from the context: if $\lambda$ is the $OZ$ spin component, $\Lambda = L$; if $\lambda$ is the helicity, $\Lambda = H$. We then define the *Fock space*, $\mathcal{F}_B$, of a set of identical particles as the space of superpositions of states with any number of such particles. We construct it by introducing the *creation*, $a^+(p,\lambda)$, and *annihilation*, $a(p,\lambda)$, operators, which satisfy the commutation relations

$$[a(p,\lambda),a(p',\lambda')] = 0,$$

$$[a(p,\lambda),a^+(p',\lambda')] = 2p_0\delta(\mathbf{p}-\mathbf{p}')\delta_{\lambda\lambda'}, \quad p_0 = \sqrt{m^2+\mathbf{p}^2}. \tag{6.5.1}$$

We assume that the particles are *bosons*; the fermionic case will be considered later. The difference between the commutation relations (6.5.1) and the usual nonrelativistic ones is that we now want *invariant* commutation relations. Nonrelativistic ones,

$$[a_{NR}(\mathbf{p}, \lambda), a_{NR}^{+}(\mathbf{p}', \lambda')] = \delta_{\lambda\lambda'}\delta(\mathbf{p} - \mathbf{p}'),$$

are obtained by defining

$$a(p, \lambda) = \sqrt{2p_0}\, a_{NR}(\mathbf{p}, \lambda).$$

We prefer to work with (6.5.1). This will mean that the $a$ do not tend, in the NR limit, to $a_{NR}$, but to $\sqrt{2m}$ times $a_{NR}$.

We postulate that $\mathcal{F}_{\mathrm{B}}$ contains a vector $|0\rangle$, the *vacuum* or state without particles, which we assume normalized to unity, $\langle 0|0\rangle = 1$. We also postulate that, for all $p, \lambda$,

$$a(p, \lambda)|0\rangle \equiv 0. \tag{6.5.2}$$

Finally, we postulate that all the states in $\mathcal{F}_{\mathrm{B}}$ can be obtained by repeatedly applying creation operators to the vacuum, or by forming linear combinations of such states. Thus, a state of $N$ particles with well-defined momenta and spins is

$$|p_1, \lambda_1, \ldots, p_N, \lambda_N\rangle = a^{+}(p_1, \lambda_1) \ldots a^{+}(p_N, \lambda_N)|0\rangle,$$

and a general vector in $\mathcal{F}_{\mathrm{B}}$ is

$$|\Phi\rangle = \alpha|0\rangle$$
$$+ \sum_{N=1}^{\infty} \int \frac{d^3 p_1}{2p_{10}} \cdots \frac{d^3 p_N}{2p_{N0}} \sum_{\lambda} \varphi_{\lambda_1 \ldots \lambda_N}(\mathbf{p}_1, \ldots, \mathbf{p}_N)|p_1, \lambda_1, \ldots, p_N, \lambda_N\rangle.$$

For fermionic particles the Fock space $\mathcal{F}_{\mathrm{F}}$ is built as in the boson case with the only modification that the creation and destruction operators, which we now denote by $b^{+}(p, \lambda)$, $b(p, \lambda)$, satisfy the *anticommutation relations*:

$$\{b(p, \lambda), b(p', \lambda')\} = 0,$$
$$\{b(p, \lambda), b^{+}(p', \lambda')\} = 2p_0\delta(\mathbf{p} - \mathbf{p}')\delta_{\lambda\lambda'}. \tag{6.5.3}$$

More generally we could build a space containing various types of particles: bosons with destruction operators $a_1, \ldots, a_n$, and fermions, with operators $b_1, \ldots, b_k$. We assume a unique vacuum $|0\rangle$, and the commutation and anticommutation relations

$$[a_i(p, \lambda), a_{i'}(p', \lambda')] = 0, \quad \{b_i(p, \lambda), b_{i'}(p', \lambda')\} = 0,$$

$$[a_i, b_j] = [a_i, b_j^{+}] = 0,$$

$$[a_i(p, \lambda), a_j^{+}(p', \lambda')] = 2p_0\delta(p - p')\delta_{\lambda\lambda'}\delta_{ij}, \tag{6.5.4}$$

$$\{b_i(p, \lambda), b_j^{+}(p', \lambda')\} = 2p_0\delta(p - p')\delta_{\lambda\lambda'}\delta_{ij}.$$

We have chosen $[a, b] = 0$ instead of $\{a, b\} = 0$; and, for $i \neq j$, $\{b_i, b_j\} = 0$, rather than $[b_i, b_j] = 0$. Both choices are equivalent from a physical point of view, and in fact it can be proved that there exists an equivalence transformation (the *Klein transformation*) connecting them; for the proof and more details, see Bogoliubov, Logunov and Todorov (1975). The choice we have made, (6.5.4), is the most natural and convenient one.

**Exercise.** (A) Verify that (6.5.4) implies that states containing different particles are orthogonal, as should be the case. (B) Verify that (6.5.1), (6.5.3) are equivalent to the normalization

$$\langle p, \lambda | p', \lambda' \rangle = 2p_0 \delta(\mathbf{p} - \mathbf{p}') \delta_{\lambda\lambda}$$

for the states. (Note that $|p, \lambda\rangle = c^+(p, \lambda)|0\rangle$, $c^+ = a^+$ or $b^+$ as the case may be.) ∎

The relativistic transformation properties of the creation–destruction operators follow from those of the states, if we *postulate* invariance of the vacuum:

$$U(a, \Lambda)|0\rangle = |0\rangle.$$

To find them, consider the bosonic case, and a Lorentz transformation $\Lambda$. We have, for any state $|\Phi\rangle$, and with $U|\Phi\rangle \equiv |\Phi_\Lambda\rangle$,

$$\langle \Phi | p, \lambda \rangle = \langle \Phi | a^+(p, \lambda)|0\rangle = \langle \Phi | U^{-1} U a^+ U^{-1} U |0\rangle$$

$$= \langle \Phi_\Lambda | U(\Lambda) a^+(p, \lambda) U^{-1}(\Lambda)|0\rangle,$$

$$\langle \Phi | p, \lambda \rangle = \langle \Phi | U^{-1} U | p, \lambda \rangle = \langle \Phi_\Lambda | U(\Lambda)| p, \lambda \rangle$$

$$= \sum D_{\lambda'\lambda}(p, \Lambda)\langle \Phi_\Lambda | \Lambda p, \lambda' \rangle = \sum D_{\lambda'\lambda}\langle \Phi_\Lambda | a^+(\Lambda p, \lambda')|0\rangle.$$

The transformation properties are found by equating these. The result, written now in general, is[6]

$$U(a, \Lambda)\hat{a}^+(p, \lambda)U^{-1}(a, \Lambda) = e^{ia \cdot \Lambda p} \sum D_{\lambda'\lambda}(\Gamma(p, \Lambda))\hat{a}^+(\Lambda p, \lambda'),$$

$$U(a, \Lambda)\hat{a}(p, \lambda)U^{-1}(a, \Lambda) = e^{-ia \cdot \Lambda p} \sum D^*_{\lambda'\lambda}(\Gamma(p, \Lambda))\hat{a}(\Lambda p, \lambda'), \qquad (6.5.5)$$

$$\Gamma(p, \Lambda) = (\Lambda(\Lambda p))^{-1} \Lambda\Lambda(p),$$

and the same holds for the fermion case if $\hat{a}, \hat{a}^+$ are replaced by $\hat{b}, \hat{b}^+$.

## 6.6 Connection with the Wave Function Formalism

As will be argued in Sects. 9.1 and 2, the construction of relativistic states with well-defined position, $|\mathbf{r}, t, a\rangle$ ($t$ is the time, and $a$ represents possible extra labels) does not make much physical sense. Therefore, the connection between the abstract ket formalism and the wave function formalism is now less straightforward than in the nonrelativistic case, where we simply have

---

[6] We distinguish the operators $\hat{a}, \hat{a}^+$ with carets to differentiate them from the four-translation $a$.

$\Psi_a(\mathbf{r}, t) = \langle \mathbf{r}, t, a | \Psi \rangle$. Now, we will connect with the *momentum space* wave functions; these can be then linked, via the appropriate Fourier transformations, to $x$-space ones.

We then want to establish the correspondence between the ket states we have constructed in the previous sections and (multicomponent) wave functions $\psi_a^{(\mathbf{k}, \lambda)}(\mathbf{p})$, corresponding to momentum $\mathbf{k}$ and spin component $\lambda$ (note that here $\mathbf{p}$ is the variable). We will work in the Heisenberg representation, so the $\psi$ are time independent. Time dependence can be introduced, if so wished, by writing

$$\Psi_a^{(\mathbf{k}, \lambda)}(\mathbf{p}, t) = e^{-ik_0 t} \psi_a^{(\mathbf{k}, \lambda)}(\mathbf{p}), \quad k_0 = \sqrt{m^2 + \mathbf{k}^2}. \tag{6.6.1}$$

Here, as in all this chapter, we work in natural units, $\hbar = c = 1$.

The case of spinless particles is simple. We just have

$$\varphi^{(\mathbf{k})}(\mathbf{p}) = \langle p | k \rangle = 2k_0 \delta(\mathbf{p} - \mathbf{k}), \tag{6.6.2}$$

but spin poses nontrivial problems. We will only consider the spin $1/2$ case; the generalization to higher spins is straightforward, for $m \neq 0$, and can be found in Moussa and Stora (1968), Weinberg (1964) and Zwanziger (1964a,b). (The latter also treat the massless case).

The wave function of a particle of spin $1/2$, with third component of covariant spin $s_3$ and momentum $\mathbf{k}$ was found in (3.5.11). Fourier transforming, extracting the $t$ dependence and slightly changing the notation, we have

$$\psi^{(k, s_3)}(\mathbf{p}) = D(L(k)) u(0, s_3) 2k_0 \delta(\mathbf{k} - \mathbf{p}). \tag{6.6.3}$$

Taking into account that

$$u(0, 1/2) = \begin{pmatrix} 1 \\ 0 \\ 0 \\ 0 \end{pmatrix}, \quad u(0, -1/2) = \begin{pmatrix} 0 \\ 1 \\ 0 \\ 0 \end{pmatrix} \ .$$

it becomes convenient for our calculations to change the labels $s_3 = \pm 1/2$ to $\tau = 1, 2$, so that $1/2 \to 1$, $-1/2 \to 2$. Then we may write $u_a(0, \tau) = \delta_{a\tau}$, and (6.6.3) adopts the simple form

$$\psi_a^{(\mathbf{k}, \tau)}(\mathbf{p}) = D_{a\tau}(L(k)) 2k_0 \delta(\mathbf{k} - \mathbf{p}), \tag{6.6.4}$$

and we then have the explicit expression

$$u_a(k, \tau) = D_{a\tau}(L(k)). \tag{6.6.5}$$

$D_{ab}(L(k))$ is the $ab$ matrix element of the matrix $D(L(k))$, whose explicit form is given in (3.5.10). For the purposes of this section, however, it is preferable to use the Weyl representation of the $\gamma$ matrices,

$$\gamma_\mu^{\mathrm{W}} = \begin{pmatrix} 0 & \tilde{\sigma}_\mu \\ \sigma_\mu & 0 \end{pmatrix}, \quad \tilde{\sigma}_i = -\sigma_i, \ \tilde{\sigma}_0 = \sigma_0 = 1.$$

We have

$$D(L(k)) \equiv D(L(\mathbf{k})) = \frac{1}{\sqrt{m}}(k_0 + \mathbf{k}\boldsymbol{\alpha})^{1/2} = \frac{1}{\sqrt{m}}(k \cdot \gamma\gamma_0)^{1/2},$$

a formula valid in any representation. In Weyl's, this becomes

$$D^{\mathrm{W}}(L(k)) = \frac{1}{\sqrt{m}} \begin{pmatrix} (k \cdot \tilde{\sigma})^{1/2} & 0 \\ 0 & (k \cdot \sigma)^{1/2} \end{pmatrix}. \tag{6.6.6a}$$

This is of course the reason why the Weyl representation is useful for us: the matrix $D^{\mathrm{W}}$ is "diagonal". Taking into account (Appendix A.4) that the matrix that leads from the Pauli to the Weyl representation is

$$\frac{1}{\sqrt{2}}(\gamma_0^{\mathrm{P}} + \gamma_5^{\mathrm{P}}) = \frac{1}{\sqrt{2}} \begin{pmatrix} 1 & 1 \\ 1 & -1 \end{pmatrix},$$

we find for the spinors $u(0, \tau)$, in the Weyl realization,

$$u^{\mathrm{W}}(0,1) = \frac{1}{\sqrt{2}} \begin{pmatrix} 1 \\ 0 \\ 1 \\ 0 \end{pmatrix}, \quad u^{\mathrm{W}}(0,2) = \frac{1}{\sqrt{2}} \begin{pmatrix} 0 \\ 1 \\ 0 \\ 1 \end{pmatrix}. \tag{6.6.6b}$$

In what follows we suppress the label "$W$".

Using (6.6.6) we may rewrite the wave function (6.6.4) as

$$\psi = \begin{pmatrix} \varphi \\ \tilde{\varphi} \end{pmatrix},$$

$$\psi_a^{(\mathbf{k},\tau)}(\mathbf{p}) = \varphi_\alpha^{(k,\tau)}(\mathbf{p}), \ a = \alpha = 1, 2; \tag{6.6.7a}$$

$$\psi_b^{(\mathbf{k},\tau)}(\mathbf{p}) = \tilde{\varphi}_{\dot{\beta}}^{(k,\tau)}(\mathbf{p}), \ b = \dot{\beta} + 2 = 3, 4,$$

with

$$\varphi_\alpha^{(k,\tau)}(\mathbf{p}) = \frac{1}{\sqrt{2m}}((k \cdot \tilde{\sigma})^{1/2})_{\alpha\tau} 2k_0\delta(\mathbf{p} - \mathbf{k}),$$

$$\tilde{\varphi}_{\dot{\beta}}^{(k,\tau)}(\mathbf{p}) = \frac{1}{\sqrt{2m}}((k \cdot \sigma)^{1/2})_{\dot{\beta}\tau} 2k_0\delta(\mathbf{p} - \mathbf{k}) \tag{6.6.7b}$$

(the notation with dotted indices, such as $\dot{\beta}$, for the components $\tilde{\varphi}_{\dot{\beta}}$ is the traditional one).

Because $\psi$ satisfies the Dirac equation, it follows that we can get $\tilde{\varphi}$ in terms of $\varphi$ (or vice versa). Indeed, we have

$$\tilde{\varphi}_{\dot{\beta}}^{(k,\tau)}(\mathbf{p}) = \sum_a \left( \frac{k \cdot \sigma}{m} \right)_{\dot{\beta}\alpha} \varphi_\alpha^{(k,\tau)}(\mathbf{p}). \tag{6.6.8}$$

**Exercise.** (A) Prove (6.6.8) by verifying that the indentity $(k \cdot \sigma)(k \cdot \tilde{\sigma}) = k \cdot k$ implies that (6.6.8) is equivalent to the Dirac equation $(\not{k} - m)\psi^{(\mathbf{k},\tau)}(\mathbf{p}) = 0$. (B) Check that

$$(k \cdot \tilde{\sigma})^{1/2} = [2(k_0 + m)]^{-1/2}(m + k_0 + \mathbf{k}\sigma) \ \blacksquare$$

Owing to this relation (6.6.8), it is sufficient to establish the connection between the states $|k, \tau\rangle$ and the wave functions $\varphi_\alpha^{(k,\tau)}(\mathbf{p})$. This is achieved by introducing the so-called *spinorial states*, $|p, \alpha\rangle$, defined to be such that

$$\varphi_\alpha^{(k,\tau)}(\mathbf{p}) \equiv \langle p, \alpha|k, \tau\rangle. \tag{6.6.9a}$$

Taking into account the explicit form of the $\varphi$ as given in (6.6.7), we obtain the formula that links the spinorial states to the familiar states with given covariant spin $|k, \tau\rangle$: it is

$$|p, \alpha\rangle = \sum_\tau \int \frac{d^3k}{2k_0} \left( \left( \frac{k \cdot \tilde{\sigma}}{2m} \right)^{1/2} \right)_{\tau\alpha} 2k_0\delta(\mathbf{p} - \mathbf{k})|k, \tau\rangle, \tag{6.6.9b}$$

and we have used the Hermiticity of the matrix $(k \cdot \tilde{\sigma})^{1/2}$.

The matrix $(k \cdot \tilde{\sigma}/m)^{1/2}$ is *not* unitary. The basis $|p, \alpha\rangle$ is therefore not orthogonal; rather one has

$$\langle p', \alpha'|p, \alpha\rangle = \frac{(p \cdot \tilde{\sigma})_{\alpha'\alpha}}{2m} 2p_0\delta(\mathbf{p} - \mathbf{p}'). \tag{6.6.10}$$

The index $\alpha$ does not correspond to any quantum number.


## Problems

**P.6.1.** Prove that $d^3p/2p_0$, $2p_0\delta(\mathbf{p} - \mathbf{p}')$ are invariant by writing, for $p_0 > 0$,

$$\delta_4(p - p') = \delta(p^2 - p'^2)2p_0\delta(\mathbf{p} - \mathbf{p}').$$

**P.6.2.** Find $R(p, \Lambda)$ in the NR limit, including corrections $O(v^2/c^2)$.

**P.6.3.** Find $R(p, I_s)$ for $\Lambda(p) = L(p)$. Find $|H(p), \lambda\rangle$ in terms of $|L(p), \eta\rangle$, and vice versa.

**P.6.4.** Let $W_\mu = \sum g_{\nu\nu}g_{\rho\rho}g_{\sigma\sigma}\epsilon_{\mu\nu\rho\sigma}P_\nu M_{\rho\sigma}$ (*Pauli–Lubanski* vector). Prove that $W^2 = $ invariant $= -m^2s(s + 1)$, $s$ the spin.

**P.6.5.** Using formulas of Appendix A.3 verify that, for any $\Lambda$,

$$U(\Lambda): \ \varphi_\alpha^{(k,\tau)}(p) \rightarrow \sum_{\alpha'} D_{\alpha\alpha'}^{(1/2)}(\Lambda)\varphi_{\alpha'}^{(k,\tau)}(\Lambda^{-1}p),$$

$$U(\Lambda): \ \varphi_\alpha^{(k,\tau)}(p) \rightarrow \sum_{\tau'} D_{\tau'\tau}^{(1/2)}(R(k, \Lambda))\varphi_\alpha^{(\Lambda k,\tau')}(p).$$

Here, $D(\Lambda) = D(L)D(R)$, for $\Lambda = LR$, with

$$D_{\alpha\beta}^{(1/2)}(L(p)) = m^{-1}(p \cdot \tilde{\sigma})_{\alpha\beta}^{1/2}, \ D_{\alpha\beta}^{(1/2)}(R(\boldsymbol{\theta})) = (e^{-i\boldsymbol{\theta}\sigma/2})_{\alpha\beta}, \text{ etc.}$$

**P.6.6.** Suppose that, for a particle, there existed a state $|\bar{p}_\perp\rangle$ *different* from all the $p = \Lambda\bar{p}$. Prove then that $\langle \bar{p}_\perp|\Lambda\bar{p}\rangle = 0$ for all $\Lambda$, and that the representation turns out to be reducible.

# 7. General Description of Relativistic Collisions: $S$ Matrix, Cross-sections and Decay Rates. Partial Wave Analyses

## 7.1 Two-Particle States. Separation of the Centre of Mass Motion. States with Well-Defined Angular Momentum

Let us consider two free particles (which for simplicity we take to be distinguishable), $A$, $B$, with masses $m_A$, $m_B$. A state of these two particles can be specified by giving the momenta $\mathbf{p}_A$, $\mathbf{p}_B$ and spin quantum numbers (for example, the helicities) to be denoted by $\alpha, \beta$: we thus write it as

$$|p_A, \alpha; p_B, \beta\rangle, \quad p_{A0} \equiv \sqrt{m_A^2 + \mathbf{p}_A^2}, \quad p_{B0} \equiv \sqrt{m_B^2 + \mathbf{p}_B^2} \tag{7.1.1a}$$

with normalization

$$\langle p_A', \alpha'; p_B', \beta' | p_A, \alpha; p_B, \beta\rangle = \delta_{\alpha\alpha'} 2p_{A0}\delta(\mathbf{p}_A - \mathbf{p}_A') \\ \times \delta_{\beta\beta'} 2p_{B0}\delta(\mathbf{p}_B - \mathbf{p}_B'). \tag{7.1.1b}$$

The same state can be specified by giving the total four-momentum, $p = p_A + p_B$, the direction of the relative three-momentum, $\mathbf{k} = (\mathbf{p}_A - \mathbf{p}_B)/2$, and the spin labels $\alpha$, $\beta$:

$$|p_A, \alpha; p_B, \beta\rangle = |p; \mathbf{k}; \alpha, \beta\rangle; \tag{7.1.2}$$

we write $\mathbf{k}$, which is redundant (just as $p_{A0}$, $p_{B0}$ are redundant in (7.1.1a)), instead of $\Omega_{\mathbf{k}}$ (the angular variables of $\mathbf{k}$) for simplicity of notation.

**Exercise.** Show that, given $p_0$, $\mathbf{p}$, $\Omega_{\mathbf{k}}$ we can reconstruct $p_A$, $p_B$ ∎

The tensor product notation is at times convenient, and we will thus write

$$|p_A, \alpha\rangle \otimes |p_B, \beta\rangle = |p_A, \alpha; p_B, \beta\rangle$$

$$= |p; \mathbf{k}; \alpha, \beta\rangle \tag{7.1.3}$$

$$= |p\rangle \otimes |\mathbf{k}; \alpha, \beta\rangle.$$

The scalar product (7.1.1b) can be easily expressed in terms of the new variables: first,

$$\delta(\mathbf{p}_A - \mathbf{p}_A')\delta(\mathbf{p}_B - \mathbf{p}_B') = \delta(\mathbf{p} - \mathbf{p}')\delta(\mathbf{k} - \mathbf{k}');$$

then, we can use the relation

$$\delta(\mathbf{k} - \mathbf{k}') = \frac{1}{\mathbf{k}^2}\delta(|\mathbf{k}| - |\mathbf{k}'|)\delta(\Omega_\mathbf{k} - \Omega_{\mathbf{k}'})$$

$$= \frac{1}{\mathbf{k}^2}J^{-1}\delta(p_0 - p_0')\delta(\Omega_\mathbf{k} - \Omega_{\mathbf{k}'}),$$

where $J$ is the Jacobian $J = \partial|\mathbf{k}|/\partial p_0$, to get

$$\delta(\mathbf{p}_A - \mathbf{p}_A')\delta(\mathbf{p}_B - \mathbf{p}_B') = (1/J\mathbf{k}^2)\delta(p_0 - p_0')\delta(\Omega_\mathbf{k} - \Omega_{\mathbf{k}'}).$$

We will only need the relative motion (described by $\mathbf{k}$) in the centre of mass (c.m.) system, $\mathbf{p} = 0$. Here, $p_0 = p_{A0} + p_{B0} = (m_A^2 + \mathbf{k}^2)^{1/2} + (m_B^2 + \mathbf{k}^2)^{1/2}$ so that

$$J = \partial|\mathbf{k}|/\partial p_0 = p_{A0}p_{B0}/p_0|\mathbf{k}|,$$

and finally we obtain

$$\langle p_A', \alpha'; p_B', \beta'|p_A, \alpha; p_B, \beta\rangle =$$

$$\langle p'; \mathbf{k}'; \alpha', \beta'|p; \mathbf{k}; \alpha, \beta\rangle = \frac{4p_0}{|\mathbf{k}|}\delta_4(p - p')\delta(\Omega_\mathbf{k} - \Omega_{\mathbf{k}'})\delta_{\alpha\alpha'}\delta_{\beta\beta'}, \qquad (7.1.4\text{a})$$

$$\delta(\Omega - \Omega') \equiv \delta(\cos\theta - \cos\theta')\delta(\varphi - \varphi'),$$

with $\theta$, $\varphi$ the polar angles corresponding to the solid angle $\Omega$. We write (7.1.4a) also as

$$\langle p'|p\rangle = \delta_4(p' - p), \quad \langle \mathbf{k}'; \alpha', \beta'|\mathbf{k}; \alpha, \beta\rangle = \frac{4p_0}{|\mathbf{k}|}\delta(\Omega_{\mathbf{k}'} - \Omega_\mathbf{k})\delta_{\alpha\alpha'}\delta_{\beta\beta'}. (7.1.4\text{b})$$

This will allow us to introduce a completeness relation once we ascertain the range of the variables $p_0$, $\mathbf{p}$. Clearly, $\mathbf{p}$ varies over all space; but $p_0$ is limited by

$$p_0 = p_{A0} + p_{B0} = \sqrt{m_A^2 + \mathbf{p}_A^2} + \sqrt{m_B^2 + \mathbf{p}_B^2} = \sqrt{p^2 + \mathbf{p}^2},$$

$$p^2 \geq (m_A + m_B)^2.$$

We can thus write the four-dimensional delta in (7.1.4) as

$$\delta_4(p - p') = 2p_0\delta(\mathbf{p} - \mathbf{p}')\delta(p^2 - p'^2), \qquad (7.1.5)$$

so that the completeness relation can be expressed separating the c.m. piece, which behaves as a composite particle with (variable) squared mass $p^2$ and momentum $\mathbf{p}$, and the relative motion, described by $\mathbf{k}$, as follows:

$$1 = \sum_{\alpha\beta} \int \frac{d^3p_A}{2p_{A0}} \int \frac{d^3p_B}{2p_{B0}} |p_A, \alpha; p_B, \beta\rangle\langle p_A, \alpha; p_B, \beta|$$

$$= \sum_{\alpha\beta} \int d^4p \int d\Omega_{\mathbf{k}} \frac{|\mathbf{k}|}{4p_0} |p; \mathbf{k}; \alpha, \beta\rangle\langle p; \mathbf{k}; \alpha, \beta|$$

$$= \int_{(m_A+m_B)^2}^{\infty} d(p^2) \int \frac{d^3p}{2p_0} |p\rangle\langle p|$$

$$\otimes \sum_{\alpha\beta} \int d\Omega_{\mathbf{k}} \frac{|\mathbf{k}|}{4p_0} |\mathbf{k}; \alpha, \beta\rangle\langle \mathbf{k}; \alpha, \beta| = 1_{\mathrm{c.m.}} \otimes 1_{\mathrm{rel}}. \tag{7.1.6}$$

In the c.m. system one can construct states with well-defined *orbital* angular momentum $l$, and third component $M$ as in the nonrelativistic case: we have

$$|l, M; \alpha, \beta\rangle = \int d\Omega_{\mathbf{k}} Y_M^l(\Omega_{\mathbf{k}}) |\mathbf{k}; \alpha, \beta\rangle. \tag{7.1.7}$$

Of course, we can (and will do so later) compose $l$, $M$ and the spins, $s_A, \alpha$; $s_B, \beta$ to obtain states with well-defined *total* angular momentum.

The completeness relation (7.1.6) can again be expressed in terms of the states $|l, M; \alpha, \beta\rangle$: separating c.m. and relative motion, we get

$$1 = 1_{\mathrm{c.m.}} \otimes 1_{\mathrm{rel}},$$

$$1_{\mathrm{c.m.}} = \int d^4p |p\rangle\langle p|,$$

$$1_{\mathrm{rel}} = \sum_{\alpha\beta} \int d\Omega_{\mathbf{k}} \frac{|\mathbf{k}|}{4p_0} |\mathbf{k}; \alpha, \beta\rangle\langle \mathbf{k}; \alpha, \beta| \tag{7.1.8}$$

$$= \frac{|\mathbf{k}|}{4p_0} \sum_{\alpha\beta} \sum_{lM} |l, M; \alpha, \beta\rangle\langle l, M; \alpha, \beta|.$$

## 7.2 Kinematics of Two-Particle Collisions

The collision of two particles, $A + B$, giving two other particles $A' + B'$, equal or different from the incoming ones (Fig. 7.2.1) is the simplest of all scattering processes. We will therefore present in some detail the kinematics in this special case. The four-momenta are denoted by $p_A$, $p_B$; $p_{A'}$, $p_{B'}$, and they satisfy the constraints

$$p_A^2 = m_A^2, \ p_B^2 = m_B^2; \ p_{A'}^2 = m_{A'}^2, \ p_{B'}^2 = m_{B'}^2,$$

where $m_A$, etc. are the masses (which can vanish), and one has

$$p_A + p_B = p_{A'} + p_{B'}$$

(four-momentum conservation). Because of this relation, and since the masses are fixed, it follows that, out of the ten invariant scalar products one can write, $p_A \cdot p_A$, $p_A \cdot p_B, \ldots$, only *two* are variables. A useful choice is the *Mandelstam variables*,

$$s = (p_A + p_B)^2 = (p_{A'} + p_{B'})^2,$$

$$t = (p_A - p_{A'})^2 = (p_B - p_{B'})^2, \tag{7.2.1a}$$

$$u = (p_A - p_{B'})^2 = (p_{A'} - p_B)^2.$$

We stated that there are only *two* independent invariants, and certainly the Mandelstam variables are related; one has

$$s + t + u = m_A^2 + m_B^2 + m_{A'}^2 + m_{B'}^2. \tag{7.2.1b}$$

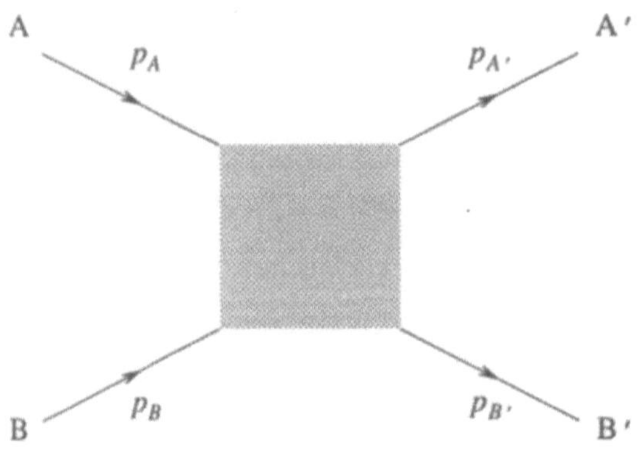


Fig. **7.2.1.** Two-body collision $A + B \to A' + B'$.

The variable $t$ is usually called *momentum transfer*; $s$ is the *c.m. energy squared.* No special name exists for $u$.

In actual scattering experiments, however, what one measures are not $s$, $t$, $u$ but energies and scattering angles. We will use two reference systems, the *centre of mass* system, where

$$\mathbf{p}_A^{\text{c.m.}} = -\mathbf{p}_B^{\text{c.m.}}, \quad \mathbf{p}_{A'}^{\text{c.m.}} = -\mathbf{p}_{B'}^{\text{c.m.}}, \tag{7.2.2}$$

and the laboratory one (lab); cf. Fig. 7.2.2. In the c.m. system we can express all quantities in terms of invariants: *energy,*

$$E^{\text{c.m.}} = p_{A0}^{\text{c.m.}} + p_{B0}^{\text{c.m.}} = p_{A'0}^{\text{c.m.}} + p_{B'0}^{\text{c.m.}} = \sqrt{s}; \tag{7.2.3a}$$

and *momenta,*

$$p_{A0}^{\text{c.m.}} = \sqrt{m_A^2 + |\mathbf{p}_A^{\text{c.m.}}|^2} = \frac{s + m_A^2 - m_B^2}{2s^{1/2}},$$

$$p_{B0}^{\text{c.m.}} = \frac{s + m_B^2 - m_A^2}{2s^{1/2}}; \tag{7.2.3b}$$

$$|\mathbf{p}_A^{\text{c.m.}}| = |\mathbf{p}_B^{\text{c.m.}}| = \frac{1}{2s^{1/2}}\sqrt{\lambda(s, m_A^2, m_B^2)}.$$

For $p_{A'0}$, $p_{B'0}$, $|\mathbf{p}_{A'}| = |\mathbf{p}_{B'}|$, change $A \to A'$, $B \to B'$. In (7.2.3b) $\lambda$ is *Källén's quadratic form*,

$$\lambda(a, b, c) = a^2 + b^2 + c^2 - 2ab - 2ac - 2bc.$$

Finally, for the *scattering angle*, $\cos\theta^{\text{cm}}$, we have

$$\cos\theta^{\text{cm}} = [\lambda(s, m_A^2, m_B^2)\lambda(s, m_{A'}^2, m_{B'}^2)]^{-1/2}$$
$$\times \left\{ 2st + [s^2 - (m_A^2 + m_{A'}^2 + m_B^2 + m_{B'}^2)s \right. \tag{7.2.3c}$$
$$\left. + m_A^2 m_{A'}^2 + m_B^2 m_{B'}^2 - m_A^2 m_{B'}^2 - m_{A'}^2 m_B^2] \right\}.$$

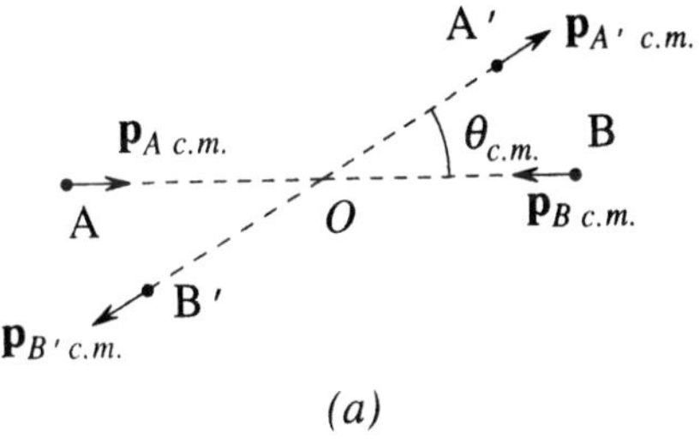


*(a)*

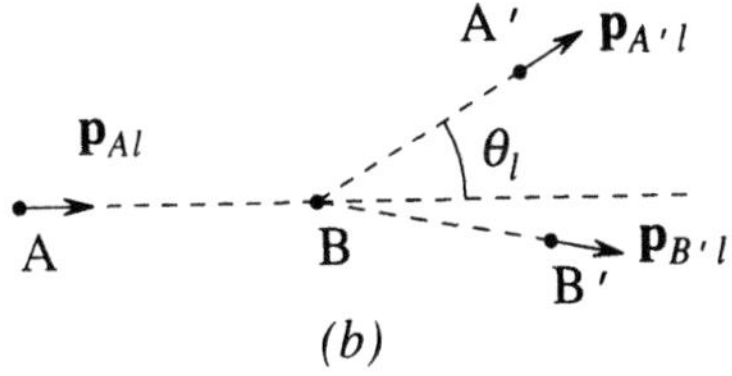


*(b)*

**Fig. 7.2.2.** Collision of two particles $A + B \to A' + B'$. (a) In the centre of mass system. (b) In the laboratory reference system.

In the laboratory reference system

$$\mathbf{p}_B = 0, \quad p_{B0} = m_B, \tag{7.2.4}$$

and we have *momenta*, and *energy*[1],

---

[1] The *lab* energy is defined by many authors to be only the energy of the *moving* particle $\tilde{E}^{\text{lab}} = E^{\text{lab}} - m_B = p_{A0}^{\text{lab}}$.

$$p_{A0}^{\text{lab}} = E^{\text{lab}} - m_B = \frac{s - m_A^2 - m_B^2}{2m_B},$$

$$p_{A'0}^{\text{lab}} = \frac{1}{m_B} p_{A'} \cdot p_B = \frac{m_{A'}^2 + m_B^2 - u}{2m_B},$$

$$p_{B'0}^{\text{lab}} = \frac{1}{m_B} p_{B'} \cdot p_B = \frac{m_B^2 + m_{B'}^2 - t}{2m_B};$$

$$|\mathbf{p}_A^{\text{lab}}| = \frac{\sqrt{\lambda(s, m_A^2, m_B^2)}}{2m_B},$$

$$|\mathbf{p}_{A'}^{\text{lab}}| = \frac{\sqrt{\lambda(u, m_{A'}^2, m_B^2)}}{2m_B},$$

$$|\mathbf{p}_{B'}^{\text{lab}}| = \frac{\sqrt{\lambda(t, m_B^2, m_{B'}^2)}}{2m_B}.$$

$$(7.2.5a)$$

For the proof of some of these relations a trick is very useful. One realizes that an invariant quantity is, of course, independent of the reference system. Therefore, in the laboratory, where

$$p_B \sim \begin{pmatrix} m_B \\ 0 \end{pmatrix},$$

the scalar product of any vector $p$ by $p_B$, on the one hand, is $p \cdot p_B = m_B p_0^{\text{lab}}$; and, on the other, it can be related to invariants. For example, $p_{A'} \cdot p_B = m_B p_{A'0}^{\text{lab}}$, on one the hand; on the other, $u = (p_{A'} - p_B)^2 = p_{A'}^2 + p_B^2 - 2p_{A'} \cdot p_B$; equating these we get the expression for $p_{A'0}^{\text{lab}}$ recorded in (7.2.5a) above.

## 7.3 The $S$ Matrix. Scattering Amplitude. Nonrelativistic Limit

The $S$ matrix is defined as the operator that transforms remote past into remote future. If $U(t_2, t_1)$ is the evolution operator from time $t_1$ to time $t_2$, one may write

$$S = \lim_{\substack{t_1 \to -\infty \\ t_2 \to +\infty}} U(t_2, t_1).$$

$$(7.3.1)$$

Because $U$ is unitary, so will $S$ be:[2]

$$S^+ S = 1.$$

$$(7.3.2)$$

---

[2] The rigorous proof of the existence of the limit (7.3.1) is far from trivial, and indeed can only be achieved in some simple cases, almost all of them nonrelativistic. These, together with references, may be found in e.g. the text of Galindo and Pascual (1978).

$S$ is also invariant under translations and Lorentz transformations: if $U(a, \Lambda)$ is the operator corresponding to the four-translation $a$ and the Lorentz transformation $\Lambda$,

$$U(a, \Lambda) S U(a, \Lambda)^{-1} = S. \tag{7.3.3}$$

The proof is not difficult. For finite $t_1, t_2$,

$$U(t_2, t_1) = \exp \frac{-i}{\hbar}(t_2 - t_1)H,$$

where $H$ is the Hamiltonian. Since this commutes with rotations and four-translations, so will $U$. The invariance under Lorentz accelerations is also simple and will be left to the reader.

If we normalize states invariantly,

$$\langle p', \alpha' | p, \alpha \rangle = \delta_{\alpha\alpha'} 2p_0 \delta(\mathbf{p} - \mathbf{p}'), \tag{7.3.4}$$

it then follows that the matrix elements of $S$ will be invariant functions. It is convenient, first, to separate a trivial[3] unity off $S$ (if there were no interactions we would have $S = 1$), letting

$$S = 1 + i\mathcal{M}; \tag{7.3.5a}$$

then to extract explicitly a delta function of four-momentum conservation from the matrix elements of $\mathcal{M}$. If we consider the matrix element of $S$ between states $|i\rangle$ ("initial") and $|f\rangle$ ("final"), we can then write

$$\begin{aligned}
\langle f|S|i \rangle &= \langle f|i \rangle + i\langle f|\mathcal{M}|i \rangle \\
&= \langle f|i \rangle + i\delta_4(p_f - p_i)F(i \to f),
\end{aligned} \tag{7.3.5b}$$

where $p_{f/i}$ is the total four-momentum of the final/initial state and, by virtue of the invariance of $S$ and the states, $F(i \to f)$, known as the *transition amplitude*, or *scattering amplitude*, will be an invariant function of four-momenta and spins.

It is possible to use a nonrelativistic normalization for the states. If we consider states normalized by

$$\overline{\langle p', \alpha' |} \overline{p, \alpha \rangle} = \delta_{\alpha\alpha'} \delta(\mathbf{p} - \mathbf{p}'), \tag{7.3.6}$$

we define

$$\langle \overline{f}|S|\overline{i} \rangle = \langle \overline{f}|\overline{i} \rangle + i\delta_4(p_f - p_i)T(i \to f). \tag{7.3.7}$$

The relation between $T$ and $F$ is easily found. If the $|i\rangle$ state contains particles with four-momenta $p_1, p_2, \ldots, p_i$, and the final one with $p'_1, p'_2, \ldots, p'_f$, one has

$$T(i \to f) = (2p_{10}2p_{20} \cdots 2p_{i0}2p'_{10}2p'_{20} \cdots 2p'_{f0})^{-1/2}F(i \to f), \tag{7.3.8}$$

as is obvious from the relation

---

[3] Actually this is less trivial than it appears at first sight: if we say $S \to 1$ for interaction $\to 0$, we are actually fixing the relative phase of states at $t \to \pm\infty$.

$$|\overline{p,\alpha}\rangle = (2p_0)^{-1/2}|p,\alpha\rangle,$$

obtained by comparing (7.3.4) and (7.3.6).

$T$ is not relativistically invariant; it is, however, useful for studying the nonrelativistic limit. Nonrelativistic states are normalized as in (7.3.6), viz.,

$$_{NR}\langle p',\alpha'|p,\alpha\rangle_{NR} = \delta_{\alpha\alpha'}\delta(\mathbf{p}-\mathbf{p}'),$$

so if we define $T_{NR}$ by

$$_{NR}\langle f|S|i\rangle_{NR} =_{NR}\langle f|i\rangle_{NR}$$

$$+i\delta(E_f - E_i)\delta(\mathbf{p}_f - \mathbf{p}_i)T_{NR}(i \to f), \tag{7.3.9}$$

comparison with (7.3.7) shows that $T_{NR}$ is just the NR limit of $T$. In terms of $F$ we then have

$$F(i \to f) \underset{NR}{\to} (2p_{10}\ldots 2p_{i0}2p'_{10}\ldots 2p'_{f0})^{1/2}T_{NR}(i \to f). \tag{7.3.10}$$


## 7.4 Cross-sections and Decay Rates. The Optical Theorem

In this section we will evaluate cross-sections (and decay rates) in terms of the $S$ matrix. For the first we will consider that the initial state, $|i\rangle$, contains two particles $A+B$ with masses $m_A$, $m_B$, four-momenta $p_A$, $p_B$ and spin quantum numbers $\alpha$, $\beta$; and the final state, $|f\rangle$, will contain $n$ particles with momenta $p'_1,\ldots,p'_n$ and spin quantum numbers $\lambda'_1,\ldots,\lambda'_n$. The spin quantum numbers will be omitted when this does not lead to confusion.

Writing the $S$-matrix elements as in (7.3.5) we thus have

$$\langle f|S|i\rangle = \langle p_A,\alpha;p_B,\beta|S|p'_1,\lambda'_1,\ldots,p'_n,\lambda'_n\rangle$$

$$= \langle f|i\rangle + i\delta_4(p_i - p_f)F(i \to f), \tag{7.4.1}$$

$$p_i \equiv p_A + p_B;\ p_f \equiv p'_1 + \ldots + p'_n.$$

We define the *cross-section* for $i \to f$ as the transition probability per unit time and per unit flux. This implies some footwork, as the states that define $F(i \to f)$, (7.4.1), are not normalized and are also defined at infinite time, cf. (7.3.1). A rigorous description goes beyond the scope of this text, and may be found, for example, in Goldberger and Watson (1965), or standard textbooks on quantum mechanics such as those by Galindo and Pascual (1978) or Gottfried (1966); but an essentially correct treatment is as follows.

A state $|p\rangle$, normalized to $\langle p'|p\rangle = 2p_0\delta(\mathbf{p} - \mathbf{p}')$, corresponds to a wave function (omitting spin that plays no role in what follows)

$$\Psi^{(p)}(\mathbf{r}) = \sqrt{2p_0}\frac{1}{(2\pi)^{3/2}}e^{i\mathbf{pr}}. \tag{7.4.2}$$

Using the nonrelativistic expression for the $x$-space scalar product of the wave functions, we get

$$\langle \Psi^{(p')} | \Psi^{(p)} \rangle = \int d^3 r \, \Psi^{(p')}(\mathbf{r})^* \Psi^{(p)}(\mathbf{r})$$

$$= 2p_0 \delta(\mathbf{p} - \mathbf{p}').$$

The $\delta(\mathbf{p}_i - \mathbf{p}_f)$ piece of (7.4.1) is due to the fact that the wave functions correspond to well-defined momentum, and are thus extended over all of space; likewise, the $\delta(p_{i0} - p_{f0})$ is due to the infinite length of the time interval $t_2 - t_1$ in (7.3.1). We will interpret the occurrence of deltas of zero argument as

$$\delta(\mathbf{p}_i - \mathbf{p}_f) \simeq \frac{1}{(2\pi)^3} \int_V d^3 r \, e^{i(\mathbf{p}_f - \mathbf{p}_i)\mathbf{r}}$$

$$\underset{\mathbf{p}_i \simeq \mathbf{p}_f}{\simeq} \frac{1}{(2\pi)^3} \int_V d^3 r \, e^{i(\mathbf{p}_f - \mathbf{p}_i)\mathbf{r}} \simeq V/(2\pi)^3,$$

$$(7.4.3\text{a})$$

where $V$ is the large but finite volume where the collision process takes place; and

$$\delta(p_{i0} - p_{f0}) \simeq \frac{1}{2\pi} \int_{t_1}^{t_2} dt \, e^{-i(p_{f0} - p_{i0})t}$$

$$\underset{p_{f0} \simeq p_{i0}}{\simeq} T/(2\pi),$$

$$(7.4.3\text{b})$$

$T = t_2 - t_1$ being the large time during which such a process occurs.

We then define the cross-section $\sigma$ as the transition probability in time $T$, per unit time, in volume $V$, per unit volume, and per unit flux (in the centre of mass) of the incoming particles, that the state $|i\rangle$ evolves to $|f\rangle$. The limit of very large $V$, $T$ is to be taken.

According to the ordinary quantum mechanical rules for computing transition probabilities we then have

$$\sigma(i \to f) = \lim_{V,T \to \infty} \frac{|\langle f | U(t_2, t_1) | i \rangle|^2}{TV\Phi_{12}},$$

$$(7.4.4)$$

where $\Phi_{12}$ is the flux. For plane waves this is equal to the product of the densities, $|\Psi_A^{(p_A)}|^2 \, |\Psi_B^{(p_B)}|^2 = 4p_{A0}^{\text{c.m.}} \, p_{B0}^{\text{c.m.}} \, (2\pi)^{-6}$, times the relative velocity in the c.m. frame,

$$v_{12} = |\mathbf{p}_A^{cm}|/|p_{A0}^{\text{c.m.}}| + |\mathbf{p}_B^{cm}|/|p_{B0}^{\text{c.m.}}|,$$

so that, using also (7.2.3),

$$\Phi_{12} = 4(2\pi)^{-6} |\mathbf{p}_A^{cm}| (p_{A0}^{\text{c.m.}} + p_{B0}^{\text{c.m.}}) = 2\lambda^{1/2}(s, m_A^2, m_B^2)/(2\pi)^6. \qquad (7.4.5)$$

Taking the limits $T, V \to \infty$ in (7.4.4) whenever possible, and substituting (7.4.1) for the $S$-matrix elements[4], we have

---

[4] We will make the assumption that the final momenta are not identical with the initial ones, so that $\langle f|i\rangle = 0$. The case $|i\rangle = |f\rangle$ is to be taken as a limiting case, at the end of the calculations.

$$\sigma(i \to f) = \lim_{V,T} \frac{1}{VT\Phi_{12}} \delta(\mathbf{p}_i - \mathbf{p}_f)|_{\mathbf{p}_i \simeq \mathbf{p}_f} \delta(p_{i0} - p_{f0})|_{p_{i0} \simeq p_{f0}}$$

$$\times |F(i \to f)|^2 \delta_4(p_i - p_f)$$

$$= \frac{1}{\Phi_{12}} \frac{1}{(2\pi)^4} |F|^2 \delta_4(p_i - p_f) = \frac{2\pi^2}{\lambda^{1/2}(s, m_A^2, m_B^2)} |F(i \to f)|^2 \delta_4(p_i - p_f),$$

where we have effected the replacements (7.4.3), and substituted (7.4.5). Because the final states are normalized to

$$\sum_{\lambda'} \int \frac{d^3p'}{2p_0'} |p', \lambda'\rangle\langle p', \lambda'| = 1,$$

we have to take the density of final states as

$$\frac{d^3p_1'}{2p_{10}'} \cdots \frac{d^3p_n'}{2p_{n0}'},$$

so that we finally obtain the *differential cross-section,*

$$d\sigma(i \to f) = \frac{2\pi^2}{\lambda^{1/2}(s, m_A^2, m_B^2)} |F(i \to f)|^2 \delta_4(p_i - p_f) \frac{d^3p_1'}{2p_{10}'} \cdots \frac{d^3p_n'}{2p_{n0}'}. \tag{7.4.6}$$

From (7.4.6) we can obtain differential cross-sections with different variables fixed. For example, the cross-section $d\sigma/d\Omega|_{\text{c.m.}}$ at given solid angle $\cos\theta, \varphi$ for quasi-elastic scattering,

$$A + B \to A' + B',$$

with $A', B'$ different or equal to $A + B$, is

$$\left.\frac{d\sigma}{d\Omega}\right|_{\text{c.m.}} = \frac{\pi^2}{4s} \frac{\lambda^{1/2}(s, m_{A'}^2, m_{B'}^2)}{\lambda^{1/2}(s, m_A^2, m_B^2)} |F(i \to f)|^2, \tag{7.4.7}$$

$$d\Omega = d\varphi d\cos\theta.$$

It is easily obtained by integrating (7.4.6) subject to the condition that $\cos \mathrm{ang}(\mathbf{p}_A, \mathbf{p}_{A'}')$ equal to $\cos\theta$, azimuthal angle $\varphi_{\mathbf{p}_A'}$ of $\mathbf{p}_{A'}'$ equal to the prescribed $\varphi$, i.e., choosing $OZ \parallel \mathbf{p}_A$, we get

$$\left.\frac{d\sigma}{d\Omega}\right|_{\text{c.m.}} = \int d\sigma(i \to f)\delta(\cos\theta_{\mathbf{p}_{A'}'} - \cos\theta)\delta(\varphi_{\mathbf{p}_{A'}'} - \varphi), \tag{7.4.8}$$

with $d\sigma(i \to f)$ given by (7.4.6), with $n = 2$.

**Exercise.** Evaluate the integrals in (7.4.8) and verify that one gets the expression (7.4.7) ∎

A differential cross-section much used for relativistic collisions is that at fixed momentum transfer, $t = (p_A - p_{A'}')^2$, which is invariant. We can make use of (cf. (7.2.3c))

$$d\cos\theta^{\text{c.m.}}_{\mathbf{p}'_{A'}}/dt = 2s/\lambda^{1/2}(s, m_A^2, m_B^2)\lambda^{1/2}(s, m_{A'}^2, m_{B'}^2)$$

and integrate $d\varphi$ in (7.4.7) so that

$$\frac{d\sigma}{dt} = \int d\varphi \left(\left.\frac{d\sigma}{d\Omega}\right|_{\text{c.m.}}\right)\frac{d\cos\theta}{dt},$$

and thus

$$\frac{d\sigma}{dt} = \frac{\pi^3}{\lambda(s, m_A^2, m_B^2)}|F(i \to f)|^2. \tag{7.4.9}$$

For *decay rates*, a similar analysis yields the *differential decay rate* of a particle $|i\rangle = |p, \alpha\rangle$ into a final state $|f\rangle = |p'_1, \lambda'; \ldots; p'_n, \lambda'_n\rangle$, in the rest system of the decaying particle:

$$d\Gamma(i \to f) = \frac{1}{4\pi m}|F(i \to f)|^2\delta(m - p_{f0})\delta(\mathbf{p}_f)\frac{d^3p'_1}{2p'_{10}}\cdots\frac{d^3p'_n}{2p'_{n0}}; \tag{7.4.10}$$

$m$ is the mass of the decaying particle.

We will next give a formula for the total cross-section, i.e., the cross section for an initial state to go into *any* final state:

$$\sigma_{\text{tot}} = \sum_f d\sigma(i \to f) = \frac{2\pi^2}{\lambda^{1/2}(s, m_A^2, m_B^2)}$$

$$\times \sum_n \sum_{\lambda'_1\cdots\lambda'_n} \int \frac{d^3p'_1}{2p'_{10}}\cdots\frac{d^3p'_n}{2p'_{n0}}\delta(p_i - p_f)|F(i \to f)|^2. \tag{7.4.11}$$

Now we turn to the unitarity of $S$. This implies that, for any states $|a\rangle$, $|b\rangle$,

$$\langle a|b\rangle = \langle a|S^+S|b\rangle = \langle a|(1 - i\mathcal{M}^+)(1 + i\mathcal{M})|b\rangle$$

$$= \langle a|b\rangle + i\langle a|\mathcal{M} - \mathcal{M}^+|b\rangle + \langle a|\mathcal{M}^+\mathcal{M}|b\rangle,$$

so that

$$i\langle a|\mathcal{M} - \mathcal{M}^+|b\rangle = i\delta_4(p_a - p_b)\{F(b \to a) - F^*(a \to b)\}$$

$$= -\langle a|\mathcal{M}^+\mathcal{M}|b\rangle = -\sum_f \langle a|\mathcal{M}^+|f\rangle\langle f|\mathcal{M}|b\rangle$$

$$= -\sum_f \delta(p_a - p_f)\delta(p_f - p_b)F(a \to f)^*F(b \to f).$$

We thus find that

$$\frac{1}{2i}\{F(b \to a) - F^*(a \to b)\} =$$

$$\frac{1}{2}\sum_f \delta(p_a - p_f)F(a \to f)^*F(b \to f). \tag{7.4.12a}$$

This is the expression of the unitarity relation in terms of the $F$. It is to be noted that the sum in $f$ stands for the sums over spins and states, and integrals on momenta:

$$\sum_f \rightarrow \sum_n \sum_{\lambda_1' \ldots \lambda_n'} \int \frac{d^3 p_1'}{2p_{10}'} \cdots \frac{d^3 p_n'}{2p_{n0}'}, \qquad (7.4.12b)$$

where it is assumed that the state $|f\rangle$ is one of $n$ particles with momenta $p_1' \ldots p_n'$, and spin quantum numbers $\lambda_1', \ldots, \lambda_n'$. Now, if we compare (7.4.12) with the particular choice $|a\rangle = |b\rangle = |i\rangle$ with (7.4.11), we find a relation between $\sigma_{\text{tot}}$ and the imaginary part of the forward elastic scattering amplitude, $F(i \rightarrow i)$:

$$\sigma_{\text{tot}} = \frac{4\pi^2}{\lambda^{1/2}(s, m_A^2, m_B^2)} \mathrm{Im} F(i \rightarrow i), \qquad (7.4.13)$$

a relation known as the *optical theorem*.

**Exercise.** (A) Write unitarity for elastic scattering. (B) Show that, in this case, (7.4.13) can also be made to follow from the phase shift analysis and the reality of the phase shifts, $\delta_l$ (see Sect. 7.5 for the latter) ∎

To finish this section, we give the pertinent formulas with nonrelativistic normalizations of states, i.e., in terms of $T$ (rather than $F$), explicitly keeping $\hbar$'s and $c$'s:

$$d\sigma = \frac{(2\pi\hbar)^2}{v_{12}} |T(i \rightarrow f)|^2 \delta(E_f - E_i) \delta(\mathbf{p}_f - \mathbf{p}_i) d^3 p_1' \ldots d^3 p_n', \qquad (7.4.14)$$

with $v_{12}$ the c.m. relative velocity of the initial particles; for initial particles with mass, $v_{12} = |\mathbf{p}_1^{\text{c.m.}}|(m_1 + m_2)/m_1 m_2$, in which case the elastic differential cross-section is

$$\left.\frac{d\sigma}{d\Omega}\right|_{\text{c.m.}} = (2\pi\hbar m)^2 |T(i \rightarrow f)|^2. \qquad (7.4.15)$$

Also,

$$\sigma_{\text{tot}} = \frac{2(2\pi\hbar)^2}{v_{12}} \mathrm{Im} T(i \rightarrow i). \qquad (7.4.16)$$

Finally, in the rest system of the decaying particle,

$$d\Gamma(i \rightarrow f) = \frac{1}{2\pi} |T(i \rightarrow f)|^2 \delta(E_f - E_i) \delta(\mathbf{p}_f) d^3 p_1' \ldots d^3 p_n' \qquad (7.4.17)$$

**Exercise.** Verify (7.4.14)–(7.4.17) ∎

## 7.5 Partial Wave Analysis and Phase Shifts. I. Spinless Elastic Scattering. Effective Range Expansion

### 7.5.1 Partial Wave Analysis

Let us consider the matrix element of $S$ between two-particle states, and use (7.1.4) to separate the c.m. piece:

$$\langle p'_A, p'_B | S | p_A, p_B \rangle = \langle p'_A, p'_B | p_A, p_B \rangle$$

$$+ i\delta_4(p' - p) F(i \to f)$$

$$= \delta_4(p' - p) \left\{ \frac{4p_0^{\text{c.m.}}}{|\mathbf{k}|^{\text{c.m.}}} \delta(\Omega_{\mathbf{k}} - \Omega_{\mathbf{k}'}) + iF(i \to f) \right\}. \tag{7.5.1}$$

The particles in the final state are assumed to be the same as those in the initial state (*elastic* scattering). We recall that

$$p_0^{\text{c.m.}} = s^{1/2}, \ k \equiv |\mathbf{k}^{\text{c.m.}}| = |\mathbf{p}_A^{\text{c.m.}}| = \frac{1}{2s^{1/2}} \lambda(s, m_A^2, m_B^2) \tag{7.5.2}$$

(cf. (7.2.3)). $F(i \to f)$, being an invariant function, will only depend upon the two invariant variables, which (omitting the c.m. label henceforth) we choose as energy, $E = s^{1/2}$, and scattering angle, $\cos\theta$:

$$F(i \to f) = F(E, \cos\theta).$$

We may expand $F$ in Legendre polynomials,

$$F(E, \cos\theta) = \sum_l (2l + 1) P_l(\cos\theta) f_l(s); \tag{7.5.3}$$

the $f_l(s)$ are called *partial wave amplitudes*. Also, by writing

$$\delta(\Omega_{\mathbf{k}} - \Omega_{\mathbf{k}'}) = \sum_{lM} Y_M^l(\Omega_{\mathbf{k}})^* Y_M^l(\Omega_{\mathbf{k}'})$$

$$= \frac{1}{4} \sum (2l + 1) P_l(\cos\theta), \tag{7.5.4a}$$

which follows by choosing $OZ$ along $\mathbf{k}$ so that

$$Y_M^l(\Omega_{\mathbf{k}})^* = \delta_{M0} \sqrt{(2l + 1)/4\pi},$$

$$Y_0^l(\Omega_{\mathbf{k}}) = \sqrt{(2l + 1)/4\pi} \ P_l(\cos\theta),$$

$$\tag{7.5.4b}$$

$$\theta = \text{ang}(\mathbf{k}, \mathbf{k}'),$$

$$Y_M^l(\cos\theta, \varphi = 0) = \sqrt{\frac{2l + 1}{4\pi} \frac{(l - M)!}{(l + M)!}} \ P_l^M(\cos\theta),$$

we can substitute these into (7.5.1) and find that

$$\langle p'_A, p'_B | S | p_A, p_B \rangle$$

$$\tag{7.5.5a}$$

$$= \delta_4(p' - p) \sum_l (2l + 1) P_l(\cos\theta) \left\{ \frac{s^{1/2}}{k\pi} + if_l(s) \right\}.$$

On the other hand, we can write the states $|p_A, p_B\rangle$, $|p'_A, p'_B\rangle$ as in (7.1.3): then,

$$\langle p'_A, p'_B | S | p_A, p_B \rangle = \langle p' | \otimes \langle \mathbf{k}' | S | \mathbf{k} \rangle \otimes | p \rangle = \delta_4(p - p') \langle \mathbf{k}' | S | \mathbf{k} \rangle.$$

Expanding the $|\mathbf{k}\rangle$, $|\mathbf{k}'\rangle$ in terms of the $|l, M\rangle$ as in (7.1.7), and letting $OZ \parallel \mathbf{k}$, we get

$$\langle p'_A, p'_B|S|p_A, p_B\rangle = \delta_4(p - p')$$

$$\times \sum_{lM} \sum_{l'M'} Y_{M'}^{l'}(\Omega_{\mathbf{k}'}) Y_M^l(\Omega_{\mathbf{k}}) \langle l'M'|S|l, M\rangle \tag{7.5.5b}$$

$$= \delta_4(p - p') \sum_l (2l + 1) P_l(\cos\theta) \frac{1}{4\pi} S_l,$$

where

$$S_l \delta_{ll'} \delta_{MM'} = \langle l'M'|S|l, M\rangle. \tag{7.5.5c}$$

**Exercise.** Show that $\langle l'M'|S|l, M\rangle$ is indeed of the form (7.5.5c), where $S_l$ may depend on $l$ (and on the energy $E$) but is independent of $M$.

*Hint.* Use rotational invariance of $S$ ∎

On comparing (7.5.5c) with (7.5.5a) we find that

$$S_l = \frac{4s^{1/2}}{k} + 4\pi i f_l(s). \tag{7.5.6}$$

More information is gained by using the unitarity of $S$. Consider $\langle l'M'|l, M\rangle$; we have

$$\frac{4s^{1/2}}{k} \delta_{ll'} \delta_{MM'} = \langle l'M'|lM\rangle = \langle l'M'|S^+ S|lM\rangle$$

$$= \sum_{l''M''} \langle l'M'|S^+ \frac{k}{4s^{1/2}}|l''M''\rangle \langle l''M''|S|lM\rangle$$

$$= \frac{k}{4s^{1/2}} \sum \langle l''M''|S|l'M'\rangle^* \langle l''M''|S|lM\rangle$$

$$= \frac{k}{4s^{1/2}} \delta_{ll'} \delta_{MM'} S_l^* S_l :$$

therefore, $S_l^* S_l = (4s^{1/2}/k)^2$, and we can write

$$S_l = \frac{4s^{1/2}}{k} e^{2i\delta_l(s)}, \tag{7.5.7}$$

where the $\delta_l$ are the *phase shifts*, and are real numbers. If we then substitute (7.5.7) into (7.5.6), it follows that

$$f_l(s) = \frac{2s^{1/2}}{\pi k} \sin \delta_l e^{i\delta_l}, \tag{7.5.8}$$

and if we further substitute this into (7.5.3), we obtain the relativistic version of the phase shift expansion:

$$F(i \to f) = \frac{2s^{1/2}}{\pi k} \sum (2l + 1) P_l(\cos\theta) \sin \delta_l(s) e^{i\delta_l(s)}. \tag{7.5.9}$$

This is to be compared with the nonrelativistic formula

$$T_{NR} = \frac{1}{2\pi mk} \sum (2l+1) P_l(\cos\theta) \sin\delta_l^{NR} e^{i\delta_l^{NR}}, \tag{7.5.10}$$

with $m$ the reduced mass of the system.

### 7.5.2 Effective Range Formalism

For completeness we will describe the effective range formalism, without going into the proofs that can be found in specialized texts like Goldberger and Watson (1965) or, for scattering by a potential, Gottfried (1966) and Galindo and Pascual (1978). There it is shown that if one defines the function $\Phi_l$ by

$$f_l(s) = \frac{4}{\pi} k^{2l} \left( \Phi_l - \frac{2i}{s^{1/2}} k^{2l+1} \right)^{-1}, \tag{7.5.11}$$

then, for interactions with finite range, $\Phi_l = \Phi_l(k^2)$ is analytic, as a function of $k^2$, in a certain neighbourhood[5] of $k^2 = 0$. Moreover, for $k^2$ real, $\Phi_l^*(k^2) = \Phi_l(k^2)$. Note that $\Phi_l$ may be written in terms of the phase shifts as

$$\Phi_l(k^2) = \frac{2k^{2l+1}}{s^{1/2}} \cot\,\delta_l. \tag{7.5.12}$$

In view of the analyticity of $\Phi$, we may expand it in a power series,

$$\Phi_l(k^2) \simeq \frac{2}{s^{1/2}} (a_l^{-1} + r_{0l}k^2 + r_{1l}k^4 + \ldots). \tag{7.5.13}$$

The definitions in (7.5.13) are dictated by custom. The quantity $a_l$ is called the $l$-wave *scattering length*; $a_l^2$ is proportional to the zero-energy scattering cross-section for the angular momentum $l$. $r_{nl}$ are the *effective range* parameters. In particular, $r_{0l}$ is related to the range of the interaction; indeed, if the interaction has infinite range, $r_{0l} = \infty$, and the development (7.5.13) fails. This may be verified explicitly for the Coulomb interaction. For other interactions, notably nuclear ones, the term in $r_{1l}k^4$ and the following ones may be neglected for low energies when the composite structure of the nucleons, or nuclei, may be ignored.

## 7.6 Partial Wave Analysis.
## II. Several Two-Body Channels

### 7.6.1 Multichannel Analysis

In increasing order of complexity, we now consider collisions of two particles without spins (or subject to interactions that do not involve the spins) giving

---

[5] The region of analyticity of $\Phi_l$ depends on the particular process under consideration.

another two, in general different, particles. Let us denote such pairs by $A_a + B_a$, $a = 1, \ldots, N$; we thus consider reactions

$$A_a + B_a \to A_b + B_b, \quad a, b = 1, \ldots, N.$$

Each such possible reaction is called a *channel*. As an example of such reactions, we have the set

$$D + D \quad \to \quad D + D,$$

$$D + D \quad \to \quad T + p,$$

$$D + D \quad \to \quad {}^3He + n,$$

$$D + D \quad \to \quad {}^4He + \gamma,$$

$$T + p \quad \to \quad D + D,$$

$$T + p \quad \to \quad {}^3He + n,$$

$$\vdots$$

$${}^4He + \gamma \quad \to \quad {}^4He + \gamma,$$

at low energy (up to a few MeV). Indeed the analysis that follows, called the multichannel formalism, is particularly useful in low-energy nuclear physics.

With self-explanatory notation we will denote states by

$$|p_1, p_2\rangle_a, \quad a = 1, \ldots, N, \tag{7.6.1a}$$

and, in particular, states with well-defined angular momentum by

$$|l, M\rangle_a, \quad a = 1, \ldots, N. \tag{7.6.1b}$$

The matrix element of an operator, in particular of the $S$ matrix, will be written as

$$\delta_{ll'}\delta_{MM'}S^l_{ab} = {}_a\langle l', M'|S|l, M\rangle_b.$$

$S^l$ was denoted $S_l$ before; the present notation is handier for the multichannel case. The meaning is clear: $S^l_{ab}$ is the matrix element of $S$ between states of particles $A_a, B_a$ and $A_b, B_b$, both states with total orbital angular momentum equal to $l$.

As is apparent, it will be convenient to use matrix notation. Denoting, in the remainder of this section, matrices by underlining them with a tilde, we then have the matrix $\underset{\sim}{S}^l$ with elements

$$\underset{\sim}{S}^l = (S^l_{ab}).$$

It is convenient also to introduce diagonal matrices whose components are momenta:

$$\underset{\sim}{k} = (k_a\delta_{ab}), \quad \underset{\sim}{k}^{1/2} = (k_a^{1/2}\delta_{ab}), \quad \text{etc.}$$

The analysis of the previous section can be repeated with simple alterations. We generalize (7.5.6) by defining

$$\underset{\sim}{S}^l = 4s^{1/2}\underset{\sim}{k}^{-1} + 4\pi i f_l(s), \quad f_l = (f_{lab}).$$
(7.6.2)

The unitarity relation is somewhat modified. We have

$$4s^{1/2}(\underset{\sim}{k}^{-1})_{ab}\delta_{ll'}\delta_{MM'} =_a \langle l'M'|lM\rangle_b$$

$$= \sum_{l''M''}\sum_c {}_a\langle l'M'|S^+|l''M''\rangle_c \frac{k_c}{4s^{1/2}} {}_c\langle l''M''|S|lM\rangle_b$$

$$= \delta_{ll'}\delta_{MM'}\frac{1}{4s^{1/2}}\sum_c (\underset{\sim}{S}^{l+})_{ac}k_c(\underset{\sim}{S}^l)_{cb},$$

so that the unitarity relation now reads

$$16s\,\underset{\sim}{k}^{-1} = \underset{\sim}{S}^{l+}\underset{\sim}{k}\underset{\sim}{S}^l,$$
(7.6.3)

i.e., the matrix

$$\underset{\sim}{U}^l \equiv \frac{1}{4s^{1/2}}\underset{\sim}{k}^{1/2}\underset{\sim}{S}^{1/2}\underset{\sim}{k}^{1/2}$$
(7.6.4a)

is unitary:

$$\underset{\sim}{U}^{l+}\underset{\sim}{U}^l = 1.$$
(7.6.4b)

Let $\underset{\sim}{C}$ be the matrix that diagonalizes $\underset{\sim}{U}$; $\underset{\sim}{C}$ itself is unitary and, if the interactions respect time reversal invariance, can be chosen real. Thus,

$$\underset{\sim}{U} = \underset{\sim}{C}_l\underset{\sim}{D}^l\underset{\sim}{D}_l^{-1},$$
(7.6.5a)

where $\underset{\sim}{D}^l$, which is diagonal and unitary, can be written as

$$\underset{\sim}{D}^l = (e^{2i\tilde{\delta}_a^l(s)}\delta_{ab}) :$$
(7.6.5b)

the scattering is described not only by the $\tilde{\delta}_a^l(s)$, called the *eigenphase-shifts*, but by the matrix elements $C_{ab}^l(s)$ of $C_l$, which are the parameters giving the mixing among the various channels.

Unwinding the thread it follows that we can write

$$\underset{\sim}{D}^l = 1 + 2i\frac{\pi}{2s^{1/2}}\underset{\sim}{C}_l^{-1}\underset{\sim}{k}^{1/2}f_l\underset{\sim}{k}^{1/2}\underset{\sim}{C}_l,$$
(7.6.6)

so that the $\tilde{f}_l$ defined by

$$F_{ab} = F(b \rightarrow a) = \sum(2l+1)P_l(\cos\theta)f_{lab}$$
(7.6.7)

satisfy

$$\underset{\sim}{f}_l = \frac{2s^{1/2}}{\pi}\underset{\sim}{C}_l\underset{\sim}{k}^{-1/2}(\sin\tilde{\delta}_a^l e^{i\tilde{\delta}_a^l}\delta_{ab})\underset{\sim}{k}^{-1/2}\underset{\sim}{C}_l^{-1},$$
(7.6.8)

which is the generalization of the phase shift expansion (7.5.9) to the multi-channel case.

**Exercise.** Show that if the matrix $\underset{\sim}{U}$ were itself diagonal there would be no mixing: $\underset{\sim}{C}_l = 1$, and (7.6.8) exactly reproduces (7.5.9), and the eigenphase-shifts $\tilde{\delta}$ coincide with ordinary phase shifts $\delta$, channel by channel, with zero amplitude for transitions between different channels ∎

This lack of mixing almost never happens, although there are many cases where the mixing is small, $\underset{\sim}{C}_l \simeq 1 + (\text{small})$.

**Exercise.** Show that, in general, unitarity implies that there exist $\delta, \eta$ with

$$f_{laa} = \frac{2s^{1/2}}{\pi k_a}\eta\sin\delta e^{i\delta}, \quad \delta = \text{real}, \ 1 \geq \eta \geq 0 \ ∎$$

### 7.6.2 Effective Range Approximation

The effective range formalism of Sect. 7.5.2 may be generalized to the quasi-elastic multichannel case. If we now write

$$\underset{\sim}{f}_l = \frac{4}{\pi}\underset{\sim}{k}^l\left(\underset{\sim}{\Phi}^l - \frac{2i}{s^{1/2}}\underset{\sim}{k}^{2l+1}\right)^{-1}\underset{\sim}{k}^l, \tag{7.6.9}$$

then the matrix elements of $\underset{\sim}{\Phi}^l$, $\Phi^l_{ab}(s)$ are analytic in $s$ around $s \simeq s_a$, where $s_a$ is the threshold for channel $a$; this is valid for short-range interactions, and provided all $s_a$ are close enough to one another. If $s_0$ is the lowest threshold, we may generalize (7.5.13) and write

$$\Phi^l_{ab}(s) \simeq \frac{2}{s^{1/2}}\left\{\Phi^l_{ab}(s_0) + (s - s_0)\Phi^{l'}_{ab}(s_0) + \ldots\right\}. \tag{7.6.10}$$

The generalization of the reality property of $\Phi$ is that now, and for $s = $ real,

$$\underset{\sim}{\Phi}^{l+} = \underset{\sim}{\Phi}^l;$$

if the interactions preserve time reversal invariance, one further has

$$\underset{\sim}{\Phi}^{l*} = \underset{\sim}{\Phi}^l,$$

so $\underset{\sim}{\Phi}^l$ is a real symmetric matrix. Its eigenvalues are related to $\cot\tilde{\delta}^l_a$, where the $\tilde{\delta}^l_a$ are the eigenphase shifts.

## 7.7 Partial Wave Analysis. III. Particles with Spin

### 7.7.1 Spin Analysis

When we constructed states with well-defined total orbital angular momentum (and third component thereof) in Sect. 7.1 we remarked that one could still compose this with spin to obtain states with well-defined total angular momentum, say $j$, and third component $\lambda$. To do so we let $s_1, s_2$ be the spins of the particles, and $\alpha, \beta$ the third components of the spins. A state can thus be labelled as

$$|p_A, s_1, \alpha\rangle \otimes |p_B, s_2, \beta\rangle$$
$$= |p = p_A + p_B\rangle \otimes |\mathbf{k}; s_1, \alpha; s_2\beta\rangle; \tag{7.7.1a}$$

cf. (7.1.3). Next, we define, as in (7.1.7),

$$|l; M; s_1, \alpha; s_2, \beta\rangle = \int d\Omega_\mathbf{k} Y_M^l(\Omega_k)|\mathbf{k}; s_1, \alpha; s_2, \beta\rangle, \tag{7.7.1b}$$

with the normalizations

$$\langle p'|p\rangle = \delta_4(p - p'), \tag{7.7.1c}$$

$$\langle l', M'; s_1, \alpha'; s_2, \beta'|l, M; s_1, \alpha; s_2, \beta\rangle$$
$$= \frac{4s^{1/2}}{|\mathbf{k}|}\delta_{ll'}\delta_{MM'}\delta_{\alpha\alpha'}\delta_{\beta\beta'}. \tag{7.7.1d}$$

The way to compose angular momenta is not unique. We will choose to combine first the spins $s_1$, $s_2$ to a total spin $s$,

$$|l, M; s_1, \alpha; s_2, \beta\rangle =$$
$$\sum_{s=|s_1-s_2|}^{s_1+s_2} (s_1, \alpha; s_2, \beta|s)|l, M; s, \alpha + \beta\rangle, \tag{7.7.2a}$$

where $(s_1, \alpha; s_2, \beta|s)$ are Clebsch–Gordan coefficients. Next we compose $s$ and $l$ to total angular momentum $j$. Because the same $j$ may be obtained from different $l, s$, we have to introduce a discrete parameter $\rho$ to distinguish these degenerate states. Thus, we have

$$|l, M; s, \alpha + \beta\rangle =$$
$$\sum_{j=|l-s|}^{l+s} (l, M; s, \alpha + \beta)|j, \lambda = \alpha + \beta + M; \rho\rangle. \tag{7.7.2b}$$

Denoting by $(s_1, \alpha; s_2, \beta; l, M|j, \rho)$ the product of the Clebsch–Gordan coefficients of (7.7.2), we can write

$$|l, M; s_1, \alpha; s_2, \beta\rangle =$$
$$\sum (s_1, \alpha; s_2, \beta; l, M|j, \rho)|j, \lambda = \alpha + \beta + M; \rho\rangle, \tag{7.7.2c}$$

so that, with $p_A + p_B = p_i, p'_A + p'_B = p_f$ denoting the total initial/final momenta,

$$\langle p'_{A'}, s'_1, \alpha'; p'_{B'}, s'_2, \beta'|S|p_A, s_1, \alpha; p_B, s_2, \beta\rangle$$
$$= \delta_4(p_f - p_i)\langle \mathbf{k}'; s'_1, \alpha'; s'_2, \beta'|S|\mathbf{k}; s_1, \alpha; s_2, \beta\rangle. \tag{7.7.3a}$$

Here

$$\langle \mathbf{k}'; s_1', \alpha'; s_2', \beta' | S | \mathbf{k}; s_1, \alpha; s_2, \beta \rangle$$

$$= \sum_{lM,l'M'} Y_{M'}^{l'}(\Omega_{\mathbf{k}'}) Y_M^l(\Omega_{\mathbf{k}})^*$$

$$\times \sum_{j\rho\rho'} \delta_{M+\alpha+\beta,M'+\alpha'+\beta'} (s_1', \alpha'; s_2', \beta'; l', M'|j, \rho')$$

$$\times (s_1, \alpha; s_2, \beta; l, M|j, \rho) S_{\rho\rho'}^j, \tag{7.7.3b}$$

and

$$\delta_{jj'} \delta_{\lambda\lambda'} S_{\rho\rho'}^j = \langle j', \lambda'; \rho' | S | j, \lambda, \rho \rangle. \tag{7.7.3c}$$

Just as in the spinless case, we could show that the matrix

$$\underset{\sim}{U}_j = \frac{k}{4s^{1/2}} \underset{\sim}{S}^j$$

is unitary; if $C_j$ is the matrix that diagonalizes it, we get the expression in terms of the phase shifts

$$S_{\rho\rho'}^j = \frac{4s^{1/2}}{k} \sum C_{\rho''\rho}^* e^{2i\bar{\delta}_{j\rho''}} C_{\rho''\rho'}. \tag{7.7.4}$$

### 7.7.2 Scattering of Spin 0 – Spin 1/2 Particles

The formulas of the previous section are very cumbersome. To show how to use them in a practical case we will consider the scattering of a spin 0 by a spin 1/2 particle, *assuming parity conservation*. This is a very important case, and not excessively complicated. To begin with, we carry out the composition of angular momentum and spin. We denote the c.m. states by $|\mathbf{k}, \lambda\rangle$, $\lambda$ being the third component of the spin of the spin 1/2 particle. The states with well-defined orbital angular momentum $l$, and third component $M$, will be

$$|l, M; \lambda\rangle = \int d\Omega_{\mathbf{k}} Y_M^l(\Omega_{\mathbf{k}}) |\mathbf{k}, \lambda\rangle, \tag{7.7.5a}$$

and those with total angular momentum $j$ are thus

$$|l, M; \lambda\rangle = \sum_{j=l-1/2}^{l+1/2} (l, M; 1/2, \lambda|j) |j, M+\lambda; \rho\rangle; \tag{7.7.5b}$$

the same value of $j$ may be obtained for $l + 1/2$, or from $l' - 1/2, l' = l + 1$; the parameter $\rho$ may here take two values. Nevertheless, if the interactions preserve parity, and since the parities of states $l$ and $l + 1$ are opposite, there is no mixture; the matrix $S_{\rho\rho'}^j$, given by

$$\delta_{jj'} \delta_{M+\lambda,M'+\lambda'} S_{\rho\rho'}^j = \langle j', M' + \lambda'; \rho' | S | j, M + \lambda, \rho \rangle, \tag{7.7.6a}$$

will be diagonal. It is customary to denote the two (diagonal) values of $S_{\rho\rho'}^j$ by $S_{l\pm}$:

$$S_{l\pm}\delta_{\rho\rho'} \equiv S^{j}_{\rho\rho'}; \tag{7.7.6b}$$

$j = l + 1/2$ for $(+)$, and $j = l - 1/2$ for the $(-)$ choice. In view of (7.7.4) we thus get

$$S_{l\pm} = \frac{4s^{1/2}}{k}\exp 2i\delta_{l\pm}, \tag{7.7.6c}$$

where we have written $\delta_{l\pm}$ for $\tilde{\delta}_{j\rho}$, also following common usage.

To obtain the detailed phase shift analysis, we take into account the results of Sect. 7.5. The analogue of (7.5.1) is now

$$\left\{\delta_{\lambda\lambda'}\frac{4s^{1/2}}{k}\delta(\Omega_{\mathbf{k}'} - \Omega_{\mathbf{k}}) + iF(i \to f)\right\}$$

$$\times\, \delta_4(p_i - p_f) = \langle p'_{A'}, \lambda'; p'_{B'}|S|p_A, \lambda; p_B\rangle$$

$$= \delta_4(p_i - p_f)\frac{4s^{1/2}}{k}\langle\mathbf{k}'|S|\mathbf{k}\rangle, \quad \mathbf{k} = \mathbf{p}^{\text{c.m.}}_A; \quad \mathbf{k}' = \mathbf{p}'^{\text{c.m.}}_{A'},$$

so that, with obvious notation, and simple manipulations,

$$F(\mathbf{k}, \lambda \to \mathbf{k}', \lambda') = -\frac{4is^{1/2}}{k}\langle\mathbf{k}', \lambda'|S|\mathbf{k}, \lambda\rangle + \frac{4is^{1/2}}{k}\delta_{\lambda\lambda'}\delta(\Omega_{\mathbf{k}} - \Omega_{\mathbf{k}'})$$

$$= -\frac{4is^{1/2}}{k}\sum_{lMl'M'}Y^{l'}_{M'}(\Omega_{\mathbf{k}'})Y^{l}_{M}(\Omega_{\mathbf{k}})^{*}\langle l', M', \lambda'|S|l, M, \lambda\rangle$$

$$+\frac{4is^{1/2}}{k}\delta_{\lambda\lambda'}\sum_{lM}Y^{l}_{M}(\Omega_{\mathbf{k}'})Y^{l}_{M}(\Omega_{\mathbf{k}})^{*}$$

$$= -\frac{4is^{1/2}}{k}\sum_{lMl'M';\pm}Y^{l'}_{M'}(\Omega_{\mathbf{k}'})Y^{l}_{M}(\Omega_{\mathbf{k}})^{*}(l', M'; 1/2, \lambda'|j, \pm)$$

$$\times (l, M; 1/2, \lambda|j, \pm)\frac{4s^{1/2}}{k}e^{2i\delta_{l\pm}} + \frac{4is^{1/2}}{k}\delta_{\lambda\lambda'}Y^{l}_{M}(\Omega_k)Y^{l}_{M}(\Omega_{\mathbf{k}})^{*}.$$

Substituting now the explicit values of the Clebsch–Gordan coefficients, choosing $\mathbf{k} \parallel OZ$ and using (7.5.4) for the values of the spherical harmonics, we obtain

$$F(\mathbf{k}, \lambda \to \mathbf{k}', \lambda') = \frac{2s^{1/2}}{\pi k}\sum_{l=0}^{\infty}(2l + 1)\sqrt{\frac{(l - \lambda + \lambda')!}{(l + \lambda - \lambda')!}}P^{\lambda-\lambda'}_{l}(\cos\theta) \tag{7.7.7}$$

$$\times\{C_-(\lambda, \lambda')\sin\delta_{l-}e^{i\delta_{l-}} + C_+(\lambda, \lambda')\sin\delta_{l+}e^{i\delta_{l+}}\},$$

where the $C$ are products of Clebsch–Gordan coefficients:

$$C_-(\lambda, \lambda) = \frac{1}{2l + 1}, \quad C_+(\lambda, \lambda) = \frac{l + 1}{2l + 1};$$

$$C_-(\lambda, -\lambda) = C_+(\lambda, -\lambda) = -\frac{\sqrt{l(l + 1)}}{2l + 1}.$$

Using this we can write (7.7.7) more explicitly as

$$F(\mathbf{k}, \lambda \to \mathbf{k}', \lambda) \equiv \frac{2s^{1/2}}{\pi} f(\cos\theta),$$

$$(7.7.8a)$$

$$F(\mathbf{k}, \lambda \to \mathbf{k}', -\lambda) \equiv -\frac{2s^{1/2}}{\pi} \sin\theta\, g(\cos\theta),$$

$$f(\cos\theta) = \frac{1}{k} \sum_{l=0}^{\infty} \{(l+1)\sin\delta_{l+}e^{i\delta_{l-}} + l\sin\delta_{l-}e^{i\delta_{l-}}\} P_l(\cos\theta),$$

$$(7.7.8b)$$

$$g(\cos\theta) = \frac{1}{k} \sum_{l=0}^{\infty} (\sin\delta_{l+}e^{i\delta_{l+}} - \sin\delta_{l-}e^{i\delta_{l-}}) \frac{dP_l(\cos\theta)}{d\cos\theta}.$$

In terms of $f, g$ we can define the spin-nonchange and spin-change cross-sections,

$$\frac{d\sigma_{\text{s.nc.}}}{d\Omega} = |f|^2, \quad \frac{d\sigma_{\text{s.c.}}}{d\Omega} = |g|^2 \sin^2\theta.$$

$$(7.7.9a)$$

When the spins are not measured, we have the unpolarized cross-section:

$$\frac{d\sigma_{\text{u.p.}}}{d\Omega} = |f|^2 + |g^2|\sin^2\theta.$$

$$(7.7.9b)$$

In our case the scattering amplitude is also usually written in another manner. If $\chi^{(\lambda)}$ are Pauli spinors,

$$\chi^{(1/2)} = \begin{pmatrix} 1 \\ 0 \end{pmatrix}, \quad \chi^{(-1/2)} = \begin{pmatrix} 0 \\ 1 \end{pmatrix},$$

we have

$$F(\mathbf{k}, \lambda \to \mathbf{k}', \lambda') = \frac{2s^{1/2}}{\pi} \sum_{ab} \chi_a^{(\lambda')*} f_{ab} \chi_b^{(\lambda)},$$

$$(7.7.10)$$

where the matrix $\underset{\sim}{f} = (f_{ab})$ is obtained in terms of the $f, g$ above:

$$\underset{\sim}{f} = f + \frac{i}{kk'}\underset{\sim}{\sigma}(\mathbf{k} \times \mathbf{k}')g.$$

**Exercise.** Verify this. Verify (7.7.9). Verify that, if the interactions are spin-independent ($\delta_{l+} = \delta_{l-}$), we recover the results of the spinless cases (7.5.9) ∎


## 7.8 Evaluation of the $S$ Matrix

### 7.8.1 The $S$ Matrix and the Interaction Picture

The *interaction picture*, also called the *Dirac* or (in quantum field theory) *Dyson picture*, was devised so that states evolve in time only with the *interaction* Hamiltonian. This will lead to an integral equation which, in turn, will allow for a nonoscillatory expression for the $S$ matrix.

Denoting by the subscripts $S$, $D$ respectively the Schrödinger and interaction pictures, we define, for states,

$$|\Psi(t)\rangle_D \equiv e^{iH_S^{(0)}t/\hbar}|\Psi(t)\rangle_S, \tag{7.8.1}$$

where $H^{(0)}$, $H^{(1)}$ will represent the free and interaction pieces of the Hamiltonian. (Indeed, the method is more general, and can be applied to the case where the total Hamiltonian can be decomposed into two pieces: one, $H^{(0)}$, containing part of the interaction, and another, $H^{(1)}$, that can be treated perturbatively.) Note that we choose the origin of time so that $S$ and $D$ pictures coincide at $t = 0$.

From (7.8.1) we find, after a simple evaluation and using the standard Schrödinger equation,

$$i\hbar\partial_t|\Psi(t)\rangle_S = (H^{(0)} + H_S^{(1)})|\Psi(t)\rangle_S,$$

the desired time evolution in the interaction picture:

$$i\hbar\partial_t|\Psi(t)\rangle_D = H_D^{(1)}(t)|\Psi(t)\rangle_D. \tag{7.8.2a}$$

Note that $H_S^{(0)}$ coincides with $H_D^{(0)}$, but

$$H_D^{(1)}(t) = e^{iH_S^{(0)}t/\hbar}H_S^{(1)}e^{-iH_S^{(0)}t/\hbar}; \tag{7.8.2b}$$

the interaction in the interaction picture is thus *different* from $H_S^{(1)}$ since $H_S^{(0)}$ and $H_S^{(1)}$ do not in general commute.

The *time evolution operator*, $U(t,t_0)$, is now more complicated to obtain than in the Schrödinger (or Heisenberg) picture. Defining $U$ by

$$U_D(t,t_0)|\Psi(t_0)\rangle_D \equiv |\Psi(t)\rangle_D, \tag{7.8.3}$$

we find from (7.8.2) that it satisfies the differential equation

$$i\hbar\partial_t U_D(t,t_0) = H_D^{(1)}U_D(t,t_0). \tag{7.8.4a}$$

This equation, together with the boundary condition

$$U_D(t,t) = 1, \tag{7.8.4b}$$

is obviously equivalent to the integral equation

$$U_D(t,t_0) = 1 - \frac{i}{\hbar}\int_{t_0}^{t} dt' H_D^{(1)}(t')U_D(t',t_0). \tag{7.8.5}$$

This equation can be solved by the Neumann–Liouville method of iterations. The zeroth order is obtained by setting $H_D^{(1)}$ to zero on the right-hand side of (7.8.5), so $U_{0D} = 1$. Substituting this back onto the right-hand side of (7.8.5), we get the first-order expression,

$$U_{1D}(t,t_0) = 1 - \frac{i}{\hbar}\int_{t_0}^{t} dt_1 H_D^{(1)}(t_1).$$

Iterating, we find the Neumann–Liouville series,

$$
U_D(t, t_0) = \sum_{n=0}^{\infty} \left(\frac{-i}{\hbar}\right)^n
$$
$$
\times \int_{t_0}^{t} dt_1 H_D^{(1)}(t_1) \int_{t_0}^{t_1} dt_2 H_D^{(1)}(t_2) \ldots \int_{t_0}^{t_{n-1}} dt_n H_D^{(1)}(t_n);
$$

$$(7.8.6)$$

the term with $n = 0$ is defined to be the unity.

Define the *time ordering* operation $T$ as follows. Let

$$
M_1(t), M_2(t), \ldots, M_n(t)
$$

be time-dependent operators. Then if the times $t_1 \ldots t_n$ are such that $t_{i_1} \geq t_{i_2} \geq \ldots t_{i_n}$, we set

$$
T\{M_1(t_1)M_2(t_2)\ldots M_n(t_n)\}
$$
$$
\equiv M_{i_1}(t_{i_1})\ldots M_{i_n}(t_{i_n}),
$$

$$(7.8.7a)$$

that is to say, $T$ orders the operators so that their time arguments grow from right to left, *as if the operators commuted*. More explicitly, if we denote by $\Pi$ the permutation

$$
\Pi(1, 2, \ldots, n) = (j_1, j_2, \ldots, j_n),
$$

then

$$
T\{M_1(t_1)\ldots M_n(t_n)\}
$$

$$
= \sum_{\Pi} \theta(t_{j_1} - t_{j_2})\ldots \theta(t_{j_{n-1}} - t_{j_n})M_{j_1}(t_{j_1})\ldots M_{j_n}(t_{j_n}),
$$

$$(7.8.7b)$$

where the sum runs over all permutations.

If all the operators $M_i$ are equal, say $M_1 = \ldots = M_n = F$, we will obviously have the equality

$$
T\{F(t_1)\ldots F(t_n)\} = \frac{1}{n!} \sum_{\Pi} T\{F(t_{j_1})\ldots F(t_{j_n})\}.
$$

Consider now an expression like that of the general term of (7.8.6), that is to say, an ordered array of integrals

$$
\int_{t_0}^{t} dt_1 F(t_1) \ldots \int_{t_0}^{t_{n-1}} dt_n F(t_n) \, .
$$

We then have

$$\int_{t_0}^{t} dt_1 F(t_1) \ldots \int_{t_0}^{t_{n-1}} dt_n F(t_n)$$

$$= \int_{t_0}^{t} dt_1 \ldots \int_{t_0}^{t} dt_n \theta(t_1 - t_2) \ldots \theta(t_{n-1} - t_n) T\{F(t_1) \ldots F(t_n)\}$$

$$= \frac{1}{n!} \int_{t_0}^{t} dt_1 \ldots \int_{t_0}^{t} dt_n \theta(t_1 - t_2) \ldots \theta(t_{n-1} - t_n) \sum_{\Pi} T\{F(t_{j_1}) \ldots F(t_{j_n})\}$$

$$= \frac{1}{n!} \int_{t_0}^{t} dt_1 \ldots \int_{t_0}^{t} dt_n \sum_{\Pi} \theta(t_{j_1} - t_{j_2}) \ldots \theta(t_{j_{n-1}} - t_{j_n}) T\{F(t_{j_1}) \ldots F(t_{j_n})\}$$

$$= \frac{1}{n!} \int_{t_0}^{t} dt_1 \ldots \int_{t_0}^{t} dt_n T\{F(t_1) \ldots F(t_n)\}$$

$$\equiv \frac{1}{n!} T \int_{t_0}^{t} dt_1 \ldots \int_{t_0}^{t} dt_n F(t_1) \ldots F(t_n),$$

where in the intermediate steps we have used the symmetry of $\sum_{\Pi}$ and that, for each value of the $t_i$, only one product of $\theta$ functions will be different from zero. Applying this to (7.8.6), we then find that we can write

$$U_D(t, t_0) = \sum_{n=0}^{\infty} \left(\frac{i}{\hbar}\right)^n \frac{1}{n!} T \int_{t_0}^{t} dt_1 \ldots \int_{t_0}^{t} dt_n H_D^{(1)}(t_1) \ldots H_D^{(1)}(t_n)$$

$$\equiv T \, \exp \frac{-i}{\hbar} \int_{t_0}^{t} dt' H^{(1)}(t'). \tag{7.8.8}$$

This is a formal expression: the right-hand side of (7.8.8) is defined through its series expansion. Nevertheless, the definition is natural as shown by the following exercise.

**Exercise.** Show that, if the operator $F(t)$ is independent of $t$, then $T\exp$ and ordinary exponential coincide:

$$T \exp \int_{t_0}^{t} dt' F \equiv e^{(t-t_0)F} \; \blacksquare$$

We thus expect (7.8.8) to converge if $H_D^{(1)}$ is not too unbounded.

Equation (7.8.8) provides a very convenient expression for the $S$ matrix:

$$S_D = \lim_{\substack{t_0 \to -\infty \\ t \to +\infty}} U_D(t, t_0) = T \exp \frac{-i}{\hbar} \int_{-\infty}^{+\infty} dt H_D^{(1)}(t). \tag{7.8.9}$$

The Born approximation is easily deduced from (7.8.9) by considering, for example, the nonrelativistic case where the interaction is given, in the Schrödinger picture, by a potential $V$. Then writing

$$|i\rangle = |\mathbf{p}_1, \mathbf{p}_2\rangle; |f\rangle = |\mathbf{p}_1', \mathbf{p}_2'\rangle,$$

and assuming that $|i\rangle \neq |f\rangle$, we find that

$$_{NR}\langle f|S|i\rangle_{NR} \simeq \frac{-i}{\hbar} \,_{NR}\langle f|H_D^{(1)}(t)|i\rangle_{NR}.$$

Dropping the label NR, and using (7.8.2b), we get

$$\langle f|Si\rangle \simeq \frac{-i}{\hbar} \int_{-\infty}^{+\infty} dt \langle \mathbf{p}_1', \mathbf{p}_2'|e^{iH_0t/\hbar} V e^{-iH_0t/\hbar}|\mathbf{p}_1, \mathbf{p}_2\rangle$$

$$= \frac{-i}{\hbar} \int_{-\infty}^{+\infty} dt \langle \mathbf{p}_1', \mathbf{p}_2'|V|\mathbf{p}_1, \mathbf{p}_2\rangle e^{i(E_1+E_2-E_1'-E_2')t/\hbar}$$

$$= \frac{-i}{\hbar} 2\pi\hbar\delta(E_f - E_i)\langle \mathbf{p}_1', \mathbf{p}_2'|V|\mathbf{p}_1, \mathbf{p}_2\rangle$$

$$= -2\pi i\delta(E_f - E_i)$$

$$\times \int d^3r_1 d^3r_2 \frac{e^{-i\mathbf{r}_1\mathbf{p}_1'/\hbar - i\mathbf{r}_2\mathbf{p}_2'/\hbar}}{(2\pi\hbar)^3} V(\mathbf{r}_1 - \mathbf{r}_2) \frac{e^{-i\mathbf{r}_1\mathbf{p}_1/\hbar - i\mathbf{r}_2\mathbf{p}_2/\hbar}}{(2\pi\hbar)^3},$$

where we have also used the fact that the NR wave function corresponding to normalization $\langle \mathbf{p}'|\mathbf{p}\rangle = \delta(\mathbf{p} - \mathbf{p}')$ is $(2\pi\hbar)^{-3/2} \exp i\mathbf{r}\mathbf{p}/\hbar$.

We may then change variables, $\mathbf{r}_1, \mathbf{r}_2 \to \mathbf{r} = \mathbf{r}_1 - \mathbf{r}_2$ and $\mathbf{R} = \frac{1}{2}(\mathbf{r}_1 + \mathbf{r}_2)$. The integral over $\mathbf{R}$ is trivial and merely gives the $\delta$ of conservation of overall three-momentum. We thus have

$$\langle f|S|i\rangle \simeq -\frac{i}{\hbar} \frac{1}{(2\pi\hbar)^2}\delta(E_f - E_i)\delta(\mathbf{p}_f - \mathbf{p}_i)$$

$$\times \int d^3r \, e^{i(\mathbf{k}-\mathbf{k}')\mathbf{r}}V(\mathbf{r}), \quad \mathbf{k} = \mathbf{p}_1^{c.m.}, \quad \mathbf{k}' = \mathbf{p}_1'^{c.m.}.$$

(7.8.10)

On comparing this with (7.3.9) and restoring the indices NR, we obtain the nonrelativistic Born approximation to the transition, or scattering, amplitude:

$$T_{NR}^{Born}(i \to f) = -\frac{1}{\hbar^3(2\pi)^2} \int d^3r \, e^{i(\mathbf{k}-\mathbf{k}')\mathbf{r}}V(\mathbf{r}).$$

(7.8.11)

### 7.8.2 The $S$ Matrix in the Lippmann–Schwinger Formalism

Consider a system governed by a Hamiltonian $H = H_0 + H_I$, $H_0$ being the free Hamiltonian. For scattering states, the energy varies continuously. The Schrödinger equation,

$$H|\psi\rangle = E|\psi\rangle,$$

(7.8.12)

is manifestly equivalent to the equations

$$|\psi^+\rangle = |\psi_0\rangle + \lim_{\substack{\epsilon \to 0 \\ \epsilon > 0}} \frac{1}{E - H_0 + i\epsilon}H_I|\psi^+\rangle,$$

(7.8.13)

or

$$|\psi^-\rangle = |\psi_0\rangle + \lim_{\substack{\epsilon \to 0 \\ \epsilon > 0}} \frac{1}{E - H_0 - i\epsilon} H_I |\psi^-\rangle, \qquad (7.8.14)$$

with $|\psi_0\rangle$ the solution of the *free* Schrödinger equation,

$$H_0|\psi_0\rangle = E|\psi_0\rangle.$$

The equations (7.8.13), (7.8.14) are known as the *Lippmann–Schwinger equations*. The state $|\psi^+\rangle$ incorporates, by virtue of the way we have circumvented the pole of $1/(E - H_0)$, the boundary conditions pertaining to a scattering state. This is easily verified in the case of a particle in a potential: from (7.8.13),

$$\psi^+(\mathbf{r}) = \langle \mathbf{r}|\psi^+\rangle = \psi_0(\mathbf{r}) + \lim_{\epsilon \to 0} \langle \mathbf{r}| \frac{1}{E - H_0 + i\epsilon} H_I |\psi^+\rangle$$

$$= \psi_0(\mathbf{r}) + \int d^3p' \langle \mathbf{r}| \frac{1}{E - H_0 + i\epsilon} |\mathbf{p}'\rangle \langle \mathbf{p}'|H_I|\psi^+\rangle.$$

Now,

$$(E - H_0 + i\epsilon)^{-1}|\mathbf{p}'\rangle = (E - \mathbf{p}'^2/2m + i\epsilon)^{-1}|\mathbf{p}'\rangle.$$

If we define the (asymptotic) wave vector $\mathbf{k}$ by

$$E = \hbar^2\mathbf{k}^2/2m,$$

and also define $\mathbf{k}' = \hbar\mathbf{p}'$, we have

$$\psi_{\mathbf{k}}^+(\mathbf{r}) = \psi_{0\mathbf{k}}(\mathbf{r}) + 2m\hbar \int d^3k' \frac{1}{k^2 - k'^2 + i\epsilon} \langle \mathbf{r}|\mathbf{p}'\rangle \langle \mathbf{p}'|H_I|\psi_{\mathbf{k}}^+\rangle,$$

and we have made explicit use of the fact that $\psi^+$, $\psi_0$ correspond to the wave vector $\mathbf{k}$, exhibited explicitely as a subscript. Using

$$\langle \mathbf{r}|\mathbf{k}'\rangle = (2\pi\hbar)^{3/2}e^{i\mathbf{k}'\mathbf{r}},$$

$$\langle \mathbf{p}'|H_I|\psi_{\mathbf{k}}^+\rangle = \int d^3r' \langle \mathbf{p}'|\mathbf{r}'\rangle\langle \mathbf{r}'|H_I|\psi_{\mathbf{k}}^+\rangle,$$

and, with $V$ the potential,

$$\langle \mathbf{r}'|H_I = \langle \mathbf{r}'|V(\mathbf{r}'),$$

we obtain the Lippmann–Schwinger equation in the wave function formulation,

$$\psi_{\mathbf{k}}^+(\mathbf{r}) = \psi_{0\mathbf{k}}(\mathbf{r})$$

$$+ \lim_{\epsilon \to 0} \frac{2m\hbar}{(2\pi\hbar)^3} \int d^3k' d^3r' \frac{e^{i\mathbf{k}'(\mathbf{r}-\mathbf{r}')}}{k^2 - k'^2 + i\epsilon} V(\mathbf{r}')\psi_{\mathbf{k}}^+(\mathbf{r}'). \qquad (7.8.15)$$

The integral over $d^3k'$ is easily performed by going to spherical coordinates and using a contour integration:

$$\lim_{\epsilon \to 0} \frac{1}{(2\pi)^3} \int d^3k' \frac{e^{i\mathbf{k}'(\mathbf{r}-\mathbf{r}')}}{k^2 - k'^2 + i\epsilon} \equiv G_{\mathbf{k}}^+(\mathbf{r} - \mathbf{r}')$$

$$= -\frac{1}{4\pi} \frac{e^{ik|\mathbf{r}-\mathbf{r}'|}}{|\mathbf{r} - \mathbf{r}'|}.$$

(7.8.16)

Taking into account the behaviour

$$|\mathbf{r} - \mathbf{r}'| \simeq r - \mathbf{r}\mathbf{r}'/r + 0(1/r),$$

(7.8.15) gives us

$$\psi_{\mathbf{k}}^+(\mathbf{r}) \underset{r\to\infty}{\simeq} \frac{1}{(2\pi\hbar)^{3/2}} \left\{ e^{i\mathbf{k}\mathbf{r}} + \frac{e^{ikr}}{r} f \right\},$$

$$f = \frac{-m}{2\pi\hbar^2} \int d^3r' e^{-ik\mathbf{r}\mathbf{r}'} V(\mathbf{r}')\psi_{\mathbf{k}}^+(\mathbf{r}'),$$

(7.8.17)

i.e., a plane wave plus outgoing spherical wave as expected of a scattering state. If we had taken (7.8.14), we would have obtained

$$\psi_{\mathbf{k}}^-(\mathbf{r}) \underset{r\to\infty}{\simeq} \frac{1}{(2\pi\hbar)^{3/2}} \left\{ e^{i\mathbf{k}\mathbf{r}} + \frac{e^{-ikr}}{r} f^* \right\},$$

(7.8.18)

or *incoming* spherical waves. This, of course, does *not* correspond to a scattering state, but $|\psi_{\mathbf{k}}^-\rangle$ is nevertheless useful for some problems.

The *scattering amplitude* is easily obtained from the $|\psi_{\mathbf{k}}^+\rangle$, for example by comparing (7.8.17) with the known expression in elementary scattering theory[6]. With the nonrelativistic normalization of (7.3.7), we have

$$T(i \to f) = -2\pi\langle\psi_{0f}|H_I|\psi_i^+\rangle;$$

(7.8.19)

we normalize the states as

$$\langle\mathbf{p}|\mathbf{p}'\rangle = \delta(\mathbf{p} - \mathbf{p}').$$

The Lippmann–Schwinger equations may be solved by iteration (*Born series*),

$$|\psi^\pm\rangle = \sum_{n=0}^{\infty} \left( \lim_{\substack{\epsilon\to 0 \\ \epsilon>0}} \frac{1}{E - H_0 \pm i\epsilon} H_I \right)^n |\psi_0\rangle.$$

(7.8.20)

The first term reproduces the familiar Born approximation, as is easily verified in potential scattering; we get (7.8.11).

---

[6] See, e.g., Landau and Lifshitz (1958); Gottfried (1966).

### 7.8.3 Scattering by Two Interactions

The Lippmann–Schwinger equations are particularly suited to solving a problem that occurs in many applications. This is when the interaction may be split into two terms,

$$H_I = H_1 + H_2,\qquad(7.8.21)$$

and it so happens that we may solve exactly the interaction $H_1$, while $H_2$ is *small*.

Let us denote by $|\varphi^\pm\rangle$ the solutions of the Lippmann-Schwinger equations with only the interaction $H_1$:

$$|\varphi^\pm\rangle = |\psi_0\rangle + \frac{1}{E - H_0 \pm i0} H_1|\varphi^\pm\rangle.\qquad(7.8.22)$$

*Ex hypothesi* we are able to solve (7.8.22), and hence $|\varphi^\pm\rangle$ are to be considered known. (Here and in all that follows we introduce the notation $\pm i0$ meaning $\pm i\epsilon, \epsilon > 0, \epsilon \to 0$.)

We then recall the perturbative solution (7.8.20). Considering the terms there which are of first order in $H_2$, we find, for $T$ given by (7.8.19), that

$$T^{(1)}(i \to f) =$$

$$-2\pi \sum_{n=1}^{\infty} \left\langle \psi_{0f} \left| (E - H_0 + i0)\left\{ \frac{1}{E - H_0 + i0}(H_1 + H_2)\right\}^n \right| \psi_{0i} \right\rangle$$

$$= -2\pi \sum_{n=1}^{\infty}\sum_{\nu=0}^{n-1} \left\langle \left( \frac{1}{E - H_0 - i0}H_1 \right)^\nu \psi_{0f} \right|$$

$$\times H_2 \left( \frac{1}{E - H_0 + i0}H_1 \right)^{n-\nu-1} \left| \psi_{0i} \right\rangle \qquad(7.8.23)$$

$$= -2\pi \left\langle \sum_{\nu=0}^{\infty}\left( \frac{1}{E - H - i0}H_1 \right)^\nu \psi_{0f} \right|$$

$$\times H_2 \left| \sum_{\mu=0}^{\infty}\left( \frac{1}{E - H_0 + i0}H_1 \right)^\mu \psi_{0i} \right\rangle.$$

We have made use of

$$\left( H_1 \frac{1}{E - H_0 + i0} \right)^+ = \frac{1}{E - H_0 - i0}H_1,$$

and, in the last step, to get (7.8.23) we have re-arranged the sum. Identifying then the states in the sandwich in (7.8.23) with the $\varphi_{i,f}^\pm$ of (7.8.22), we finally obtain

$$T(i \to f) = -2\pi(\langle\psi_{0f}|H_1|\varphi_i^+\rangle + \langle\varphi_f^-|H_2|\varphi_i^+\rangle).\qquad(7.8.24)$$

For more details and applications, see Goldberger and Watson (1965).

## Problems

**P.7.1.** Establish the connection between (7.5.9) and (7.4.10).

*Solution.* Recalling (7.3.10), we have

$$F \underset{NR}{\to} (2^4 p_{A0}^2 p_{B0}^2)^{1/2} T_{NR},$$

and, from (7.5.9),

$$T_{NR} = (16 m_A^2 m_B^2)^{1/2} \frac{2(m_A + m_B)}{\pi k} \sum (2l + 1) P_l \sin \delta_l e^{i\delta_l}$$

$$= \left( \frac{1}{m_A} + \frac{1}{m_B} \right) \frac{1}{2\pi k} \sum (2l + 1) P_l \sin \delta_l e^{i\delta_l},$$

which indeed coincides with (7.5.10).

**P.7.2.** Calculate the scattering length and effective range paremeter, $a$, $r_0$, for $S$-wave scattering by a constant spherical well.

*Solution.*

$$\cot \delta_0 = \frac{k' + k \tan kL \tan k'L}{k \tan k'L - k' \tan kL},$$

with $L$ the radius of the well, and

$$k = \frac{1}{\hbar} \sqrt{2mE}, \quad k' = \frac{1}{\hbar} \sqrt{2m(E - v_0)},$$

$v_0$ being the height of potential inside the well. Then, for example,

$$a = \frac{\hbar \tan(L\hbar^{-1} \sqrt{2mv_0})}{\sqrt{2mv_0}} - L.$$

# 8. Quantization of the Electromagnetic Field. Interaction of Radiation with Matter

## 8.1 Normal, or Wick, Products

A very useful technique when considering the quantized version of field theory (in particular, the quantum theory of electromagnetic fields) is that of *normal, or Wick, products* of operators, which will now be described.

Consider two operators $\hat{f}$, $\hat{g}$ (in this section carets will distinguish operators) which are polynomials in creation–destruction operators,

$$\hat{f} = p_f(\hat{a}, \hat{a}^+; \hat{b}, \hat{b}^+),$$

$$\hat{g} = p_g(\hat{a}, \hat{a}^+; \hat{b}, \hat{b}^+),$$

where the $\hat{a}$, $\hat{a}^+$ refer to bosons and the $\hat{b}$, $\hat{b}^+$ to fermions. The *normal, or Wick, product* of $\hat{f}$ and $\hat{g}$ will be denoted by

$$: \hat{f}\hat{g} :,$$

and will be defined starting from the simplest case, i.e., when we have Wick products of creation–destruction operators. In this case we set

$$: \hat{a}(k,\eta)\hat{a}(k',\eta') :\equiv \hat{a}(k,\eta)\hat{a}(k',\eta'),$$

$$: \hat{a}^+(k,\eta)\hat{a}^+(k',\eta') :\equiv \hat{a}^+(k,\eta)\hat{a}^+(k',\eta'),$$

$$: \hat{a}^+(k,\eta)\hat{a}(k',\eta') :\equiv \hat{a}^+(k,\eta)\hat{a}(k',\eta'), \tag{8.1.1}$$

$$: \hat{a}(k,\eta)\hat{a}^+(k',\eta') :\equiv \hat{a}^+(k',\eta')\hat{a}(k,\eta),$$

that is to say, the Wick product orders the product by putting creators to the left of annihilators, *as if they commuted.* Equation (8.1.1) is valid for boson operators; for fermionic ones, the Wick product orders them as if they *anticommuted.* Thus, all relations except the last of (8.1.1) also hold for fermionic operators; the last relation is to be replaced by

$$: \hat{b}(k,\eta)\hat{b}^+(k',\eta') :\equiv -\hat{b}^+(k',\eta')\hat{b}(k,\eta). \tag{8.1.2}$$

For polynomials in creation–annihilation operators the definition is similar: the Wick product is like the ordinary product but orders creators to the left of annihilators as if they commuted (anticommuted for fermions). Formally, and for bosons, for example,

$$: \hat{a}(: \hat{f}_1 \ldots \hat{f}_j :) := (: f_1 \ldots f_j :)\hat{a},$$

$$: (: \hat{f}_1 \ldots \hat{f}_j :)\hat{a} := (: \hat{f}_1 \ldots \hat{f}_j :)\hat{a},$$

$$: \hat{a}^+(: \hat{f}_1 \ldots \hat{f}_j :) := \hat{a}^+(: \hat{f}_1 \ldots \hat{f}_j :),$$

$$: (: \hat{f}_1 \ldots \hat{f}_j :)\hat{a}^+ := \hat{a}^+(: \hat{f}_1 \ldots \hat{f}_j :);$$

$$: (\alpha\hat{f} + \beta\hat{g}) := \alpha : \hat{f} : +\beta : \hat{g} :,$$

from which, by iteration, we get the rules for any polynomial. We will use the colons also to bracket expressions. Thus,

$$: \hat{f} + \hat{g} :\equiv: (\hat{f} + \hat{g}) :,$$

$$: (\hat{f}\hat{g}) :\equiv: \hat{f}\hat{g} :,$$

etc.

The normal product is easily seen to satisfy the *distributive* property,

$$: \hat{f}(\hat{g} + \hat{h}) :=: \hat{f}\hat{g} : + : \hat{f}\hat{h} :,$$

and, inside a normal product, bosonic operators may be taken to commute, fermionic ones to anticommute:

$$: \ldots \hat{a}_1\hat{a}_2^+ \ldots :=: \ldots \hat{a}_2^+\hat{a}_1 \ldots :,$$

$$: \ldots \hat{b}_1^+\hat{b}_2 \ldots := - : \ldots \hat{b}_2\hat{b}_1^+ \ldots :$$

etc.


## 8.2 Quantization of the Electromagnetic Field (Coulomb Gauge). The Casimir Effect

### 8.2.1 Quantization of the Electromagnetic Field

Besides the reasons advanced in the previous sections for constructing a quantum theory of fields, there is, for the electromagnetic field, another and very compelling one: the electric and magnetic fields, $\mathcal{E}$, $\mathcal{B}$ are measurable quantities. Therefore, in quantized theory they should be replaced by operators

$$\hat{\mathcal{E}}, \ \hat{\mathcal{B}},$$

in such a way that the classical fields will be identified with the expectation values:

$$\mathcal{E}_{\text{cl}} \simeq \langle\Psi|\hat{\mathcal{E}}|\Psi\rangle, \ \mathcal{B}_{\text{cl}} \simeq \langle\Psi|\hat{\mathcal{B}}|\Psi\rangle.$$

We will continue to distinguish operators by putting a caret over the corresponding symbol.

As proved spectacularly by the existence of the Aharonov–Bohm[1] effect or by the properties of bremsstrahlung radiation (see here, Sect. 8.4 below) the basic quantities in quantum electromagnetism are *not* $\mathcal{E}$, $\mathcal{B}$, but the four-potential $A_\mu$. We have to construct an operator for it, $\hat{A}_\mu(\mathbf{r}, t)$ (in the Heisenberg picture), which immediately raises the matter of gauge indeterminacy. In this section we will work in the Coulomb gauge and will thus set

$$\hat{A}_0(\mathbf{r}, t) \equiv 0, \ \operatorname{div} \hat{\mathbf{A}}(\mathbf{r}, t) \equiv 0. \tag{8.2.1}$$

We want to recover the standard theory in the classical limit, so we have to require that $\hat{A}$ satisfy the Maxwell equations, which in this gauge are just

$$\partial^2 \hat{A}_\mu(x) = 0. \tag{8.2.2}$$

Moreover, because we want $\mathcal{E}$, $\mathcal{B}$ *real*, we assume $\hat{A}_\mu$ to be a self-adjoint operator, $\hat{A}_\mu^+(x) = \hat{A}_\mu(x)$.

The more general solution to these conditions can be written as[2]

$$\hat{\mathbf{A}}(\mathbf{r}, t) = \frac{\sqrt{\hbar c}}{2\pi} \int \frac{d^3 k}{\sqrt{|\mathbf{k}|}} \sum_{\eta = \pm 1} \left\{ e^{i(\mathbf{kr} - \omega t)} \boldsymbol{\epsilon}(\mathbf{k}, \eta) \hat{a}_{\mathrm{n.r.}}(\mathbf{k}, \eta) \right.$$
$$\left. + e^{-i(\mathbf{kr} - \omega t)} \boldsymbol{\epsilon}^*(\mathbf{k}, \eta) \hat{a}_{\mathrm{n.r.}}(\mathbf{k}, \eta) \right\}, \tag{8.2.3a}$$

$$\omega = \omega(\mathbf{k}) = c|\mathbf{k}|.$$

The polarization vectors $\boldsymbol{\epsilon}$ are those defined in Sect. 5.2. The reasons for the constant $\sqrt{\hbar c}/2\pi$ and the factor $1/\sqrt{|\mathbf{k}|}$ will be given in a moment. The success of the classical interpretation of the electromagnetic field as a set of oscillators suggests that the passage to the quantum theory should be effected by postulating harmonic oscillator commutation relations for the operators $\hat{a}_{\mathrm{n.r.}}$, $\hat{a}_{\mathrm{n.r.}}^+$; thus we set

$$[\hat{a}_{\mathrm{n.r.}}(\mathbf{k}, \eta), \hat{a}_{\mathrm{n.r.}}^+(\mathbf{k}', \eta')] = \delta_{\eta\eta'} \delta(\mathbf{k} - \mathbf{k}') f(k).$$

The normalization function $f$ can be found by requiring that the expression for the energy for the set of oscillators and that obtained from the classical one by the correspondence principle agree. With the choice of (8.2.3a) this will be the case (as we shall verify presently) if $f(k) \equiv 1$. Therefore, we postulate

$$[\hat{a}_{\mathrm{n.r.}}(\mathbf{k}, \eta), \hat{a}_{\mathrm{n.r.}}(\mathbf{k}', \eta')] = 0,$$
$$\tag{8.2.3b}$$
$$[\hat{a}_{\mathrm{n.r.}}(\mathbf{k}, \eta), \hat{a}_{\mathrm{n.r.}}^+(\mathbf{k}', \eta')] = \delta_{\eta\eta'} \delta(\mathbf{k} - \mathbf{k}').$$

The energy corresponding to the (continuous) set of oscillators will be

[1] See Galindo and Pascual (1978); Sakurai (1967); Ynduráin (1988).

[2] Although the discussion here is reasonably self-contained, familiarity with the elementary treatment of the quantized radiation field would help.

$$\hat{H}^{\text{osc.}}_{\text{rad}} = \sum_\eta \int d^3 k\, \hbar\omega(k) \hat{a}^+_{\text{n.r.}}(\mathbf{k},\eta) \hat{a}_{\text{n.r.}}(\mathbf{k},\eta)$$
$$+ C^{\text{osc}},$$

(8.2.4a)

and we will show that it agrees with that obtained from the correspondence principle,

$$\hat{H}^{\text{c.p.}}_{\text{rad}} = \frac{1}{8\pi} \int d^3 r (\hat{\mathcal{E}}^2 + \hat{\mathcal{B}}^2) + C^{\text{c.p.}}.$$

(8.2.4b)

The constants $C^{\text{osc}}$, $C^{\text{c.p.}}$ need not (and will not) be the same; recall that the energy is only defined up to an additive constant. $C^{\text{osc}}$ would correspond, in the oscillator interpretation, to the sum of the zero-mode energies,

$$C^{\text{osc}} = \sum_\eta \int d^3 k \frac{1}{2} \hbar\omega(k),$$

and is actually divergent.

The normalization of (8.2.3) is appropriate to study the interaction of the radiation with slowly moving particles, being nonrelativistic (hence the label n.r. in $\hat{a}_{\text{n.r.}}$). This does *not* mean that (8.2.3) is not relativistically acceptable; but relativistic invariance is more apparent if we introduce, in lieu of the $\hat{a}_{\text{n.r.}}$, the operators $\hat{a}$ with

$$\hat{a}(k,\eta) \equiv \sqrt{2k_0}\, \hat{a}_{\text{n.r.}}(\mathbf{k},\eta), \quad k_0 \equiv |\mathbf{k}|,$$

so that, in natural units, $c = \hbar = 1$, (8.2.3) become

$$\hat{\mathbf{A}}(x) = \frac{\sqrt{4\pi}}{(2\pi)^{3/2}} \int \frac{d^3 k}{2k_0} \sum_\eta \left\{ e^{-ik\cdot x} \epsilon(\mathbf{k},\eta) \hat{a}(k,\eta) \right.$$
$$\left. + e^{ik\cdot x} \epsilon^*(\mathbf{k},\eta) \hat{a}^+(k,\eta) \right\};$$

(8.2.5a)

$$[\hat{a}(k,\eta), \hat{a}^+(k',\eta')] = 2k_0 \delta(\mathbf{k} - \mathbf{k}') \delta_{\eta\eta'}, \quad k_0 \equiv |\mathbf{k}|.$$

(8.2.5b)

Moreover, we have to admit that Lorentz accelerations must be accompanied by gauge transformations as in Sect. 5.2, to restore the Coulomb conditions (8.2.1). (In Sect. 9.6 we will describe a manifestly covariant treatment of the electromagnetic field.)

The above expressions for $\hat{\mathbf{A}}$ hold in the Gauss system of electromagnetic units, which have been used up till now, and which will go on being used unless explicitly stated otherwise. In fully relativistic calculations, however, it is customary to use the *Heaviside rationalized* system. Denoting it by the suffix $R$, we have

$$e_R = \sqrt{4\pi}e, \quad A_R = (1/\sqrt{4\pi})A,$$

so that the creation–destruction operators $\hat{a}$, $\hat{a}^+$ and the minimal substitution (the product $eA = e_R A_R$) do not change; (8.2.5b) is thus unaltered, and (8.2.5a) changes to

$$\hat{\mathbf{A}}_R(x) = \frac{1}{(2\pi)^{3/2}} \int \frac{d^3k}{2k_0} \sum_{\eta=\pm 1} \left\{ e^{-ik\cdot x} \boldsymbol{\epsilon}(\mathbf{k},\eta)\hat{a}(k,\eta) \right.$$
$$\left. + e^{ik\cdot x} \boldsymbol{\epsilon}^*(\mathbf{k},\eta)\hat{a}^+(k,\eta) \right\}. \tag{8.2.6}$$

The *fine-structure constant* $\alpha$ is defined to be the same in both systems, $\alpha = e^2 = e_R^2/4\pi \simeq 1/137.036$.

After these disgressions, let us return to the basic problems. If we had a system of oscillators, the operators $\hat{a}/\hat{a}^+$ would destroy/create energy excitations; now we interpret them as destroying/creating *photons*, particles which are the quanta of the radiation field. Thus,

$$\hat{a}^+(k_1,\eta_1)\ldots\hat{a}^+(k_n,\eta_n)|0\rangle \tag{8.2.7a}$$

will be interpreted as a state of $n$ photons with momenta (or wave vectors) $\mathbf{k}_1,\ldots,\mathbf{k}_n$, energies $k_{10},\ldots,k_{n0}$ (in natural units) and helicities $\eta_1,\ldots,\eta_n$. In (8.2.7a) $|0\rangle$ is the state without photons, the *vacuum*; it is the state of minimum energy and it is assumed to satisfy

$$\hat{a}(k,\eta)|0\rangle = 0. \tag{8.2.7b}$$

If we define the operators $\hat{H}_{\mathrm{rad}}$, $\hat{\mathbf{P}}_{\mathrm{rad}}$ by

$$\hat{H}_{\mathrm{rad}} = \sum_{\eta} \int \frac{d^3k}{2k_0} k_0 \hat{a}^+(k,\eta)\hat{a}(k,\eta), \tag{8.2.8a}$$

$$\hat{\mathbf{P}}_{\mathrm{rad}} = \sum_{\eta} \int \frac{d^3k}{2k_0} \mathbf{k}\,\hat{a}^+(k,\eta)\hat{a}(k,\eta), \tag{8.2.8b}$$

then these operators give respectively the correct value for the (total) energy and momentum of the state (8.2.7a). Indeed, straightforward use of the commutation relations (8.2.5b) shows that the four-vector operators $\hat{P}_\mu$ with $\hat{P}_0 = \hat{H}_{\mathrm{rad}}$, $\hat{P}_i = \hat{P}_{\mathrm{rad}\,i}$ satisfy

$$\hat{P}_\mu \left\{ \hat{a}^+(k_1,\eta_1)\ldots\hat{a}^+(k_n,\eta_n)|0\rangle \right\}$$
$$= (k_{1\mu} + \ldots + k_{n\mu}) \left\{ \hat{a}^+(k_1,\eta_1)\ldots\hat{a}^+(k_n,\eta_n)|0\rangle \right\}. \tag{8.2.8c}$$

As stated at the begining of this section, we then have to show the correspondence with the classical expressions,

$$E_{\mathrm{cl}} = \frac{1}{8\pi} \int d^3r (\boldsymbol{\mathcal{E}}_{\mathrm{cl}}^2 + \boldsymbol{\mathcal{B}}_{\mathrm{cl}}^2),$$
$$\mathbf{P}_{\mathrm{cl}} = \frac{1}{4\pi c} \int d^3r (\boldsymbol{\mathcal{E}}_{\mathrm{cl}} \times \boldsymbol{\mathcal{B}}_{\mathrm{cl}}). \tag{8.2.9}$$

We consider the first in detail, and leave the second as an exercise. Integrating by parts, using the gauge conditions and the relations

$$\boldsymbol{\mathcal{E}} = \frac{-1}{c}\partial_t \mathbf{A}, \quad \boldsymbol{\mathcal{B}} = \nabla \times \mathbf{A},$$

we write, in terms of $\mathbf{A}$,

$$E_{\mathrm{cl}} = \frac{1}{8\pi} \int d^3r \left\{ \frac{1}{c^2} (\partial_t \mathbf{A}_{\mathrm{cl}})^2 - \mathbf{A}_{\mathrm{cl}} \triangle \mathbf{A}_{\mathrm{cl}} \right\}.$$

We then expect the quantum expression (in natural units)

$$\hat{H}'_{\mathrm{rad}} = \frac{1}{8\pi} \int d^3r \left\{ (\partial_0 \hat{\mathbf{A}})^2 - \hat{\mathbf{A}} \triangle \hat{\mathbf{A}} \right\} + C$$

to coincide with (8.2.8a), conveniently adjusting the constant $C$.

To see this, substitute (8.2.5) into $\hat{H}'_{\mathrm{rad}}$ here; we find that

$$\hat{H}'_{\mathrm{rad}} = \frac{1}{2(2\pi)^3} \sum_{\eta\eta'} \sum_j \int \frac{d^3k\, d^3k'}{2k_0 2k_0'} \int d^3r$$

$$\left\{ -k_0 k_0' \left[ e^{-ik\cdot x} \epsilon_j(\mathbf{k},\eta)\hat{a}(k,\eta) + e^{ik\cdot x} \epsilon_j^*(\mathbf{k},\eta)\hat{a}^+(k,\eta) \right] \right.$$

$$\times \left[ -e^{-ik'\cdot x} \epsilon_j(\mathbf{k}',\eta')\hat{a}(k',\eta') + e^{ik'\cdot x} \epsilon_j^*(\mathbf{k}',\eta')\hat{a}^+(k,\eta) \right] \qquad (8.2.10a)$$

$$+\mathbf{k}'^2 \left[ e^{-ik\cdot x} \epsilon_j(\mathbf{k},\eta)\hat{a}(k,\eta) + e^{ik\cdot x} \epsilon_j^*(\mathbf{k},\eta)\hat{a}^+(k,\eta) \right]$$

$$\left. \times \left[ e^{-ik'\cdot x} \epsilon_j(\mathbf{k}',\eta')\hat{a}(\mathbf{k}',\eta') + e^{ik'\cdot x} \epsilon_j^*(\mathbf{k}',\eta')\hat{a}^+(k',\eta') \right] \right\} + C.$$

There are here three types of term: (i) Terms with $\hat{a}\ldots\hat{a}'$ or $\hat{a}^+\ldots\hat{a}'^+$. Integrating $d^3r$ with the corresponding exponentials, $\exp(\pm i(\mathbf{k}+\mathbf{k}')\mathbf{r})$ we get $(2\pi)^3\delta(\mathbf{k}+\mathbf{k}')$. Using $k_0 = |\mathbf{k}|$, direct inspection shows all such terms to actually vanish. (ii) Terms containing $\hat{a}^+\ldots\hat{a}'$. These terms annihilate the vacuum, as was to be expected: we would like $\hat{H}_{\mathrm{rad}}|0\rangle = 0$. (iii) Terms containing $\hat{a}\ldots\hat{a}'^+$. We can use the commutation relations to write then as terms of class (ii) plus a constant:

$$\hat{a}(k,\eta)\hat{a}^+(k',\eta') = \hat{a}^+(k',\eta')\hat{a}(k,\eta) + 2\delta_{\eta\eta'} k_0 \delta(\mathbf{k}-\mathbf{k}'). \qquad (8.2.10b)$$

Integrating $\int d^3r \exp i\mathbf{r}(\mathbf{k}'-\mathbf{k}) = (2\pi)^3\delta(\mathbf{k}-\mathbf{k}')$, then $d^3k$ with this $\delta$, and using the relation

$$\sum_j \epsilon_j^*(k,\eta')\epsilon_j(k,\eta) = \delta_{\eta\eta'},$$

we finally obtain

$$\hat{H}'_{\mathrm{rad}} = \int \frac{d^3k}{2k_0} k_0 \hat{a}^+(k,\eta)\hat{a}(k,\eta) + C' + C$$

$$= \hat{H}_{\mathrm{rad}} + C' + C,$$

where $C'$ is the constant[3]

---

[3] This constant is actually divergent. We can give sense to these types of expression by integrating up to a fixed $|\mathbf{k}| = K$, large but finite, and allowing $K \to \infty$ at the end. We will have to admit that $C', C$ also depend on $K$.

$$C' \sim \sum_{\eta} \int d^3k$$

coming from the terms in the commutator of (8.2.10b). Adjusting $C$ to cancel this exactly, we obtain precisely $\hat{H}'_{\text{rad}} = \hat{H}_{\text{rad}}$, with $\hat{H}_{\text{rad}}$ given by (8.2.8a), as desired.

**Exercise.** Verify that

$$\hat{H}_{\text{rad}} = \frac{1}{8\pi} \int d^3r : \hat{\boldsymbol{\mathcal{E}}}^2 + \hat{\boldsymbol{B}}^2 :,$$

$$\hat{\mathbf{P}}_{\text{rad}} = \frac{1}{4\pi} \int d^3r : \hat{\boldsymbol{\mathcal{E}}} \times \hat{\boldsymbol{B}} : \; \blacksquare$$

(8.2.11)

For $\hat{\mathbf{P}}_{\text{rad}}$, the same result is obtained by symmetrization (no need to use the Wick product):

$$\hat{\mathbf{P}}_{\text{rad}} = \frac{1}{8\pi} \int d^3r \left( \hat{\boldsymbol{\mathcal{E}}} \times \hat{\boldsymbol{B}} - \hat{\boldsymbol{B}} \times \hat{\boldsymbol{\mathcal{E}}} \right).$$

### 8.2.2 Multipole Expansion

Expressions like (8.2.5), where the electromagnetic field is developed in plane waves, are particularly useful for scattering problems. There are, however, situations where photons are emitted or absorbed with definite angular momentum. For these cases it is better to use an expansion into multipoles, given in (5.3.23). We thus write

$$\hat{\mathbf{A}}(\mathbf{r},t) = c\sqrt{2\pi\hbar} \sum_{IlM} \int_0^\infty d\omega \frac{1}{\omega^{3/2}} \left\{ e^{-i\omega t} \mathbf{X}_M^{(I,\omega,l)}(\mathbf{r}) \hat{a}_I(\omega,l,M) \right.$$

$$\left. + e^{i\omega t} \mathbf{X}_M^{(I,\omega,l)*}(\mathbf{r}) \hat{a}_I^+(\omega,l,M) \right\}.$$

(8.2.12)

The commutation relations of the $\hat{a}_I(\omega,l,M)$ can be obtained from those of the $\hat{a}(\mathbf{k},\eta)$ by expresing the $\hat{a}_I(\omega,l,M)$ in terms of $\hat{\mathbf{A}}$ (using the orthogonality condition of the multipoles, (5.3.22b)), and $\hat{\mathbf{A}}$ in turn in terms of the $\hat{a}(\mathbf{k},\eta)$. We get

$$[\hat{a}_I(\omega,l,M), \hat{a}_I^+,(\omega',l',M')]$$

$$= \frac{1}{\omega^2} \sum_{\eta\eta'} \int \frac{d^3k\, d^3k'}{|\mathbf{k}|} \sum_{jj'} \epsilon_j(\mathbf{k},\eta) \epsilon_{j'}^*(\mathbf{k}',\eta')$$

$$\times X_{Mj}^{(I,\omega,l)}(\mathbf{k})^* X_{Mj'}^{(I',\omega',l')}(\mathbf{k}') \delta(\mathbf{k}-\mathbf{k}') \delta_{\eta\eta'}.$$

Using

$$\sum_{\eta} \epsilon_j(\mathbf{k},\eta) \epsilon_{j'}^*(\mathbf{k},\eta) = \delta_{jj'} - k_j k_{j'}/k^2,$$

(8.2.13)

and the fact that the **X** are transverse, as well as their orthogonality properties, we finally obtain the desired commutation relations:

$$[\hat{a}_I(\omega,l,M),\hat{a}_{I'}^{+}(\omega',l',M')] = \delta(\omega - \omega')\delta_{II'}\delta_{ll'}\delta_{MM'}, \tag{8.2.14}$$

whose simplicity is, of course, the reason for the choice of factor $c\sqrt{2\pi\hbar}/\omega^{3/2}$ in (8.2.12).

### 8.2.3 The Casimir Effect

The equivalence between the replacement of product by Wick product and the adjustment of a constant for the expression of $\hat{H}_{\mathrm{rad}}$ in (8.2.11a), say, holds true only in unbounded vacuum. For electromagnetic fields, for example that are confined, there may be boundary conditions that Wick's product does not take into account. A striking consequence of this is the *Casimir effect*, where an attraction develops between two conducting plates in the absence of any charge, owing to a mismatch of the boundary conditions.

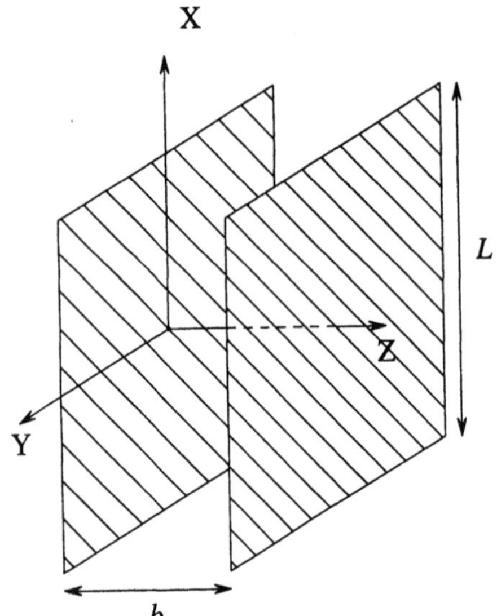


**Fig. 8.2.1.** Conducting plates for the Casimir effect.

The arrangement is like that of Fig. 8.2.1: two conducting plates, which we take to be square with side $L$ very large, are located at a distance $b$, $b \ll L$, but still $b \gg$ (interatomic distances), so that we can take the plates to be smooth. Typically, $b \sim 1$ $\mu$m. In the presence of the plates, the electromagnetic field must satisfy boundary conditions; classically, these are that the nodes of the oscillators fall on the plates. We take the same for the quantum case (since $\mathcal{E}_{\mathrm{cl}} \sim \langle \Phi|\hat{\mathcal{E}}|\Phi\rangle$, etc.) so for the vector potential in the presence of the plates we replace (8.2.3a) by $\hat{\mathbf{A}}^{(b)}$, given by

$$\hat{\mathbf{A}}^{(b)}(\mathbf{r},t) =$$

$$\frac{\sqrt{\hbar c}}{2\pi} \int dk_x dk_y \frac{\pi}{b}$$

$$\times \sum_{n=-\infty}^{+\infty} \frac{1}{\sqrt{|\mathbf{k}(b)|}} \left\{ e^{-i\omega t} e^{i\mathbf{k}(b)\mathbf{r}} \boldsymbol{\epsilon}(\mathbf{k}(b),\eta) \hat{a}_{\text{n.r.}}^{(b)}(\mathbf{k}(b),\eta) \right.$$

$$\left. + e^{i\omega t} e^{-i\mathbf{k}(b)\mathbf{r}} \boldsymbol{\epsilon}^*(\mathbf{k}(b),\eta) \hat{a}_{\text{n.r.}}^{(b)+}(\mathbf{k}(b),\eta) \right\},$$

(8.2.15a)

where $\omega = c|\mathbf{k}(b)|$ and $\mathbf{k}_x(b) = k_x$, $k_y(b) = k_y$ and

$$k_z(b) = \frac{\pi n}{b}, \quad n = \text{integer}.$$

Moreover, the commutation relations (8.2.3b) are to be replaced by

$$\left[ \hat{a}_{\text{n.r.}}^{(b)}(\mathbf{k}(b),\eta), \hat{a}_{\text{n.r.}}^{(b)}(\mathbf{k}'(b),\eta')^+ \right] = \delta_{\eta\eta'} \delta(k_x - k_x') \delta(k_y - k_y') \frac{b}{\pi} \delta_{nn'}. \quad (8.2.15b)$$

We use nonrelativistic normalization, and write $\hbar$ and $c$ explicitly.

The energy inside the volume $V$ limited by the plates, in the presence of the plates, is now

$$\hat{H}_{\text{rad}}^{(b)} = \frac{1}{8\pi} \int_V d^3r \left\{ \left( \partial_0 \hat{\mathbf{A}}^{(b)} \right)^2 - \hat{\mathbf{A}}^{(b)} \triangle \hat{\mathbf{A}}^{(b)} \right\} + C;$$

without the plates we would have had

$$\hat{H}_{\text{rad}} = \frac{1}{8\pi} \int_V d^3r \left\{ \left( \partial_0 \hat{\mathbf{A}} \right)^2 - \hat{\mathbf{A}} \triangle \hat{\mathbf{A}} \right\} + C.$$

The point is that the constant must of course be the same in both cases. This means that the presence of the plates will introduce an energy shift. Even in the vacuum, and without charges, there will be an effect. The corresponding energy shift will be

$$\Delta E(b) = \left\langle 0 \left| \left( \hat{H}_{\text{rad}}^{(b)} - \hat{H}_{\text{rad}} \right) \right| 0 \right\rangle$$

$$= \frac{1}{8\pi} \int_V d^3r \left\langle 0 \left| \left\{ \left( \partial_0 \hat{\mathbf{A}}^{(b)} \right)^2 - \left( \partial_0 \hat{\mathbf{A}} \right)^2 \right. \right. \right.$$

(8.2.16)

$$\left. \left. \left. + \hat{\mathbf{A}} \triangle \hat{\mathbf{A}} - \hat{\mathbf{A}}^{(b)} \triangle \hat{\mathbf{A}}^{(b)} \right\} \right| 0 \right\rangle.$$

If we substitute (8.2.3a), (8.2.15a) for $\hat{\mathbf{A}}$, $\hat{\mathbf{A}}^{(b)}$, to get expressions like (8.2.10a) inside the curly bracket of (8.2.16), only the terms in the commutators of expressions with $\hat{a} \ldots \hat{a}^+$ will not give zero when sandwiched between the vacuum. A simple calculation then gives

$$\Delta E(b) = \frac{\hbar c}{4(2\pi)} \left(\frac{\pi}{b}\right)^2 \frac{b}{\pi} \sum_\eta \epsilon^* \epsilon$$

$$\times \int dk_x dk_y \sum_{n=-\infty}^{+\infty} 2|\mathbf{k}(b)|^2 \int_V d^3r$$

$$- \frac{\hbar c}{4(2\pi)^3} \sum_\eta \epsilon^* \epsilon$$

$$\times \int dk_x dk_y \int dk_z 2|\mathbf{k}|^2 \int_V d^3r$$

$$= \frac{\hbar c L^2}{8\pi^2} \int dk_x dk_y \left\{ \sum_{n=-\infty}^{+\infty} \sqrt{k_x^2 + k_y^2 + \left(\frac{\pi n}{b}\right)^2} \right.$$

$$\left. - \int_{-\infty}^{+\infty} d\nu \sqrt{k_x^2 + k_y^2 + \left(\frac{\pi \nu}{b}\right)^2} \right\},$$

where we have changed the variable of integration $dk_z$ to $d\nu = (b/\pi)dk_z$. Because of the symmetry of the last expression with respect to $n \to -n$, $\nu \to -\nu$, and by also introducing polar coordinates for the $k_x, k_y$ integration, we obtain the energy shift per unit area,

$$\Delta\epsilon(b) \equiv \frac{\Delta E(b)}{L^2} = -\frac{\hbar c}{4\pi} \int_0^\infty dx \left\{ \int_0^\infty d\nu \sqrt{x + \left(\frac{\nu\pi}{b}\right)^2} \right.$$

$$\left. - \sum_{n=0}^\infty \sqrt{x + \left(\frac{n\pi}{b}\right)^2} + \frac{1}{2}\sqrt{x} \right\}. \tag{8.2.17}$$

This expression is undefined. This should not be too surprising: for large values of the wave vector, $\mathbf{k}$, we are sensitive to the detailed structure of the plates. Thus (8.2.17) must fail for $x \sim k^2 \sim 1/a^2$, where $a \sim 1$ Å is of the order of the interatomic distances. We may take this into account by introducing a cut-off in (8.2.17). It will happen that the large $x, \nu(n)$ contributions to (8.2.17) do actually cancel, and one obtains a finite, cut-off independent result. Because of this we will *not* write the cut-off explicitly.

Exchanging sums and integrals we can rewrite (8.2.17) as

$$\Delta\epsilon(b) = -\frac{\hbar c}{4\pi} \left\{ \int_0^\infty d\nu f(\nu) - \sum_{n=0}^\infty f(n) + \frac{1}{2}f(0) \right\}, \tag{8.2.18}$$

where we have defined

$$f(\nu) \equiv \int_0^\infty dx \sqrt{x + \left(\frac{\nu\pi}{b}\right)^2}.$$

(With an eventual cut-off.) The next step is the use of the Euler formula[4], valid for any smooth function $f$,

---

[4] See, e.g., Abramowicz and Stegun (1965).

$$\int_0^\infty d\nu\, f(\nu) = \sum_{n=0}^\infty f(n) - \frac{1}{2} f(0) + \sum_{l=1}^\infty \frac{B_{2l}}{(2l)!} f^{(2l-1)}(0), \qquad (8.2.19)$$

where the $B_{2l}$ are the Bernouilli numbers:

$$B_2 = 1/6, \; B_4 = 1/30 \ldots.$$

We see that the divergent pieces $\frac{1}{2} f(0)$ in (8.2.18) and (8.2.19) cancel one another. Moreover,

$$f'(\nu) = \frac{\pi^2 \nu}{b^2} \int_0^\infty dx \left( x + \left( \frac{\nu \pi}{b} \right)^2 \right)^{-1/2},$$

which vanishes at $\nu = 0$ for any reasonable cut-off. Higher derivatives are *convergent*,

$$f'''(\nu) = -4 \frac{\pi^3}{b^3},$$

and indeed vanish from $f^{\mathrm{IV}}$ onwards. Thus,

$$\Delta\epsilon(b) = -\frac{\hbar c}{4\pi} \frac{B_4}{4!} \frac{-4\pi^3}{b^3} = \frac{\pi^2 \hbar c}{720\, b^3}.$$

This energy depends on $b$, so it will generate a force (per unit plaque surface) $\mathcal{F}$,

$$\mathcal{F} = \frac{\partial \Delta\epsilon(b)}{\partial b} = -\frac{\pi^2 \hbar c}{240\, b^4}.$$

The force is *attractive* and small but measurable (and measured); for $b = 1\mu$m, $\mathcal{F} = -1.3 \times 10^{-2}$ dyn/cm$^2$.

## 8.3 Interaction of the Radiation with Slowly Moving Particles

In this section we consider the interaction of photons with matter. Our treatment will be semi-relativistic and partially quantal; that is to say, the photons are of course relativistic particles, but we consider their interactions with particles that move slowly, with average speeds much less than the speed of light; so they will be treated nonrelativistically. Likewise, full quantum consistency would force us to describe not only photons, but other particles by quantum fields. Since the probability of creation of massive particles is small[5], because the velocities, and hence the kinetic energies, are small compared with the rest mass, we can dispense with a full quantum description of matter that will be discussed later on. This means that we will consider systems in which,

---

[5] Even if there is no possibility of creating *real* particles, there is always a small probability of producing virtual fluctuations. This we also neglect for the moment.

typically, we have a slowly moving particle, say an electron in a classical potential, whose state is described by a wave function $\psi_e(\mathbf{r})$ subject to the Schrödinger equation; and an (in general) variable number $n$ of photons with wave vectors and helicities $\mathbf{k}_1, \eta_1; \ldots; \mathbf{k}_n, \eta_n$. The corresponding state will be denoted by

$$|\psi_e; \mathbf{k}_1, \eta_1; \ldots; \mathbf{k}_n, \eta_n\rangle = \frac{1}{\sqrt{n!}} \hat{a}^+(\mathbf{k}_1, \eta_1) \ldots \hat{a}^+(\mathbf{k}_n, \eta_n)|\psi_e\rangle. \tag{8.3.1a}$$

We will again systematically use carets to denote operators. Because the state $|\psi_e\rangle$ contains no photons, it behaves like the vacuum for the photon field, and thus we postulate that

$$\hat{a}(\mathbf{k}, \eta)|\psi_e\rangle = 0. \tag{8.3.1b}$$

The Hamiltonian of the system is obtained as follows. If there were no interaction between matter and radiation, we would have the Hamiltonian

$$\hat{H}_1 = \hat{H}_e + \hat{H}_{\text{rad}}, \tag{8.3.2a}$$

where, if we use nonrelativistic normalization,

$$\hat{H}_{\text{rad}} = \sum_{\eta=\pm 1} \int d^3 k \, \hbar\omega(k) \hat{a}^+_{\text{n.r.}}(\mathbf{k}, \eta) \hat{a}_{\text{n.r.}}(\mathbf{k}, \eta), \tag{8.3.2b}$$

and

$$\hat{H}_e = \frac{1}{2m}\hat{\mathbf{P}}_e^2 + V; \quad \hat{\mathbf{P}}_e = -i\hbar\boldsymbol{\nabla}_r, \tag{8.3.2c}$$

with $V$ a potential (not quantized) to which the electron may possibly be subjected.

The matter–radiation interaction is implemented by postulating the minimal replacement

$$\hat{\mathbf{P}}_e \to \hat{\mathbf{P}}_e - \frac{e}{c}\hat{\mathbf{A}}_S(\mathbf{r}).$$

Here $e$ is the charge (and $m$ the reduced mass) of the particle and $\hat{A}_S(\mathbf{r}) = \hat{A}(\mathbf{r}, t = 0)$ the Schrödinger picture electromagnetic operator; we choose the origin of times as the moment in which Schrödinger and Heisenberg pictures coincide. So, with nonrelativistic normalization and in the Gauss system of units,

$$\hat{\mathbf{A}}_S(\mathbf{r}) = \frac{\sqrt{\hbar c}}{2\pi} \int \frac{d^3 k}{\sqrt{k}} \sum_{\eta=\pm 1} \left\{ e^{i\mathbf{kr}} \boldsymbol{\epsilon}(\mathbf{k}, \eta) \hat{a}_{\text{n.r.}}(\mathbf{k}, \eta) \right. \tag{8.3.3}$$
$$\left. + e^{-i\mathbf{kr}} \boldsymbol{\epsilon}^*(\mathbf{k}, \eta) \hat{a}^+_{\text{n.r.}}(\mathbf{k}, \eta) \right\}.$$

The index $S$ in $\hat{\mathbf{A}}_S$ will be dropped, as will n.r. in $\hat{a}_{\text{n.r.}}$, in the remainder of this section. The full Hamiltonian is now

$$\hat{H} = \frac{1}{2m}\left(\hat{\mathbf{P}}_e - \frac{e}{c}\hat{\mathbf{A}}\right)^2 + V + \hat{H}_{\text{rad}} - \frac{2\mu}{\hbar}\hat{\mathbf{S}}\hat{\mathbf{B}} + C. \tag{8.3.4a}$$

We have added a term $-(2\mu/\hbar)\hat{\mathbf{S}}\hat{\boldsymbol{\mathcal{B}}}$ of spin interaction with the magnetic field, as suggested by our relativistic corrections to the Schrödinger equation (cf. Sect. 3.4). For elementary particles, say again electrons, $\mu_e = -|e|\hbar/2m_e c$. An arbitrary constant, $C$, has also been taken into account.

Because we will be studying slowly moving particles, we should be able to work perturbatively, considering that the terms in (8.3.4a) that contain $c$ in the denominator are a small perturbation of the rest. It will thus be useful to split $\hat{H}$ as

$$\hat{H} = \hat{H}^{(0)} + \hat{H}_I,$$

$$\hat{H}^{(0)} = \frac{1}{2m}\hat{\mathbf{P}}_e^2 + V + \hat{H}_{\mathrm{rad}},$$

$$\hat{H}_I = \frac{e^2}{2mc^2}\hat{\mathbf{A}}^2(\mathbf{r}) - \frac{e}{mc}\hat{\mathbf{P}}_e\hat{\mathbf{A}}(\mathbf{r})$$

$$-\frac{2\mu}{\hbar}\hat{\mathbf{S}}\hat{\boldsymbol{\mathcal{B}}} + C,$$

(8.3.4b)

and we have used the fact that, in the Coulomb gauge in which we will work in the remainder of this section,

$$\hat{\mathbf{P}}_e\hat{\mathbf{A}} = -i\hbar\boldsymbol{\nabla}\hat{\mathbf{A}} = \hat{\mathbf{A}}\hat{\mathbf{P}}_e - i\hbar\,\mathrm{div}\,\hat{\mathbf{A}} = \hat{\mathbf{A}}\hat{\mathbf{P}}_e,$$

so $\hat{\mathbf{P}}_e$ and $\hat{\mathbf{A}}$ can be taken to commute.

In what remains of this section we will present the evaluation of some processes[6] of interaction of matter with radiation, using (8.3.4); apart from the intrinsic interest of these processes, the calculations will serve to help build intuition (and evidence) for a quantum field-theoretic treatment of interactions. Because only slowly moving particles shall be considered here, *nonrelativistic normalization will be employed throughout*; the labels nr, NR will be *omitted*.

### 8.3.1 Radiative Decays, and Absorption of Radiation

In this subsection we consider the decay of the excited state of a system, emitting one photon, written symbolically as

$$\mathcal{M}^* \rightarrow \mathcal{M}' + \gamma. \tag{8.3.5a}$$

The related process, absorption of a photon

$$\gamma + \mathcal{M} \rightarrow \mathcal{M}'^*, \tag{8.3.5b}$$

can be treated in full parallel with (8.3.5a), and we will not give explicit formulas for it.

---

[6] A very comprehensive set of processes may be found in the classic textbooks of Akhiezer and Berestetskii (1963) and Sakurai (1967).

For definiteness we will consider the matter system $\mathcal{M}^*$, $\mathcal{M}'$ to be electrons, bound by some potential, with well-defined energy, orbital angular momentum and third component thereof. Thus we write the electron wave functions before ($\psi_e$) and after ($\psi'_e$) the emission as

$$\psi_e = f_{nl}(r)Y_M^l(\theta, \varphi),$$
$$\psi'_e = f_{n'l'}(r)Y_{M'}^{l'}(\theta, \varphi). \tag{8.3.6}$$

We will neglect the interaction with the electron spin, in general of higher order in $1/c$, and work to lowest order in $1/c$. Thus, of (8.3.4) we retain as effective perturbation the piece (in the Coulomb gauge)

$$\hat{H}_{I\,\text{eff}} = -\frac{e}{mc}\hat{\mathbf{A}}\hat{\mathbf{P}}_e. \tag{8.3.7}$$

According to (7.4.17) the (lowest-order) decay width for emission of a photon with wave vector $\mathbf{k}$ and helicity $\eta$ can be written as

$$d\Gamma(i \to f) = \frac{1}{2\pi}|T(i \to f)|^2\delta(E_f - E_i)d^3k\,d^3p',$$

with $\mathbf{k}$ the wave vector of the final photon, and $\mathbf{p}'$ the momentum of the final atom, usually not measured. Because of this, we integrate it, so that, still denoting the rate by $d\Gamma$, we get

$$d\Gamma(i \to f) = \frac{\hbar}{2\pi}|T(i \to f)|^2\delta(E_f - E_i)d^3k.$$

(These equations assume the photon state to be normalized according to $\langle\mathbf{k}|\mathbf{k}'\rangle = \delta(\mathbf{k} - \mathbf{k}')$, which is the reason why we write $d^3k$ for the differential density of final photon states.)

Now, to lowest order,

$$\langle f|S|i\rangle \simeq \frac{-i}{\hbar}\int_{-\infty}^{+\infty} dt\langle f|\hat{H}_{ID}(t)|i\rangle$$
$$= \frac{-i}{\hbar}\int_{-\infty}^{+\infty} dt\, e^{-it(E_i - E_f)/\hbar}\langle f|\hat{H}_{IS}|i\rangle$$
$$= -\frac{2\pi\hbar i}{\hbar}\delta(E_i - E_f)\langle f|\hat{H}_{IS}|i\rangle,$$

where $\hat{H}_{ID}$ is the Dirac picture interaction Hamiltonian, and $\hat{H}_{IS}$ is the same in the Schrödinger picture. Extracting a momentum conservation delta from $\langle f|\hat{H}_{IS}|i\rangle$, and still writing $\langle f|\hat{H}_{I\,\text{eff}}|i\rangle$, with the understanding that $|i\rangle$, $|f\rangle$ now refer to the c.m. motion, we obtain

$$\langle f|S|i\rangle = -2\pi i\delta(E_i - E_f)\delta(\mathbf{p}_i - \mathbf{p}_f)\langle f|\hat{H}_{I\,\text{eff}}|i\rangle,$$

so that

$$T(i \to f) \simeq -2\pi\langle|\hat{H}_{I\,\text{eff}}|i\rangle,$$

and thus

$$d\Gamma(i \to f) = 2\pi|\langle\psi'_e; \mathbf{k}, \eta|\hat{H}_{I\,\text{eff}}|\psi_e\rangle|^2\delta(E_f - E_i)d^3k. \tag{8.3.8}$$

We will now evaluate the matrix element in (8.3.8). After simple manipulations, and using (8.3.3), we have

$$\langle f|\hat{H}_{I\,\text{eff}}|i\rangle = \langle\psi'_e|\hat{a}(\mathbf{k}, \eta)\hat{H}_{I\,\text{eff}}|\psi_e\rangle$$

$$= \frac{-e}{mc}\frac{\sqrt{\hbar c}}{2\pi}\sum_{\eta'}\int\frac{d^3k'}{\sqrt{k'}}\boldsymbol{\epsilon}^*(\mathbf{k}', \eta')\langle\psi'_e|e^{-i\mathbf{k}'\mathbf{r}}\hat{\mathbf{P}}_e|\psi_e\rangle$$

$$\times\delta_{\eta\eta'}\delta(\mathbf{k} - \mathbf{k}') \tag{8.3.9}$$

$$= \frac{-e}{mc}\frac{\sqrt{\hbar c}}{2\pi k^{1/2}}\langle\psi'_e|e^{-i\mathbf{k}\mathbf{r}}\hat{\mathbf{P}}_e|\psi_e\rangle\boldsymbol{\epsilon}^*(\mathbf{k}, \eta).$$

Consider now the case in which the electron is that in a hydrogen atom. The mean value of $r$ is then of the order of the Bohr radius, $a_B = \hbar^2/m_ee^2$; whereas the energy of the emitted photon is $c\hbar k$, equal to the difference of energies between the initial and final electron, of the order of the Rydberg, $me^4/2\hbar^2$. Therefore, $|\mathbf{rk}| \simeq e^2/\hbar c = \alpha \simeq 1/137$, and, to the order at which we are working, we can replace $\exp(-i\mathbf{kr}) \to 1$. This is the so-called *dipole approximation*.

In some cases the dipole approximation vanishes, and we have to go to higher orders in the expansion of $\exp(-i\mathbf{kr})$. The ensuing terms are called *multipole terms*, and for higher orders it is actually much simpler *not* to proceed as we are doing; but replace in $\hat{A}$ *not* the plane wave expansion (8.2.3) but the *multipole expansion* (8.2.12). Details, together with many applications, may be found in the books of Galindo and Pascual (1978), or Condon and Shortley (1967); we will now continue with the case where the dipole approximation is valid, so that

$$\langle f|\hat{H}_{I\,\text{eff}}|i\rangle = \frac{-e\sqrt{\hbar c}}{2\pi mck^{1/2}}\boldsymbol{\epsilon}^*(\mathbf{k}, \eta)\langle\psi'_e|\hat{\mathbf{P}}_e|\psi_e\rangle.$$

We have thus reduced the problem to the simple one of evaluating matrix elements of the momentum operator. This can be simplified still further by noting that

$$\hat{P}_{ej} = m\hat{\dot{Q}}_{ej} = \frac{im}{\hbar}\left[\hat{H}_e, \hat{Q}_{ej}\right],$$

where $\hat{H}_e$ is the electron Hamiltonian, and $\hat{Q}_e$ its position operator $\hat{Q}_{ej}\psi_e(\mathbf{r}) = r_j\psi_e(\mathbf{r})$. If $E_n, E_{n'}$ are the electron energies in states $\psi_e, \psi'_e$, then

$$\langle\psi'_e|\hat{\mathbf{P}}_e|\psi_e\rangle = \frac{mi}{\hbar}(E_{n'} - E_n)\int d^3r\,\psi'^*_e\mathbf{r}\psi_e.$$

The last integral can be evaluated by relating $\mathbf{r}$ to the spherical harmonic $Y^1_\lambda(\Omega_\mathbf{r})$: from (5.3.4), (5.3.13),

$$r_j = r\sqrt{\frac{4\pi}{3}} \sum_\lambda U_{j\lambda}^* Y_\lambda^1(\Omega_{\mathbf{r}}), \ U_{j\lambda} = \chi_j(\lambda).$$

Thus, also using (8.3.6) and assuming the $f_{nl}$ to be real,

$$\int d^3r \, \psi_e'^* r_j \psi_e = \sqrt{\frac{4\pi}{3}} \sum_\lambda U_{j\lambda}^* \int_0^\infty dr \, r^3 f_{n'l'}(r) f_{nl}(r)$$

$$\times \int d\Omega Y_{M'}^{l'}(\Omega) Y_M^l(\Omega) Y_\lambda^1(\Omega).$$

The evaluation is finished using standard formulas for products of spherical harmonics. We find that

$$\langle f|\hat{H}_{I\,\text{eff}}|i\rangle = -i\frac{e\sqrt{E_\gamma}}{2\pi}\sqrt{\frac{2l+1}{2l'+1}}\,(1,0;l,0|l')$$

$$\times \sum_\lambda \epsilon_\lambda^*(\mathbf{k},\eta)(1,\lambda;l,M|l') \int_0^\infty dr \, r^3 f_{n'l'}(r) f_{nl}(r), \tag{8.3.10}$$

$$\epsilon_\lambda \equiv \sum_j U_{k\lambda}\epsilon_j,$$

for $l' = l \pm 1$, and zero if $l' \neq l \pm 1$.

**Exercise.** Use invariance under parity to prove generally that the matrix element must vanish if $l = l'$ ∎

It is not difficult, using (8.3.10), to evaluate widths for various cases, or to generalize the method. Here we will present the details for the evaluation of the radiative decay of the $2p \to 1s + \gamma$ levels of hydrogen. To obtain the *total* width, and if we do not know the third component of angular momentum in the initial state, $M$, or if we measure the wave vector or helicity of the emitted photon, we will have to average over $M, (1/3)\Sigma_M$, and sum over $\mathbf{k}, \eta$, $\int d^3k \Sigma_\eta$. Thus, from (8.3.10),

$$\Gamma(2\mathrm{p}) \equiv \Gamma(2\mathrm{p} \to 1\mathrm{s} + \gamma) = \frac{1}{3}\sum_M \sum_\eta \int d^3k \, 2\pi |\langle f|\hat{H}_{I\,\text{eff}}|i\rangle|\delta(E_f - E_i).$$

The only remaining step that is not totally trivial is the evaluation of the integral, that appears after substituting $\langle f|\hat{H}_{I\,\text{eff}}|i\rangle$, $\int d^3k \, k^{-1}\delta(E_f - E_i)$.

We have

$$E_f = E_{1\mathrm{s}} + E_\gamma = -\frac{me^4}{2\hbar^2} + c\hbar|\mathbf{k}|,$$

$$E_i = E_{2\mathrm{p}} = -\frac{me^4}{8\hbar^2},$$

so

$$\int \frac{d^3k}{k}\delta(E_f - E_i) = \int d\Omega_{\mathbf{k}} \int_0^\infty dk\, k\, \delta\left(\frac{-3me^4}{8\hbar^2} + c\hbar k\right)$$

$$= 2\pi \frac{3me^4}{4c^2\hbar^4}.$$

Finally,

$$\Gamma(2\mathrm{p}) = \left(\frac{2}{3}\right)^8 \frac{e^{10}}{c^3\hbar^5}m = \left(\frac{2}{3}\right)^8 \alpha^5 mc^2.$$

The corresponding mean life is then

$$\tau(2\mathrm{p}) = \hbar/\Gamma(2\mathrm{p}) = (3/2)^8 \hbar/\alpha^5 mc^2 \simeq 1.596 \times 10^{-9} \text{ seconds,}$$

to be compared with the experimental value

$$\tau_{\exp}(2\mathrm{p}) = (1.60 \pm 0.01) \times 10^{-9} \text{ seconds.}$$

### 8.3.2 Low-Energy Compton Scattering

We consider here the scattering (*Compton scattering*) of a photon by a charged particle, which we choose to be an electron; the extension to similar situations presents no difficulty. The process is $e(\mathbf{p}) + \gamma(\mathbf{k}, \eta) \to e(\mathbf{p}') + \gamma(\mathbf{k}', \eta')$, with obvious notation. The initial and final states are thus

$$|i\rangle = \hbar^{-3/2}|\mathbf{p}; \mathbf{k}, \eta\rangle; \quad |f\rangle = \hbar^{3/2}|\mathbf{p}'; \mathbf{k}', \eta'\rangle,$$

where the factors $\hbar^{3/2}$ are introduced so that the normalization of $|i\rangle$, $|f\rangle$ is $\delta(\mathbf{p} - \mathbf{p}')\delta(\mathbf{p}_\gamma - \mathbf{p}_\gamma')$ with $\mathbf{p}_\gamma = \hbar\mathbf{k}$, $\mathbf{p}_\gamma' = \hbar\mathbf{k}'$; but photon states are still normalized to $\langle\mathbf{k}'|\mathbf{k}\rangle = \delta(\mathbf{k} - \mathbf{k}')$.

The new feature of the present process is that we have to work to second order in perturbation theory. In fact, writing the perturbation as

$$\hat{H}_I = \hat{H}'_{IS} + \hat{H}''_{IS},$$

$$\hat{H}'_{IS} = -\frac{e}{mc}\hat{\mathbf{P}}_e\hat{\mathbf{A}}_S(\mathbf{r}); \quad \hat{H}''_{IS} = \frac{e^2}{2mc^2}\hat{\mathbf{A}}_S^2(\mathbf{r}),$$

(8.3.11)

and neglecting the spin-magnetic moment interaction, which gives effects of higher order in $1/c$, we will have to go to second order in $\hat{H}'_{IS}$. (The subscript $S$ reminds us that expression (8.3.11) is written in the Schrödinger picture.) The reason is that the matrix element $\langle f|\hat{H}_{IS}|i\rangle$ vanishes, as $\hat{H}'_{IS}$ changes the number of photons by one unit. Therefore, the first nonzero term is of order $e^2/c^2$, either from twice $\hat{H}'_{IS}$ or once $\hat{H}''_{IS}$:

$$\langle f|\hat{S}|i\rangle = \left(\frac{-i}{\hbar}\right)^2 \int_{-\infty}^{+\infty} dt \int_{-\infty}^t dt'\langle f|\hat{H}'_{ID}(t)\hat{H}'_{ID}(t')|i\rangle$$

$$+ \frac{-i}{\hbar}\int_{-\infty}^{+\infty} dt\langle f|\hat{H}''_{ID}(t)|i\rangle + \text{higher orders}$$

(8.3.12)

$$\equiv \langle f|\hat{S}'|i\rangle + \langle f|\hat{S}''|i\rangle + \text{higher orders,}$$

with self-explanatory notation. The $\hat{H}_{ID}$ are the operators in the Dirac picture,

$$\hat{H}_{ID}(t) = e^{i\hat{H}_0 t/\hbar}\hat{H}_{IS}e^{-i\hat{H}_0 t/\hbar}.$$

We first evaluate $\langle f|\hat{S}''|i\rangle$; as it will turn out, this will be the dominant term at low energies. We have

$$\langle f|\hat{S}''|i\rangle = -\frac{i}{\hbar}\int_{-\infty}^{+\infty} dt\; e^{i(E_f-E_i)t/\hbar}\langle f|\hat{H}''_{IS}|i\rangle$$

$$= \frac{2\pi i e^2}{2mc^2}\delta(E_f - E_i)\langle p'|\hat{a}(\mathbf{k}',\eta')\hat{A}_S^2(\mathbf{r})\hat{a}^+(\mathbf{k},\eta)|\mathbf{p}\rangle\hbar^{-2}.$$

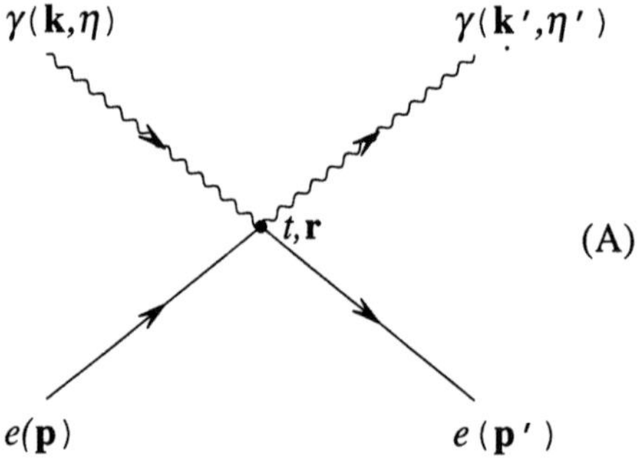


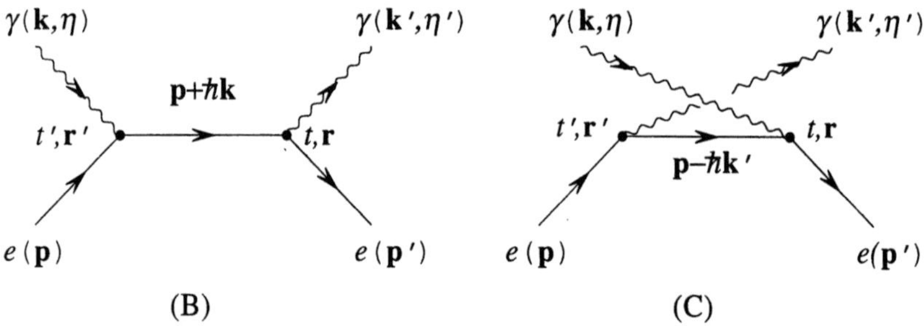


Fig. 8.3.1. Feynman diagrams for nonrelativistic Compton scattering.

We can represent this piece by the diagram (a so-called Feynman diagram) of Fig. 8.3.1 (A); the two photons are joined as they are emitted and absorbed at the same time $t$, and at the same location, $\mathbf{r}$. Substituting $\hat{A}_S$,

$$\langle f|\hat{S}''|i\rangle = -\frac{ie^2}{4\pi mc\hbar^2}\,\delta(E_f - E_i)$$

$$\times \sum_{\eta_1\eta_2}\int \frac{d^3k_1 d^3k_2}{\sqrt{k_1 k_2}}\,\langle \mathbf{p}|\hat{a}(\mathbf{k}',\eta')$$

$$\times \left\{ \boldsymbol{\epsilon}(\mathbf{k}_1,\eta_1)e^{i\mathbf{k}_1\mathbf{r}}\hat{a}(\mathbf{k}_1,\eta_1)\boldsymbol{\epsilon}^*(\mathbf{k}_2,\eta_2)e^{-i\mathbf{k}_2\mathbf{r}}\hat{a}^+(\mathbf{k}_2,\eta_2) \right.$$

$$\left. + (k_1,\eta_1 \leftrightarrow k_2,\eta_2) \right\} \hat{a}^+(\mathbf{k},\eta)|\mathbf{p}\rangle$$

$$= -\frac{ie^2}{2\pi mc\hbar^2}\,\frac{1}{(2\pi\hbar)^3}\,\frac{1}{\sqrt{kk'}}\boldsymbol{\epsilon}^*(\mathbf{k}',\eta')\boldsymbol{\epsilon}(\mathbf{k},\eta)$$

$$\times\delta(E_f - E_i)\langle \mathbf{p}'|e^{i(\mathbf{k}-\mathbf{k}')\mathbf{r}}|\mathbf{p}\rangle,$$

where we have repeatedly used the fact that $\langle \mathbf{p}|\hat{a}^+ = \hat{a}|\mathbf{p}\rangle = 0$. Now, with our normalization, the wave function for the state $|\mathbf{p}\rangle$ is $(2\pi\hbar)^{-3/2}e^{i\mathbf{p}\mathbf{r}/\hbar}$, so

$$\langle \mathbf{p}'|e^{i(\mathbf{k}-\mathbf{k}')\mathbf{r}}|\mathbf{p}\rangle = (2\pi\hbar)^3 \int d^3r\; e^{i(\mathbf{p}+\hbar\mathbf{k}-\mathbf{p}'-\hbar\mathbf{k}')\mathbf{r}/\hbar}$$

$$= \delta(\mathbf{p} + \hbar\mathbf{k} - \mathbf{p}' - \hbar\mathbf{k}').$$

Finally,

$$T''(i \to f) = -\frac{e^2}{2\pi mc\hbar^2\sqrt{kk'}}\boldsymbol{\epsilon}^*(\mathbf{k}',\eta')\boldsymbol{\epsilon}(\mathbf{k},\eta)$$

$$= -\frac{\alpha}{2\pi m\hbar\sqrt{kk'}}\boldsymbol{\epsilon}^*(\mathbf{k}',\eta')\boldsymbol{\epsilon}(\mathbf{k},\eta). \tag{8.3.13a}$$

In the c.m. frame, $k = k' = p_\gamma/\hbar$, and this simplifies to

$$T'' = \frac{-\alpha}{2\pi m p_\gamma}\boldsymbol{\epsilon}'^*\boldsymbol{\epsilon}. \tag{8.3.13b}$$

Let us now turn to $\langle f|\hat{S}'|i\rangle$. We have

$$\langle f|\hat{S}'|i\rangle = -\frac{1}{\hbar^2}\,\frac{1}{\hbar^3}\int_{-\infty}^{+\infty}dt\int_{-\infty}^{t}dt'\langle f|e^{i\hat{H}_0 t/\hbar}\hat{H}'_{IS}e^{-i\hat{H}_0 t/\hbar}$$

$$\times e^{i\hat{H}_0 t'/\hbar}\hat{H}'_{IS}e^{-i\hat{H}_0 t'/\hbar}|i\rangle.$$

To evaluate this we have to introduce a complete sum of states, $\sum_n |n\rangle\langle n|$, so that

$$\langle f|\hat{S}'|i\rangle = -\frac{1}{\hbar^5}\int_{-\infty}^{+\infty}dt\int_{-\infty}^{t}dt'\; e^{i(E_f-E_n)t/\hbar}$$

$$\times\sum_n\langle f|H'_{IS}|n\rangle\langle n|H'_{IS}|i\rangle e^{i(E_n-E_i)t'/\hbar}$$

$$= -\frac{1}{\hbar^5}\frac{e^2}{m^2c^2}\sum_n\int_{-\infty}^{+\infty}dt\; e^{i(E_f-E_n)t/\hbar}\frac{-\hbar}{i(E_f-E_n)}e^{i(E_n-E_i)t/\hbar}$$

$$\times\langle\mathbf{p}'|\hat{a}(\mathbf{k}',\eta')\hat{\mathbf{A}}_S(\mathbf{r})\hat{\mathbf{P}}|n\rangle\langle n|\hat{\mathbf{A}}_S(\mathbf{r}')\hat{\mathbf{P}}\hat{a}^+(\mathbf{k},\eta)|\mathbf{p}\rangle. \tag{8.3.14}$$

It is easy to become convinced that the only states that do not give zero in (8.3.14) are states with one electron, with momentum $\mathbf{p}_n$; or with one electron and *two* photons:

$$\sum|n\rangle\langle n| = \int d^3p_e|\mathbf{p}_e\rangle\langle\mathbf{p}_e| + \int d^3p_e\frac{d^3k_1 d^3k_2}{2!}\sum_{\eta_1\eta_2}|\mathbf{p}_e,\mathbf{k}_1,\eta_1;\mathbf{k}_2,\eta_2\rangle$$

$$\times\langle\mathbf{p}_e,\mathbf{k}_1,\eta_1;\mathbf{k}_2\eta_2|.$$

We can describe the process by the diagram of Fig. 8.3.1 (B): the incoming photon is absorbed at time $t'$, at location $\mathbf{r}'$; the electron then propagates to time $t$, location $\mathbf{r}$, where the final photon is emitted; or with Fig. 8.3.1 (C) where the emission occurs *before* the absorption.

Let us continue the calculation. We write

$$\langle f|\hat{S}'|i\rangle = -\frac{2\pi ie^2}{m^2c^2\hbar^3}\delta(E_f - E_i)\left\{\mathcal{M}_0 + \mathcal{M}_2\right\}, \tag{8.3.15a}$$

where the index $N$ in $\mathcal{M}_N$ refers to the number of photons in the intermediate state, and

$$\mathcal{M}_0 = \int d^3p_e\frac{1}{E_f - \mathbf{p}_e^2/2m}\langle\mathbf{p}'|\hat{a}(\mathbf{k}',\eta')\mathbf{p}_e\hat{\mathbf{A}}_S(\mathbf{r})|\mathbf{p}_e\rangle$$

$$\times\langle\mathbf{p}_e|\mathbf{p}_e\hat{\mathbf{A}}_S(\mathbf{r}')\hat{a}^+(\mathbf{k},\eta)|\mathbf{p}\rangle, \tag{8.3.15b}$$

$$\mathcal{M}_2 = \frac{1}{2!}\int d^3p_e d^3k_1 d^3k_2\sum_{\eta_1\eta_2}\frac{1}{E_f - \mathbf{p}_e^2/2m - c\hbar k_1 - c\hbar k_2}$$

$$\times\langle\mathbf{p}'|\hat{a}(\mathbf{k}',\eta')(\mathbf{p}_e + \hbar\mathbf{k}_1 + \hbar\mathbf{k}_2)\hat{\mathbf{A}}_S(\mathbf{r})\hat{a}^+(\mathbf{k}_1,\eta_1)\hat{a}^+(\mathbf{k}_2,\eta_2)|\mathbf{p}_e\rangle \tag{8.3.15c}$$

$$\times\langle\mathbf{p}_e|\hat{a}(\mathbf{k}_1,\eta_1)\hat{a}(\mathbf{k}_2,\eta_2)(\mathbf{p}_e + \hbar\mathbf{k}_1 + \hbar\mathbf{k}_2)\hat{\mathbf{A}}_S(\mathbf{r}')\hat{a}^+(\mathbf{k},\eta)|\mathbf{p}\rangle.$$

The evaluation of the $\mathcal{M}$ is straightforward. Substituting $\hat{\mathbf{A}}_S$, we get

$$\mathcal{M}_0 = \frac{\hbar c}{(2\pi)^2} \int d^3 p_e \frac{1}{E_f - \mathbf{p}_e/2m}$$

$$\times \sum_{\eta_1 \eta_2} \int \frac{d^3 k_1 d^3 k_2}{\sqrt{k_1 k_2}} \mathbf{p}_e \boldsymbol{\epsilon}^*(\mathbf{k}_1, \eta_1) \mathbf{p}_e \boldsymbol{\epsilon}(\mathbf{k}_2, \eta_2)$$

$$\times \langle \mathbf{p}' | \hat{a}(\mathbf{k}', \eta') \hat{a}^+(\mathbf{k}_1, \eta_1) e^{-i\mathbf{k}_1 \mathbf{r}} | \mathbf{p}_e \rangle \langle \mathbf{p}_e | \hat{a}(\mathbf{k}_2, \eta_2) \hat{a}^+(\mathbf{k}, \eta) e^{i\mathbf{k}_2 \mathbf{r}'} | \mathbf{p} \rangle$$

$$= \frac{\hbar c}{(2\pi)^2} \frac{1}{\sqrt{kk'}} \int d^3 p_e \frac{1}{E_f - \mathbf{p}_e^2/2m} \mathbf{p}_e \boldsymbol{\epsilon}^*(\mathbf{k}', \eta') \mathbf{p}_e \boldsymbol{\epsilon}(\mathbf{k}, \eta)$$

$$\times \langle \mathbf{p}' | e^{-i\mathbf{k}'\mathbf{r}} | \mathbf{p}_e \rangle \langle \mathbf{p}_e | e^{i\mathbf{k}\mathbf{r}'} | \mathbf{p} \rangle \tag{8.3.16a}$$

$$= \frac{\hbar c}{(2\pi)^2} \frac{1}{\sqrt{kk'}} \frac{1}{(2\pi\hbar)^6} \mathbf{p}_f \boldsymbol{\epsilon}'^* \mathbf{p}_e \boldsymbol{\epsilon}$$

$$\times \int d^3 r \; e^{i(\mathbf{p}_e - \hbar \mathbf{k}' - \mathbf{p}')\mathbf{r}/\hbar} \int d^3 r \; e^{i(\mathbf{p} + \hbar \mathbf{k} - \mathbf{p}_e)\mathbf{r}/\hbar} =$$

$$\frac{\hbar c}{(2\pi)^2 \sqrt{kk'}} \frac{1}{E_f - (\mathbf{p} + \hbar \mathbf{k})^2/2m} \mathbf{p}_f \boldsymbol{\epsilon}^*(\mathbf{k}', \eta') \mathbf{p}_i \boldsymbol{\epsilon}(\mathbf{k}, \eta) \delta(\mathbf{p}_i - \mathbf{p}_f).$$

The evaluation of $\mathcal{M}_2$, which is slightly more cumbersome, follows along the same lines. The result is

$$\mathcal{M}_2 = \frac{\hbar c}{(2\pi)^2} \frac{1}{\sqrt{kk'}} \frac{1}{E_f - ((\mathbf{p}' - \hbar \mathbf{k})^2/2m + c\hbar(k + k'))} \tag{8.3.16b}$$

$$\times \mathbf{p}_f \boldsymbol{\epsilon}^*(\mathbf{k}', \eta') \mathbf{p}_i \boldsymbol{\epsilon}(\mathbf{k}, \eta) \delta(\mathbf{p}_i - \mathbf{p}_f).$$

The terms $1/(E_f - E_n)$ in (8.3.16) are called *nonrelativistic propagators*, because they are connected with the propagation of the intermediate electron. The energy $E_n$ is that of the intermediate state, a simple electron for $\mathcal{M}_0$, an electron and two photons for $\mathcal{M}_2$. The momentum of the electron is obtained by requiring momentum conservation at each vertex (Fig. 8.3.1). Note also that in the scalar products of (8.3.16) $\mathbf{p}_f \boldsymbol{\epsilon}^*(\mathbf{k}', \eta')$, $\mathbf{p}_i \boldsymbol{\epsilon}(\mathbf{k}, \eta)$ we can if so wished replace $\mathbf{p}_f$ by $\mathbf{p}'$, $\mathbf{p}_i$ by $\mathbf{p}$ because $\mathbf{k}' \boldsymbol{\epsilon}^*(\mathbf{k}', \eta') = \mathbf{k} \boldsymbol{\epsilon}(\mathbf{k}, \eta) = 0$.

Substituting (8.3.16) into (8.3.15), and also adding (8.3.13), we obtain the final result:

$$T(i \to f) = -\frac{\alpha}{2\pi m\hbar \sqrt{kk'}} \sum_{ij} \epsilon_i^*(\mathbf{k}', \eta') \epsilon_j(\mathbf{k}, \eta) \tag{8.3.17a}$$

$$\times \left\{ \delta_{ij} + \frac{p_i' p_j}{m} \left( \frac{1}{D_0} + \frac{1}{D_2} \right) \right\},$$

$$D_0 = E_f - (\mathbf{p} + \hbar \mathbf{k})^2/2m,$$

$$D_2 = E_f - ((\mathbf{p}' - \hbar \mathbf{k})^2/2m + c\hbar(k + k')). \tag{8.3.17b}$$

The expression for the $D_N$ is particularly simple in the c.m. frame. There $\mathbf{p} + \hbar\mathbf{k} = 0$; writing $E_\gamma = c\hbar k$, $E'_\gamma = c\hbar k'$, and letting $\mathbf{v} = \mathbf{p}/m$, $\mathbf{v}' = \mathbf{p}'/m$ be the velocity of the electron before and after the collision, we find that

$$D_0 = E_\gamma(1 + v/2c), \quad D_2 = -E_\gamma\left(1 + v/2c + \frac{v}{c}\cos\theta\right),$$

$\theta$ being the angle between $\mathbf{p}, \mathbf{p}'$. Thus,

$$\frac{1}{D_0} + \frac{1}{D_2} \simeq \frac{1}{E_\gamma}\frac{v}{c}\cos\theta + \text{higher orders in } v/c$$

so that, in the low velocity limit in which we are working,

$$T(i \to f) \simeq -\frac{\alpha}{2\pi m\hbar\sqrt{kk'}} \sum_{ij} \epsilon_i^*(\mathbf{k}',\eta')\epsilon_j(\mathbf{k},\eta)$$

$$\times \left\{\delta_{ij} + \frac{v_j v_i'}{c^2}\cos\theta\right\} \underset{vv'\to 0}{\longrightarrow} \frac{-\alpha}{2\pi m\hbar\sqrt{kk'}}\epsilon^*(\mathbf{k}',\eta')\epsilon(\mathbf{k},\eta). \tag{8.3.18}$$

**Exercise.** Evaluate the c.m. cross-section. Verify that the nonpolarized zero-energy one is

$$\frac{d\sigma_{\text{n.p.}}}{d\Omega}\underset{k\to 0}{=}(\alpha r_e)^2, \quad r_e \equiv \hbar/mc \tag{8.3.19}$$

(*Thompson's formula*), where $r_e$ is the so-called *classical electron radius*. Numerically, $r_e \simeq 3 \times 10^{-11}$ cm, $\alpha r_e \simeq 2 \times 10^{-13}$ cm, $(\alpha r_e)^2 \simeq 5 \times 10^{-26}$ cm $= 0.05$ barn $\blacksquare$

The Thompson formula is important, among other reasons, because it is exact. Although deduced nonrelativistically, and to second order in perturbation theory, it can be proved (Thirring, 1950) that (8.3.19) is *exact* in the limit of zero photon energy.

## 8.4 Bremsstrahlung

The process in which a charged particle radiates when accelerated is called *bremsstrahlung* (German for "braking radiation"). The situation is as follows. Consider a particle of spin 1/2, with charge $Qe$, mass $m$ and magnetic moment $\mu = ge/4mc$; $g$ is called the gyromagnetic ratio, and our results for the Dirac equation show that, for elementary particles, $g = 2$. The particle is scattered by an interaction, described by the operator $\hat{V}$ (usually a nonquantized potential) that we will suppose weak enough to be treated to first order. Being scattered, the particle undergoes an acceleration and it will therefore radiate. We take the interaction of the radiation with matter to be

$$\hat{H}_1 = \frac{-eQ}{mc}\hat{\mathbf{A}}\hat{\mathbf{P}} - (2\mu/\hbar)\hat{\mathbf{B}}\mathbf{S}, \tag{8.4.1}$$

where, as stated,

$$\mu = ge/4mc$$

and we have not written the term $(e^2Q^2/2m^2c^2)\hat{A}^2$ since we will consider the lowest order in $e$. We will also start by by neglecting the spin-magnetic moment interaction; later we will take it into account.

The initial and final states will be respectively

$$|i\rangle = |\psi_{0\mathbf{k}}\rangle, \quad |f\rangle = |\psi_{0\mathbf{k}'}; k_\gamma, \eta\rangle = \hat{a}(\mathbf{k}_\gamma, \eta)|\psi_{0\mathbf{k}'}\rangle;$$

the initial and final particles are identified by their wave vectors $\mathbf{k}$ and $\mathbf{k}'$. We choose them normalized by

$$\langle \psi_{0\mathbf{k}}|\psi_{0\mathbf{k}'}\rangle = (2\pi)^3\delta(\mathbf{k} - \mathbf{k}').$$

With this normalization the scattering amplitude for scattering *without radiation*, which we denote by $T^{(0)}(\mathbf{k}_1 \to \mathbf{k}_2)$ is given, in the Born approximation, by

$$T^{(0)}(\mathbf{k}_1 \to \mathbf{k}_2) = -\frac{1}{(2\pi)^2\hbar^3}\langle \psi_{0\mathbf{k}_2}|\hat{V}|\psi_{0\mathbf{k}_1}\rangle. \tag{8.4.2}$$

When the particle radiates, we have to take into account $\hat{H}_1$ as well. Expanding $\hat{S}$, we easily see that the mixed second-order term is the first that gives a nonzero contribution:

$$\langle f|S|i\rangle = \frac{1}{2!}\int_{-\infty}^{+\infty}dt\int_{-\infty}^{t}dt'\Big\langle f\Big|\Big\{\hat{H}_{1D}(t)\hat{V}_D(t')$$
$$+ \hat{V}_D(t)\hat{H}_{1D}(t')\Big\}\Big|i\Big\rangle + \text{higher orders.} \tag{8.4.3}$$

The calculation is very similar to that of the previous subsection and we will not repeat the details. We get two terms, each associated with one of the terms of (8.4.3):

$$T_B(\mathbf{k} \to \mathbf{k}' + \gamma(\mathbf{k}_\gamma, \eta)) = \frac{Qe}{(2\pi)^3m\hbar^2\sqrt{E_\gamma}}$$
$$\times \boldsymbol{\epsilon}^*(\mathbf{k}_\gamma, \eta)\{\boldsymbol{\Phi}_1 + \boldsymbol{\Phi}_2\}, \tag{8.4.4a}$$

where $E(\mathbf{k}) \equiv \hbar^2\mathbf{k}^2/2m$ and

$$\boldsymbol{\Phi}_1 = \hbar(\mathbf{k}' + \mathbf{k}_\gamma)\frac{(2\pi)^2\hbar^3}{E(\mathbf{k}) - E(\mathbf{k}' + \mathbf{k}_\gamma)}T_B^{(0)}(\mathbf{k} \to \mathbf{k}' + \mathbf{k}_\gamma),$$
$$\boldsymbol{\Phi}_2 = \hbar(\mathbf{k}' - \mathbf{k}_\gamma)\frac{(2\pi)^2\hbar^3}{E(\mathbf{k}) - E(\mathbf{k} - \mathbf{k}_\gamma)}T_B^{(0)}(\mathbf{k} - \mathbf{k}_\gamma \to \mathbf{k}'). \tag{8.4.4b}$$

Equation (8.4.4) is a remarkable expression which relates the amplitude for scattering plus radiation to that without radiation. It can be visualized, as in the previous case, in a a very pictorial way (Fig. 8.4.1): first $\hat{V}$ acts, and then the photon is emitted (term in $\boldsymbol{\Phi}_1$) or conversely (term in $\boldsymbol{\Phi}_2$).

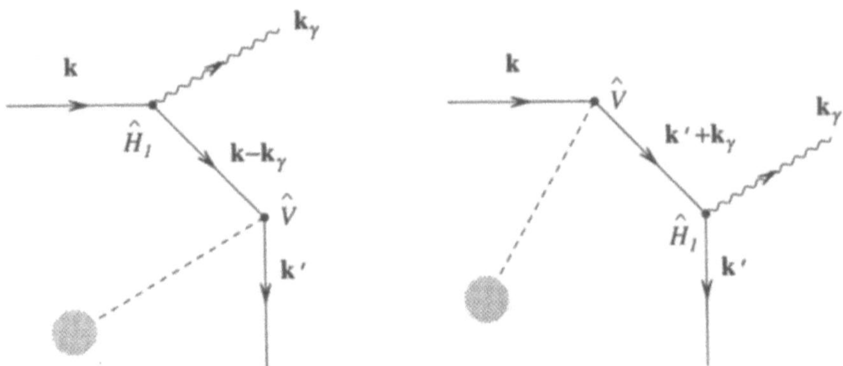

**Fig. 8.4.1.** Bremsstrahlung diagrams.

Now, both $T_B^{(0)}$ in (8.4.4b) are actually equal since, in the Born approximation, $T_B^{(0)}(\mathbf{k}_1 \to \mathbf{k}_2)$ only depends upon the difference $\mathbf{k}_1 - \mathbf{k}_2$. Using this and some simple algebra we get the overall result,

$$T_B(\mathbf{k} \to \mathbf{k}' + \gamma(\mathbf{k}_\gamma, \eta)) = T_B^{(0)}(\mathbf{k}' + \mathbf{k}_\gamma - \mathbf{k})\Phi_{\mathrm{rad}},$$

$$\Phi_{\mathrm{rad}} = \frac{eQ\hbar^2}{2\pi m E_\gamma^{3/2}}\epsilon^*(\mathbf{k}_\gamma, \eta)\left\{\frac{\mathbf{k}}{D} - \frac{\mathbf{k}'}{D'}\right\}, \tag{8.4.5}$$

$$D' = 1 - \hbar(k_\gamma + 2k'\cos\theta'_\gamma)/2mc, \quad D = 1 + \hbar(k_\gamma - 2k\cos\theta_\gamma)/2mc.$$

$\theta_\gamma$, $\theta'_\gamma$ are the angles of $\mathbf{k}_\gamma$ with $\mathbf{k}$, $\mathbf{k}'$.

The differential cross-section for scattering of the particle into the momentum interval $d^3\hbar k'$, and the photon into the wave interval $d^3k_\gamma$ is then

$$d\sigma\,(\mathbf{k} \to \mathbf{k}' + \gamma(\mathbf{k}_\gamma, \eta))$$

$$= \left\{\frac{(2\pi\hbar)^2}{v_i}|T_B^{(0)}(\mathbf{k}' + \mathbf{k}_\gamma - \mathbf{k})|^2\delta(E(\mathbf{k}') + E_\gamma - E(\mathbf{k}))\hbar d^3k'\right\} \tag{8.4.6}$$

$$\times |\Phi_{\mathrm{rad}}|^2 d^3k_\gamma;$$

$v_i$ is the initial velocity of the particle, $v_i = \hbar k/m$, and we have disposed of the $\delta$ of three-momentum conservation by integrating over the momentum of the scattering centre. For long-wavelength ("*soft*") photons with $k_\gamma$ very small, the term in curly brackets in (8.4.5) coincides with the cross-section for scattering of the particles by the interaction $\hat{V}$, *without emitting photons*, $d\sigma^{(0)}(\mathbf{k} \to \mathbf{k}')/d\Omega$: the cross-section factorizes. Working with small velocities (compared with $c$), so that we may approximate $D \simeq 1$, $D' \simeq 1$, we get

$$\frac{d\sigma(\mathbf{k} \to \mathbf{k}' + \gamma(\mathbf{k}_\gamma, \eta))}{d\Omega_{k'} dE_\gamma} \simeq \frac{d\sigma^{(0)}(\mathbf{k} \to \mathbf{k}')}{d\Omega_{k'}}$$

$$\times \frac{Q^2 e^2 \hbar^4}{(2\pi)^2 m E_\gamma^3} \sum_{ij} \epsilon_i^*(\mathbf{k}_\gamma, \eta) \epsilon_j(\mathbf{k}_\gamma, \eta)(k_i - k_i')(k_j - k_j') d^3 k_\gamma. \tag{8.4.7}$$

If we are not interested in the direction or polarization of the radiated photon, we have to sum over $\eta$ and integrate in $d\Omega_\gamma$, the angular variables of $\mathbf{k}_\gamma$. We get

$$\frac{d\sigma(\mathbf{k} \to \mathbf{k}' + \gamma(E_\gamma))}{d\Omega_{k'} dE_\gamma} \underset{\substack{v \ll c \\ k_\gamma \ll k}}{\simeq} \frac{d\sigma^{(0)}(\mathbf{k} \to \mathbf{k}')}{d\Omega_{k'}}$$

$$\times \frac{8}{3} \frac{Q^2 \alpha}{\pi E_\gamma} \left(\frac{v_i}{c}\right)^2 \sin^2 \frac{\theta_p}{2}, \tag{8.4.8}$$

$\theta_p$ being the angle of deflection of the particle, i.e., that formed by $\mathbf{k}$ and $\mathbf{k}'$. Clearly, the more the particle trajectory is bent, the larger $\sin^2 \theta_p/2$ becomes, and the larger the cross-section for radiation becomes, as could be expected. Equation (8.4.8) has the important feature of being *divergent* when integrated in $dE_\gamma$, owing to the $E_\gamma$ in the denominator. This is the so-called *infrared catastrophe*.

The situation is not really so catastrophic. We should take into account that *detectors* have a finite resolution power. Photons with energies below a certain threshold, $E_{\min}$, will not be detected. The measured cross-section will therefore *not* be the integral in $dE_\gamma$, from a given maximum energy to zero, but to $E_{\min}$. The observable quantity is then

$$\int_{E_{\min}}^{E_{\max}} dE_\gamma \frac{d\sigma}{d\Omega_{k'} dE_\gamma} = \frac{d\sigma^{(0)}}{d\Omega} \frac{8\alpha Q^2}{3\pi} \left(\frac{v_i}{c}\right)^2 \left(\log \frac{E_{\max}}{E_{\min}}\right) \left(\sin^2 \frac{\theta_p}{2}\right). \tag{8.4.9}$$

When $E_{\min}$ is very small, the product $\alpha \log E_{\max}/E_{\min}$ becomes of order unity. Because we can iterate the procedure that gives $\sigma(\mathbf{k} \to \mathbf{k}' + \gamma)$ in terms of $\sigma^{(0)}(\mathbf{k} \to \mathbf{k}')$, we can obtain $\sigma(\mathbf{k} \to \mathbf{k}' + n\gamma)$ in terms of $\sigma(\mathbf{k} \to \mathbf{k}' + (n-1)\gamma)$. When $\alpha \log E_{\max}/E_{\min} \sim 1$, all these cross-sections are comparable, so we expect that many photons will be radiated. Perturbation theory breaks down; the problem becomes almost classical and can actually be solved exactly (in the limit $E_\gamma \to 0$) by a method due to Bloch and Nordsiek (1937). The calculations, with details, may be found in the text of Akhiezer and Berestetskii (1963), and in Chap. 12 below.

Let us next take into account, for bremsstrahlung by particles of spin 1/2, the spin-magnetic interaction piece in (8.4.1), where the spin operator is $\mathbf{S} = \frac{1}{2}\hbar\boldsymbol{\sigma}$, and the spin states of the particle are described by Pauli spinors $\chi^{(s_3)}$, $s_3$ being the third component of spin. $\Phi_{\mathrm{rad}}$ is now replaced by

$$\Phi_{\text{rad}}^{(s_3',s_3)} = \frac{\hbar^2}{2\pi m E_\gamma^{3/2}} \boldsymbol{\epsilon}^*(\mathbf{k}_\gamma,\eta)$$

$$\times \underset{\sim}{\chi}^{(s_3')+}\left\{\left(\frac{\mathbf{k}}{D'}-\frac{\mathbf{k}'}{D}\right)eQ-\frac{i}{4}e\left(\frac{1}{D}-\frac{1}{D'}\right)g\mathbf{k}_\gamma\times\boldsymbol{\sigma}\right\}\underset{\sim}{\chi}^{(s_3)}; \tag{8.4.10}$$

the correction due to the spin tends to zero as $\mathbf{k}_\gamma \to 0$. If we do not measure photon or electron polarizations, or photon orientation,

$$\frac{d\sigma_{\text{n.p.}}(\mathbf{p}\to\mathbf{k}'+\gamma(E_\gamma))}{d\Omega_{\mathbf{k}'}dE_\gamma} \underset{v\ll c}{\simeq} \frac{d\sigma_{\text{n.p.}}^{(0)}(\mathbf{k}\to\mathbf{k}')}{d\Omega_{\mathbf{k}'}}$$

$$\times\frac{8}{3}\frac{\alpha}{\pi E_\gamma}\left\{Q^2+\frac{1}{4}g^2\left(\frac{E_\gamma}{mc^2}\right)^2+0(E_\gamma^4)\right\}\left(\frac{v_i}{c}\right)^2\sin^2\frac{\theta_p}{2}. \tag{8.4.11}$$

These results are also interesting in another respect. We have found that mainly photons of low energy are radiated by charged particles, in agreement with experiment, indeed everyday experience. But this is because we have followed the minimal prescription, and we have thus written interactions with $\hat{A}$. Interactions with $\hat{\mathcal{E}}$ or $\hat{\mathcal{B}}$ (like the term $-(2\mu/\hbar)\hat{\mathcal{B}}\hat{\mathbf{S}}$ in (8.4.1)) would have given terms vanishing as $E_\gamma \to 0$ (like the $(g^2/4)(E_\gamma/mc^2)^2$ term in (8.4.11)). This, together with the Aharonov–Bohm effect,[7] is a powerful indication that the basic quantities in quantum mechanics are the components of the vector potential $A_0$, $\mathbf{A}$ and that the $\mathcal{E}$, $\mathcal{B}$ are to be considered derived ones, although they are the ones that are directly measurable. Moreover, the interaction is to be implemented by the procedure of minimal substitution also for the quantized electromagnetic field.

## 8.5 The Classical Limit. Coherent States

The classical limit in the theory of radiation corresponds to states with very large numbers of photons, $|\Phi\rangle$, such that the average of the electromagnetic field in them[8]

$$\mathcal{E}_{\text{av}} = \langle\Phi|\hat{\mathcal{E}}|\Phi\rangle, \quad \mathcal{B}_{\text{av}} = \langle\Phi|\hat{\mathcal{B}}|\Phi\rangle \tag{8.5.1}$$

obeys the classical Maxwell equations and, moreover, there are (relatively) small fluctuations.

The problem may be split into two parts. First, we have to describe the states $|\Phi\rangle$ which fulfil the properties just mentioned. Second, we have to ascertain under what conditions assemblies of photons with this classical behaviour are produced. The first part will be the subject of the rest of the present section; the second will be postponed until Chap. 12, because it is

---

[7] Which may be found in e.g. Galindo and Pascual (1978) and Ynduráin (1988).

[8] In this section we will distinguish operators by placing carets over them.

there that we will have developed the tools that allow us to present a complete solution.

To study the "classical" states $|\Phi\rangle$ it will be convenient to consider the radiation field to be enclosed in a finite square box with volume $V$; the physics will of course be independent of this, in the limit $V \to \infty$, but this confinement will be very helpful for solving a number of technical difficulties.

The boundary conditions which simplify the treatment most are periodic boundary conditions. If $L$ is the length of the cube with volume $V$, $V = L^3$, we then require

$$\hat{\mathbf{A}}_L(\mathbf{r} + L\mathbf{j}, t) = \hat{\mathbf{A}}_L(\mathbf{r}, t); \quad \hat{\mathbf{A}}_L(\mathbf{r}, t) = 0 \text{ outside } V. \tag{8.5.2}$$

Here $\mathbf{j}$ is a unit vector along the $j$ axis.

The expression of $\hat{A}$ in term of creators and annihilators, (8.2.3a), now becomes

$$\hat{\mathbf{A}}_L(\mathbf{r}, t) = \frac{\sqrt{\hbar c}}{2\pi} \rho^3 \sum_{\mathbf{n}} \frac{1}{k_{\mathbf{n}}^{1/2}} \sum_{\eta = \pm 1} \left\{ e^{i(\mathbf{k_n r} - \omega(k_{\mathbf{n}})t)} \boldsymbol{\epsilon}(\mathbf{k}_n, \eta) \hat{a}(\mathbf{k}_n, \eta) \right.$$
$$\left. + e^{-i(\mathbf{k_n r} - \omega(k_{\mathbf{n}})t)} \boldsymbol{\epsilon}^*(\mathbf{k_n}, \eta) \hat{a}^+(\mathbf{k_n}, \eta) \right\}, \tag{8.5.3a}$$

and the commutation relations corresponding to (8.2.3b) are now

$$[\hat{a}(\mathbf{k_n}, \eta), \hat{a}^+(\mathbf{k_{n'}}, \eta')] = \rho^{-3} \delta_{\eta\eta'} \delta_{\mathbf{nn'}}. \tag{8.5.3b}$$

Here

$$(\mathbf{k_n})_j = 2\pi n_j/L, \ \rho = 2\pi/L; \ n_j = \text{integer}, \ j = 1, 2, 3.$$

In the limit $L \to \infty$, one has $\rho \to 0$ and

$$\rho^3 \sum_{\mathbf{n}} \to \int d^3k, \ \rho^{-3} \delta_{\mathbf{nn'}} \to \delta(\mathbf{k} - \mathbf{k}'), \text{ etc.},$$

and we would recover the expressions without the confinement.

The commutation relations (8.5.3b) are identical to those of a set of harmonic oscillators. Therefore the states which will reproduce the classical limit will be, as for ordinary harmonic oscillators, the *coherent states*. Their construction may be found in any standard textbook on quantum mechanics (Galindo and Pascual, 1978; Ynduráin, 1988, etc.). We write

$$|\Phi(\mathbf{k}_n, \eta)\rangle = \sum_{N=0}^{\infty} C_N \frac{1}{\sqrt{N!}} \rho^{3N/2} [\hat{a}^+(\mathbf{k_n}, \eta)]^N |0\rangle, \tag{8.5.4}$$

i.e., the state is an assembly of photons all of which have the same polarization $\eta$ and wave vector $\mathbf{k_n}$. The coefficients $C_N$ are found by requiring $|\Phi\rangle$ to be an eigenstate of the annihilation operator with eigenvalue $\alpha_0$:

$$\hat{a}(\mathbf{k_{n'}}, \eta') |\Phi(\mathbf{k}_n, \eta)\rangle = \delta_{\mathbf{nn'}} \delta_{\eta\eta'} \alpha_0 |\Phi(\mathbf{k}_n, \eta)\rangle.$$

After a simple calculation, one verifies that this implies that

$$C_N = \frac{\rho^{3N/2}\alpha_0^N}{\sqrt{N!}}C; \tag{8.5.5}$$

$C$ is a constant, which may be fixed (up to a phase) by requiring e.g.

$$\langle\Phi(\mathbf{k_n},\eta)|\Phi(\mathbf{k_n},\eta)\rangle = 1.$$

It is easy to verify that these coherent states possess the desired properties. Using (8.5.3a) we immediately find that

$$\langle\Phi(\mathbf{k_n},\eta)|\hat{\mathbf{A}}_L(\mathbf{r},t)|\Phi(\mathbf{k_n},\eta)\rangle$$

$$= \frac{\sqrt{\hbar c}}{2\pi}\frac{\rho^3}{k_\mathbf{n}^{1/2}}2\mathrm{Re}\left\{e^{i(\mathbf{k_n r}-\omega(k_n)t)}\boldsymbol{\epsilon}(\mathbf{k_n},\eta)\alpha_0\right\}, \tag{8.5.6}$$

which is indeed identical with the expression for the potential of a classical electromagnetic wave $\mathbf{A}_\mathrm{cl}^{(\mathbf{k_n},\eta)}(\mathbf{r},t)$ with wave vector $\mathbf{k_n}$, polarization $\eta$ and amplitude $\alpha_0$. Likewise, one can check that the expectation value in our states of the radiation Hamiltonian, given, for example, in (8.2.8a),

$$\langle\Phi(\mathbf{k_n},\eta)|\hat{H}_\mathrm{rad}|\Phi(\mathbf{k_n},\eta)\rangle,$$

coincides, up to corrections of order $\hbar$, with the classical energy

$$E_\mathrm{cl} = \frac{1}{8\pi}\int_V d^3r\left\{\frac{1}{c^2}\left(\partial_t\mathbf{A}_\mathrm{cl}^{(\mathbf{k_n},\eta)}\right)^2\right.$$

$$\left. - \mathbf{A}_\mathrm{cl}^{(\mathbf{k_n},\eta)}\triangle\mathbf{A}_\mathrm{cl}^{(\mathbf{k_n},\eta)}\right\}$$

with $\mathbf{A}_\mathrm{cl}^{(\mathbf{k_n},\eta)}$ given by the right-hand side of (8.5.6). Finally, the equations of motion (the Maxwell equations) for the average $\langle\Phi|\hat{\mathbf{A}}_L|\Phi\rangle$ follow immediately from (8.5.6).

An expression for the coherent state (8.5.4) that will be useful when we discuss emission of radiation is

$$|\Phi(\mathbf{k_n},\eta)\rangle = \left\{\exp\left(\rho^3\alpha_0\hat{a}^+(\mathbf{k_n},\eta)\right)\right\}|0\rangle. \tag{8.5.7}$$

This may be verified easily by expanding the exponential and comparing the result with (8.5.4), using the expression (8.5.5) for the constants $C_N$.

## 8.6 Uncertainty Relations for Field Variables

The operators $\hat{\boldsymbol{\mathcal{E}}}$, $\hat{\boldsymbol{\mathcal{B}}}$ represent *observables*. A complete description of the quantization of the electromagnetic field then requires a study of their commutation relations and the ensuing compatibility and/or uncertainty relations.

In the Gauss system of units, but taking $\hbar = c = 1$, we can use the expression for the electromagnetic field potential, and the commutation relations, (8.2.5), to obtain immediately

$$[\hat{\mathcal{B}}_j(\mathbf{r},t),\hat{\mathcal{B}}_l(\mathbf{r}',t')] = [\hat{\mathcal{E}}_j(\mathbf{r},t),\hat{\mathcal{E}}_l(\mathbf{r}',t')]$$
$$= -4\pi i(\delta_{jl}\partial_t^2 - \partial_j\partial_l)D(\mathbf{r}-\mathbf{r}',t-t'), \tag{8.6.1a}$$

$$[\hat{\mathcal{E}}_j(\mathbf{r},t),\hat{\mathcal{B}}_l(\mathbf{r}',t')] = 4\pi i\sum \epsilon_{jls}\partial_t\partial_s D(\mathbf{r}-\mathbf{r}',t-t'). \tag{8.6.1b}$$

The function $D$ is

$$\begin{aligned}
D(\mathbf{r},t) &= -\frac{1}{(2\pi)^3}\int d^3k\, e^{i\mathbf{k}\mathbf{r}}\,\frac{\sin|\mathbf{k}|t}{|\mathbf{k}|}\\
&= \frac{1}{4\pi r}(\delta(r+t)-\delta(r-t)).
\end{aligned} \tag{8.6.2}$$

The commutation relations (8.6.1) were first derived by Jordan and Pauli. Because $D$ is different from zero on the light cone, they imply that a measurement of $\mathcal{E}$ or $\mathcal{B}$ at a point $\mathbf{r}, t$ influences subsequent measurements if they can be connected by a light signal. However, one also has

$$[\hat{F}_{\mu\nu}(x),\hat{F}_{\alpha\beta}(y)] = 0 \quad\text{if}\quad (x-y)^2 < 0. \tag{8.6.3}$$

This is an expression of *causality*, since it means that measurements at space-time points that cannot be connected by a signal are independent.

The reason for these uncertainty relations, as well as for the singularity of the function $D$, was given in two classical papers by Bohr and Rosenfeld (1933, 1950). To measure $\mathcal{E}$ (say) we require as a probe a charged particle, necessarily extended: thus, the really observable quantities are averages over a small volume $V$, for example

$$\int_V d^3r\,\varphi(r)\hat{\mathcal{E}}(\mathbf{r},t), \tag{8.6.4}$$

$\varphi$ being the density of charge of the probe. These quantities (8.6.4) would satisfy commutation relations with a nonsingular commutator. On the other hand, the probe satisfies the Heisenberg uncertainty relations: thus, it also has an uncertainty in location and momentum. The last creates a current, and the uncertainty in $\mathbf{r}$ gives a dipole. This unknown dipole and current *of the probe* create the uncertainty in the electromagnetic field in a region $\mathbf{r}', t'$ which may be causally influenced by $V$.

Two more commutation relations are of interest:

$$[\hat{A}_j(\mathbf{r},t),\hat{\pi}_l(\mathbf{r}',t)] = i\delta_{jl}\delta(\mathbf{r}-\mathbf{r}'),$$
$$\hat{\pi}_l(\mathbf{r},t) \equiv \frac{1}{4\pi}\partial_t\hat{A}_j(\mathbf{r},t), \tag{8.6.5}$$

which may be connected with a canonical commutation relation in a Lagrangian formulation of the quantization of the electromagnetic field.

Finally, we mention an uncertainty relation between the number of photons and the values of the $\mathcal{E}$, $\mathcal{B}$. If we have zero photons, then

$$(\Delta\mathcal{E}_j)^2 = \langle 0|\hat{\mathcal{E}}^2|0\rangle - \langle 0|\hat{\mathcal{E}}|0\rangle^2 = \infty,$$

or, if we take the average over a small volume,

$$(\Delta \mathcal{E}_j)_V^2 = \left\langle 0 \left| \int_V d^3 r \, \hat{\mathcal{E}}^2 \right| 0 \right\rangle \sim \hbar c / V^{4/3}. \tag{8.6.6}$$

This is no surprise: the coherent states discussed in Sect. 8.5 are the states that correspond to classical electromagnetic fields, and thus contain an undetermined number of photons.

## Problems

**P.8.1.** Calculate the cross-section for the photoelectric effect in which a photon hits an electron in the fundamental state of the hydrogen atom and kicks it out of the atom (a) when the energy of the photon is large compared to the ionization energy of the electron; (b) when it is close to it.

*Solution.* Consider case (b). We split the interactions into two pieces: the interaction of the electron with the Coulomb field of the nucleus, $V = -e^2/r$, which we treat exactly, and the interaction with the radiation field, which we treat to first order,

$$\hat{H}_2 = (e/m)\hat{\mathbf{A}}\hat{\mathbf{P}}_e.$$

The scattering amplitude is, with the formalism of Sect. 7.8.3,

$$T = -2\pi \langle \varphi_{\mathbf{k}'}^- | \hat{H}_2 \hat{a}^+(\mathbf{k}, \eta) | \psi_{1s} \rangle.$$

Here $\psi_{1s}(r) = 2a_B^{-3/2} \exp(-r/a_B)$, with $a_B$ the Bohr radius, is the electron wave function; $\mathbf{k}, \eta$ are the wave vector and helicity of the incoming photon; and $\varphi_{\mathbf{k}'}^-(\mathbf{r})$ is the wave function of the outgoing electron, with momentum $\hbar k'$, in the field of the leftover proton. In the dipole approximation only the $P$ wave contributes. The details may be found in the texts of Gottfried (1966) and Ynduráin (1988). One gets

$$\frac{d\sigma}{d\Omega_{k'}} = \left( 64\pi \alpha a_B^2 \sin^2 \theta \right) \left( \frac{\mathrm{Ry}}{E_\gamma} \right)^4$$

$$\times \frac{\exp(-(4/a_B k') \arctan a_B k')}{1 - \exp\left(-2\pi/k' a_B\right)}, \quad \mathrm{Ry} = \frac{1}{2} m_e c^2 \alpha^2.$$

**P.8.2.** Calculate the average number of photons, and the uncertainty in this quantity, in the coherent states of (8.5.4).

# 9. Quantum Fields: Spin 0, 1/2, 1. Covariant Quantization of the Electromagnetic Field

## 9.1 Generalities

As we have stated several times, it is a fact that a theory based upon the wave function formalism, and with interactions given by potentials, does not provide a consistent description of physical reality. There are a number of reasons for this. Some are empirical: in any process at high energy, particles are created; therefore a wave function formalism, where the number of particles stays constant in time, will not be appropriate. Moreover, even if particles cannot be created because the energy is not sufficient, they may appear as quantum fluctuations provided the time they are present, $\Delta t$, and the energy fluctuation, $\Delta E$, satisfy the Bohr–Heisenberg relation

$$\Delta t \; \Delta E \sim \hbar.$$

In particular, for photons, $\Delta E$ may be as small as wished, and indeed any process involving photons necessitates (as shown in the previous chapter) the introduction of *field operators* that describe a system with a variable number of particles.

Internal difficulties also abound, for example those already discussed in connection with strong Coulomb fields or the Klein paradox; but more interesting are difficulties of *principle*. In fact, the description of a quantum system by a wave function $\Phi(\mathbf{r}, t)$ obeying a Schrödinger-type equation involving a potential, $V(\mathbf{r})$, implies two assumptions. First, it should be possible to localize particles so that a meaning can be attached to expressions like "the value of the potential at the location $\mathbf{r}$ of the particle". Second, the very concept of interaction carried by a potential is suspect in relativity: since $V(\mathbf{r})$ is time-independent, the interaction is instantaneous, in flagrant contradiction with the very principles of relativity. We will discuss both these matters in turn.

## 9.2 Localization of Particles in Relativistic Quantum Mechanics

Let us start by considering a spinless particle with nonzero mass, $m \neq 0$. The invariant scalar product of wave functions is given by (2.3.9), (2.3.12),

$$\langle \Phi_1 | \Phi_2 \rangle = i\hbar \int d^3 r \, \Phi_1^*(\mathbf{r}, t) \, \overleftrightarrow{\partial}_t \, \Phi_2(\mathbf{r}, t)$$

$$= \int \frac{d^3 p}{2E(\mathbf{p})} \Phi_1^*(\mathbf{p}, t) \Phi_2(\mathbf{p}, t), \quad E(\mathbf{p}) = c\sqrt{m^2 c^2 + \mathbf{p}^2}, \tag{9.2.1}$$

in position and momentum space. We will now try to find an operator $\mathbf{Q}$ that represents the location of the particle (*position operator*).

What first comes to mind is the expression (in $\mathbf{p}$ space)

$$\mathbf{Q}\Phi(\mathbf{p}, t) = i\hbar \nabla_p \Phi(\mathbf{p}, t),$$

which is identical to the nonrelativistic one: but it does not work. Indeed, the operator is not Hermitean:

$$\int \frac{d^3 p}{2E(\mathbf{p})} \Phi_1^*(\mathbf{p}, t)(i\hbar \nabla_p \Phi_2(\mathbf{p}, t))$$

$$\neq \int \frac{d^3 p}{2E(\mathbf{p})} (i\hbar \nabla_p \Phi_1(\mathbf{p}, t))^* \Phi_2(\mathbf{p}, t).$$

Newton and Wigner (1949) have shown, under very general conditions[1], that the only acceptable position operator is given in momentum space by

$$\mathbf{Q}_{NW} = i\hbar \nabla_\mathbf{p} - \frac{i\hbar}{2} \frac{\mathbf{p}}{\mathbf{p}^2 + m^2 c^2}. \tag{9.2.2}$$

In the nonrelativistic limit, this Newton–Wigner position operator coincides with the ordinary one.

**Exercise.** Verify that $\mathbf{Q}_{NW}$ is Hermitean with respect to the scalar product (9.2.1) ∎

In spite of the nice properties $\mathbf{Q}_{NW}$ enjoys, it is not quite as satisfactory a position operator as one would desire. In fact, if we consider those of its eigenfunctions that should correspond to particles located at the origin, i.e., verifying

$$\mathbf{Q}_{NW} \Phi_0(\mathbf{p}, t) = 0, \tag{9.2.3a}$$

one finds that

$$\Phi_0(\mathbf{p}, t) = (\text{constant}) \sqrt{E(\mathbf{p})}, \tag{9.2.3b}$$

or, in $x$ space,

$$\Phi_0(\mathbf{r}, t) = (\text{constant}) \left(\frac{m}{r}\right)^{5/4} H_{5/4}^{(1)}\left(\frac{imcr}{\hbar}\right). \tag{9.2.3c}$$

---

[1] Essentially these are that the set of eigenfunctions $\Phi_0$ with $\mathbf{Q}_{NW}\Phi_0 = 0$ should be invariant under rotations and that the $\Phi_\mathbf{y}$ such that $\mathbf{Q}_{NW}\Phi_\mathbf{y} = \mathbf{y}\Phi_\mathbf{y}$ should be orthogonal to $U(\mathbf{a})\Phi_\mathbf{y}$, for any translation $\mathbf{a} \neq 0$.

This does *not* coincide with $\delta(\mathbf{r})$ and thus a particle that, according to (9.2.3a), should be localized at the origin instead has an extended wave function, spread over a region $r \sim \hbar/mc$ (the Hankel function $H_{5/4}^{(1)}(i\xi)$ decreases exponentially for $\xi \gg 1$). For a Dirac particle the situation is similar; the corresponding operator is

$$\mathbf{Q}_{NW} = i\hbar\boldsymbol{\nabla}_p + i\hbar c\frac{\beta\boldsymbol{\alpha}}{2E(\mathbf{p})} - i\hbar\frac{\beta(\boldsymbol{\alpha}\mathbf{p})\mathbf{p} + i|\mathbf{p}|\boldsymbol{\Sigma}\times\mathbf{p}}{2E(\mathbf{p})(E(\mathbf{p})+mc^2)|\mathbf{p}|}. \tag{9.2.4}$$

For the proof of (9.2.4), see problem 9.6. A more detailed discussion of position operators, and further references, may be found in Schweber's (1961) textbook.

Equations (9.2.3) and (9.2.4) become *singular* for massless particles; but the situation is particularly hopeless for photons. Because of gauge, and relativistic invariance, it so happens that $A_\mu(\mathbf{r},t)$ does not make much sense as a probability amplitude to find a photon in $\mathbf{r}$ at time $t$. When one detects a photon, it is localized through its interactions with charged particles: the relevant functions are then $\boldsymbol{\mathcal{E}}(\mathbf{r},t)$ and $\boldsymbol{\mathcal{B}}(\mathbf{r},t)$. However, it is easy to verify that there is no bilinear in $\boldsymbol{\mathcal{E}}$, $\boldsymbol{\mathcal{B}}$ with the desirable *relativistic* transformation properties for a candidate to represent the probability of finding a photon in $\mathbf{r}$ at time $t$.

All these pitfalls strongly suggest that there must exist a physical reason why it is so difficult to localize relativistic particles. This is indeed the case. To localize a particle in a region of extension $\Delta r$ we have to concentrate an energy $\Delta E$ associated with the momentum $\Delta p \sim \hbar/\Delta r$. As soon as $\Delta p$ is of order $\hbar/mc$, where $m$ is the mass of the particle, $\Delta E$ becomes of order $mc^2$; in the case of an electron, for example, this is enough to extract electrons from the Dirac sea, leaving the corresponding holes and thus creating a cloud of electrons–positrons which surround the particle we tried to localize. It follows that, at small distances, a description in terms of a *single*, localized particle makes little sense. For massless particles, such as the photon, the accompanying cloud appears for any precision, $\Delta r$. In view of this, it is not surprising if position operators possess peculiar properties, become singular or just do not exist.

## 9.3 Retardation and Consistency

In the previous subsection we have argued that the localization of a relativistic quantum particle is not a very well defined concept; in this section we will discuss how relativity forces us to abandon, except as an approximation for low velocities, the concept of potential. To do so, consider two particles, distinguished by the indices 1, 2 (Fig. 9.3.1A). If at the instant of time $t$ as measured, for example, by an observer at 0, the particles are located at $\mathbf{r}_1, \mathbf{r}_2$, the potential that the existence of particle 1 creates and acts on particle 2 *cannot* depend only on $\mathbf{r}_1 - \mathbf{r}_2$. In fact, imagine that particle 1 disappears at time $t$. The finite speed with which relativity forces any disturbance to travel means that particle 2 will only notice the disappearance of 1 after a time lapse of $\Delta t = |\mathbf{r}_1 - \mathbf{r}_2|c^{-1}$. Therefore, it follows that particle 1 at time $t$ can only affect particle 2 when it arrives at $\mathbf{r}_2'$ at time $t'$, with $t' - t = |\mathbf{r}_2' - \mathbf{r}_1|c^{-1}$. This effect (*retardation*) may be simulated with the help of so-

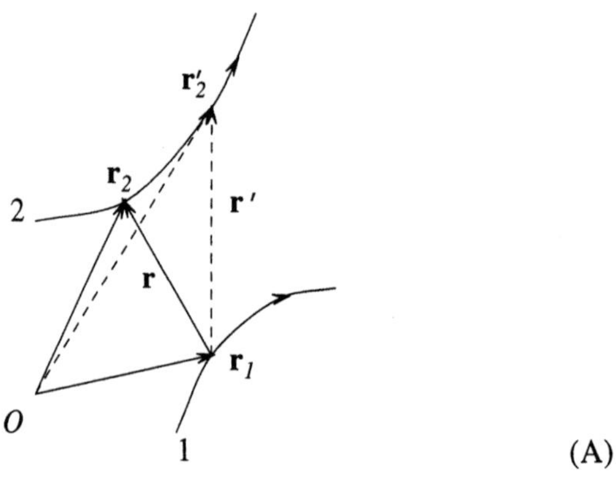


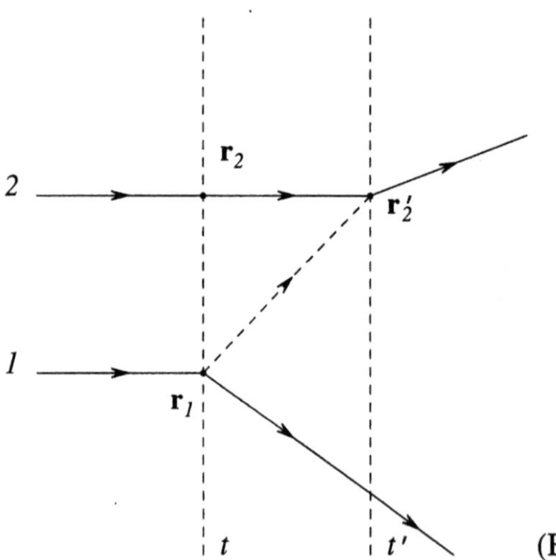


Fig. 9.3.1. Interaction between two relativistic particles. (A) Potential picture. (B) Exchange of a particle (discontinuous line).

called "retarded potentials"; but the solution is not satisfactory. The reason is that the potential (field) must have an existence of its own, for it will still affect particle 2 when the source (particle 1) has disappeared. In view of quantum duality, we expect that the oscillation of the field will have a particle aspect: so we will be led to visualize interactions as the exchange of these particles (Fig. 9.3.1B). As we shall see, retardation will then be automatic.

This exchange implies the possibility of creating and annihilating these field quanta, and from this direction we are thus also led to consider a theory of assemblies of particles.

A last argument for quantum fields is one of consistency. It appears reasonable to assume that, since a kind of particle (photons) are described by quantum field operators, all particles should correspond to quantum fields. This is made more plausible by the fact that the Maxwell equations for the electromagnetic field in the presence of matter, say (1.8.1), link $\mathcal{B}$ and $\mathcal{E}$ with $\rho$ and $\mathbf{j}$; if we quantize the former, then the latter should also become operators, $\hat{\rho}$, $\hat{\mathbf{j}}$. These operators will then be constructed from field operators for charged particles.

The process of going over from a one-particle theory to one of assemblies of particles is called *second quantization*; and the resulting theory goes by the name of *quantum field theory*.

## 9.4 Quantization of Scalar Fields and of Massive Vector Fields

### 9.4.1 Second Quantization for Spinless Particles

By analogy with the electromagnetic case we postulate the Klein–Gordon equation. If we denote by $\hat{\varphi}(\mathbf{r}, t)$ the field operator for a spinless particle, then we have

$$(\partial \cdot \partial + m^2)\hat{\varphi}(x) = 0; \tag{9.4.1}$$

here and henceforth we use natural units. Next, we Fourier-expand, writing

$$\hat{\varphi}(x) = \frac{1}{(2\pi)^{3/2}} \int d^4p \, e^{-ip \cdot x} \hat{\varphi}(p); \tag{9.4.2a}$$

the factor $(2\pi)^{-3/2}$ is the customary one. Equation (9.4.1) implies that

$$(-p \cdot p + m^2)\hat{\varphi}(p) = 0,$$

and hence

$$\hat{\varphi}(p) = \delta(p \cdot p - m^2)\hat{\alpha}(p),$$

where, in $\hat{\alpha}(p)$, $p$ is taken to satisfy $p \cdot p = m^2$. Substituting this into (9.4.1),

$$\hat{\varphi}(x) = \frac{1}{(2\pi)^{3/2}} \int d^3p \int_{-\infty}^{+\infty} dp_0 \, \delta(p_0^2 - \mathbf{p}^2 - m^2)e^{-ip \cdot x}\hat{\alpha}(p).$$

Let us split the $p_0$ integration into two pieces, from 0 to $\infty$ and from $-\infty$ to 0; and, in the second, change the variable of integration $p \to p' = -p$. We get

$$\hat{\varphi}(x) = \frac{1}{(2\pi)^{3/2}} \left\{ \int d^3p \int_0^\infty dp_0 \; \delta(p_0^2 - \mathbf{p}^2 - m^2) e^{-ip\cdot x} \hat{\alpha}(p) \right.$$

$$\left. + \int d^3p' \int_0^\infty dp_0' \; \delta(p_0'^2 - \mathbf{p}'^2 - m^2) e^{-ip'\cdot x} \hat{\alpha}(-p') \right\}.$$

The change of variables $p_0 \to p_0^2 - \mathbf{p}^2 - m^2$ is now uniform in both integrals, and the $dp_0$ integral is thus performed trivially, using the $\delta$. The result is

$$\hat{\varphi}(\mathbf{r}, t) = \frac{1}{(2\pi)^{3/2}} \int \frac{d^3p}{2E(\mathbf{p})} \left\{ e^{-iE(\mathbf{p})x_0 + i\mathbf{p}\mathbf{r}} \hat{\alpha}(E(\mathbf{p}), \mathbf{p}) \right.$$

$$\left. + e^{iE(\mathbf{p})x_0 - i\mathbf{p}\cdot\mathbf{r}} \hat{\alpha}(-E(\mathbf{p}), \mathbf{p}) \right\}, \tag{9.4.2b}$$

$$E(\mathbf{p}) \equiv +\sqrt{m^2 + \mathbf{p}^2};$$

in the second integral we have changed the name of the variable of integration, from $d^3p'$ to $d^3p$. Equation (9.4.2b) may be rewritten in a manifestly covariant manner by using the notation

$$p_0 \equiv \sqrt{m^2 + \mathbf{p}^2},$$

$$\hat{\alpha}(E(\mathbf{p}), \mathbf{p}) \equiv \hat{a}(p),$$

$$\hat{\alpha}(-E(\mathbf{p}), -\mathbf{p}) \equiv \hat{a}_c^+(p),$$

so that we have

$$\hat{\varphi}(x) = \frac{1}{(2\pi)^{3/2}} \int \frac{d^3p}{2p_0} \left\{ e^{-ip\cdot x} \hat{a}(p) + e^{ip\cdot x} \hat{a}_c^+(p) \right\}. \tag{9.4.3}$$

Note that, in general, $\hat{a}$ and $\hat{a}_c$ are *different* operators; because $E(p) > 0$, $\hat{\alpha}(E, \mathbf{p})$ and $\hat{\alpha}(-E, \mathbf{p})$ are different entities.

Equation (9.4.3) looks very much like the corresponding expression for the electromagnetic case, (8.2.6). Following the analogy, we then make the connection with the particle formalism by interpreting the operators $\hat{a}, \hat{a}_c^+$ as destruction–creation operators for particles. Before continuing, however, we have to distinguish two possibilities. (i) Particles[2] without charges, in particular without electric charge. These particles will be taken to be their own antiparticle. (One such particle is the neutral pion, $\pi^0$). (ii) Particles with some charge, for example the electric charge, different from zero. This is the case of the $\pi^+$, with charge equal to the proton's.

We start with the second case. If we now exchange $\hat{\varphi}(x) \to \hat{\varphi}(x)^+$, we obtain an equation for the latter like (9.4.3), but with $\hat{a}$ replaced by $\hat{a}_c$. On comparing this with (3.7.11), we *postulate* that the operators $\hat{a}, \hat{a}^+$ refer to

---

[2] Elementary spinless particles have not been found in nature. There exist, however, composite particles whose substructure is apparent only at very high energies (like the pions or kaons) and which can be considered elementary, to a good approximation, at small and medium energies.

*particles* and the $\hat{a}_c$, $\hat{a}_c^+$ to *antiparticles*. We require the commutation relations (cf. (6.5.1))

$$[\hat{a}(p), \hat{a}(p')] = [\hat{a}_c(p), \hat{a}_c(p')] = 0,$$

$$[\hat{a}(p), \hat{a}_c^+(p')] = 0,$$

$$[\hat{a}(p), \hat{a}^+(p')] = 2p_0\delta(\mathbf{p} - \mathbf{p}'),$$

$$[\hat{a}_c(p), \hat{a}_c^+(p')] = 2p_0\delta(\mathbf{p} - \mathbf{p}').$$

$$(9.4.4)$$

The vacuum is characterized by

$$\hat{a}|0\rangle = \hat{a}_c|0\rangle = 0; \quad \langle 0|0\rangle = 1,$$

and a state with $N$ particles and $N'$ antiparticles with well-defined momenta is

$$\hat{a}^+(p_1)\dots\hat{a}^+(p_N)\hat{a}_c^+(p'_1)\dots\hat{a}_c^+(p'_{N'})|0\rangle. \tag{9.4.5}$$

Here we interpret $\mathbf{p}$ as the three-momentum, and $p_0$ as the energy of each particle.

The Hamiltonian, and momentum operators corresponding to the system, can be written as

$$\hat{H}_0 = \int \frac{d^3p}{2p_0} \left\{ p_0\hat{a}^+(p)\hat{a}(p) + p_0\hat{a}_c^+(p)\hat{a}_c(p) \right\}, \tag{9.4.6a}$$

$$\hat{\mathbf{P}} = \int \frac{d^3p}{2p_0} \left\{ \mathbf{p}\,\hat{a}^+(p)\hat{a}(p) + \mathbf{p}\,\hat{a}_c^+(p)\hat{a}_c(p) \right\}. \tag{9.4.6b}$$

They can also be expressed directly in terms of the field operators:

$$\hat{H}_0 = \int d^3r : (\partial_0\hat{\varphi}^+(x))\partial_0\hat{\varphi}(x)$$
$$+ (\boldsymbol{\nabla}\hat{\varphi}^+(x))(\boldsymbol{\nabla}\hat{\varphi}(x)) + m^2\hat{\varphi}^+(x)\hat{\varphi}(x) :, \tag{9.4.7a}$$

$$\hat{\mathbf{P}} = -\frac{1}{2}\int d^3r : (\partial_0\hat{\varphi}^+(x))\boldsymbol{\nabla}\hat{\varphi}(x) + (\boldsymbol{\nabla}\hat{\varphi}^+(x))\partial_0\hat{\varphi}(x) : . \tag{9.4.7b}$$

**Exercise.** (A) Check that expressions (9.4.6) and (9.4.7) do indeed coincide. (B) Rewrite (9.4.7) in a manifestly covariant manner for the four-vector $\hat{P}_\mu$, $\hat{P}_0 = \hat{H}$. (C) Use (9.4.6) to verify that

$$\hat{P}_\mu|N, N'\rangle = \left(\sum p_\mu + \sum p'_\mu\right)|N, N'\rangle,$$

where $|N, N'\rangle$ is the state in (9.4.5) ∎

If the particles associated with the field $\hat{\varphi}$ have charge, we have to define, using the $\hat{\varphi}$, the operators *charge density*, $\hat{\rho}$, and current, $\hat{\mathbf{j}}$, or, in covariant notation, the four-vector current operator, $\hat{J}_\mu(x)$.

We can guess the form of $\hat{J}_\mu$ using the following argument: for a state $|p\rangle = \hat{a}^+(p)|0\rangle$, the quantity $\langle 0|\hat{\varphi}(x)\hat{a}^+(p)|0\rangle$ coincides with the wave function

$$\frac{1}{(2\pi)^{3/2}} e^{-ip\cdot x}$$

of a particle with momentum $p$, at the spacetime point $x$. We may then identify $\hat{\varphi}^+(x)$ as an operator that creates particles at $x$. We then try an expression for the current, in terms of $\hat{\varphi}(x)$, similar to that found in Sect. 2.3 (2.3.7) in terms of $\Phi$. Thus we write[3]

$$\hat{J}_\mu(x) = ie : \hat{\varphi}^+(x) \overleftrightarrow{\partial}_\mu \hat{\varphi}(x) :, \tag{9.4.8}$$

and the notation $\overleftrightarrow{\partial}$ is defined, in general, as

$$f \overleftrightarrow{\partial} g = f(\partial g) - (\partial f)g.$$

We will now check that the expectation values of (9.4.8) coincide with what is desirable for a current operator. Considering for simplicity the case of a one-particle state, we have

$$\langle p_1 | \hat{J}_\mu(x) | p_2 \rangle = \langle 0 | \hat{a}(p_1) \hat{J}_\mu \hat{a}^+(p_2) | 0 \rangle$$

$$= e(p_{1\mu} + p_{2\mu}) \frac{1}{(2\pi)^3} \exp i(p_1 - p_2) \cdot x,$$

after a simple calculation. Integrating in $d^3r$,

$$\left\langle p_1 \left| \int d^3 r \hat{J}_\mu(x) \right| p_2 \right\rangle = 2e p_{1\mu} \delta(\mathbf{p}_1 - \mathbf{p}_2)$$

$$= e \frac{p_{1\mu}}{p_{10}} \langle p_1 | p_2 \rangle.$$

Indeed the zeroth component coincides with the charge, and the space components with charge $\times$ velocity (relativistic velocity, in units of $c$, being $\mathbf{p}/p_0$).

If we had considered antiparticle states,

$$|p\rangle_c = \hat{a}_c^+(p)|0\rangle,$$

we would have obtained

$$_c\left\langle p_1 \left| \int d^3 r \hat{J}_\mu(x) \right| p_2 \right\rangle_c = -e \frac{p_{1\mu}}{p_{10}} {}_c\langle p_1 | p_2 \rangle_c;$$

that is, antiparticles appear with a charge opposite to that of particles.

**Exercise.** (A) Verify that a state $|N, N'\rangle$ given by (9.4.5) is an eigenstate of the charge operator,

$$\hat{Q} \equiv \int d^3 r \hat{J}_0(\mathbf{r}, t) = ie \int d^3 r : \hat{\varphi}^+(x) \overleftrightarrow{\partial}_0 \hat{\varphi}(x) :, \tag{9.4.9}$$

---

[3] Expression (9.4.8) is actually valid only in the absence of *photons*. If these are present, a term $-2e^2 : \hat{\varphi}^+(x)\hat{\varphi}(x) : \hat{A}_\mu(x)$ should be added to the right-hand side of (9.4.8).; cf. Bjorken and Drell (1965).

with eigenvalue $Ne - N'e$. (B) Check the commutations relations

$$[\hat{Q}, \hat{a}^+] = e\hat{a}^+, \ [\hat{Q}, \hat{a}_c^+] = -e\hat{a}_c^+,$$

$$[\hat{Q}, \hat{\varphi}(x)] = -e\hat{\varphi}(x). \tag{9.4.10}$$

(C) Verify that, in terms of creation–annihilation operators,

$$\hat{Q} = e \int \frac{d^3p}{2p_0} \left( \hat{a}^+(p)\hat{a}(p) - \hat{a}_c^+(p)\hat{a}_c(p) \right) \ \blacksquare$$

It is noteworthy that, in spite of the existence of particles and antiparticles, the Hamiltonian (as given in (9.4.6a)) is manifestly positive definite.

The transformation properties of $\hat{\varphi}(x)$ follow from the expression of $\hat{\varphi}$ in terms of creation–destruction operators, and of the transformation properties of these, (6.5.5). We have, denoting by $b$ a four-translation, to avoid confusion with the $\hat{a}$,

$$\hat{U}(b, \Lambda)\hat{\varphi}(x)\hat{U}(b, \Lambda)^{-1}$$

$$= \frac{1}{(2\pi)^{3/2}} \int \frac{d^3p}{2p_0} \left\{ e^{-ip\cdot x} e^{-ib\cdot \Lambda p} \hat{a}(\Lambda p) \right.$$

$$\left. + e^{ip\cdot x} e^{ib\cdot \Lambda p} \hat{a}_c^+(\Lambda p) \right\}.$$

Changing the variable, $p' = \Lambda p$, and taking into account that $d^3p/2p_0$ is invariant, we find that the right-hand side here becomes

$$\frac{1}{(2\pi)^{3/2}} \int \frac{d^3p'}{2p_0'} \left\{ e^{-ip'\cdot(\Lambda x + b)} \hat{a}(p') + e^{ip\cdot(\Lambda x + b)} \hat{a}_c^+(p') \right\},$$

and therefore

$$\hat{U}(b, \Lambda)\hat{\varphi}(x)\hat{U}(b, \Lambda)^{-1} = \hat{\varphi}(\Lambda x + b). \tag{9.4.11}$$

For charge conjugation we define

$$\hat{C}\hat{a}(p)\hat{C}^{-1} = \eta_C \hat{a}_c(p), \ \hat{C}\hat{a}_c(p)\hat{C}^{-1} = \eta_C \hat{a}(p), \eta_C^2 = 1,$$

and then the transformation properties of $\hat{\varphi}$ under the discrete transformations parity ($\hat{P}$), time reversal ($\hat{T}$) and $\hat{C}$ are

$$\hat{P}\hat{\varphi}(x)\hat{P}^{-1} = \eta_P \hat{\varphi}(I_s x), \ \hat{T}\hat{\varphi}(x)T^{-1} = \eta_T \hat{\varphi}(I_t x),$$

$$\hat{C}\hat{\varphi}(x)\hat{C}^{-1} = \eta_C \hat{\varphi}^+(x). \tag{9.4.12}$$

$\hat{P}, \hat{C}$ are *unitary*, $\hat{T}$ is *antiunitary*.

*Neutral scalar particles*, that is to say, spinless particles identical to their antiparticles, are described by this same formalism, but we now identify $\hat{a}(p) \equiv \hat{a}_c(p)$. The field is thus written as

$$\hat{\varphi}_0(x) = \frac{1}{(2\pi)^{3/2}} \int \frac{d^3p}{2p_0} \left\{ e^{-ip\cdot x} \hat{a}(p) + e^{ip\cdot x} \hat{a}^+(p) \right\}. \tag{9.4.13}$$

### 9.4.2 Massive Vector Particles

Particles with spin 1 are called, for obvious reasons, *vector particles*. The case $m = 0$ is the photon case, which we have studied (and will go on studying) in detail. Massive spin 1 particles exist that are neutral, and are thus their own antiparticles (like the $Z$ particle); others like the $W^\pm$ carry electric charge. We will describe them briefly, concentrating on the second case. For neutral particles, put $e = 0$ and identify $\hat{a} = \hat{a}_c$ below.

We write the field operator,

$$\hat{V}_\mu(x) = \frac{1}{(2\pi)^{3/2}} \sum_\lambda \int \frac{d^3 p}{2p_0} \left\{ e^{-ip\cdot x} \epsilon_\mu(p, \lambda) \hat{a}(p, \lambda) \right.$$
$$\left. + e^{ip\cdot x} \epsilon_\mu^*(p, \lambda) \hat{a}_c^+(p, \lambda) \right\}, \; p_0 = \sqrt{m^2 + \mathbf{p}^2}. \tag{9.4.14}$$

Here the index $\lambda$ can be the helicity, $\lambda = 0, \pm 1,$ or another of the indices introduced in Sect. 5.1; and the $\epsilon$ are the polarization vectors introduced there. $\hat{V}$ satisfies the equations

$$(\partial \cdot \partial + m^2)\hat{V}_\mu(x) = 0, \; \partial \cdot V(x) = 0. \tag{9.4.15}$$

The nonzero commutators of the creation–annihilation operators are

$$[\hat{a}(p, \lambda), \hat{a}^+(p', \lambda')] = 2p_0 \delta(\mathbf{p} - \mathbf{p}') \delta_{\lambda\lambda'},$$
$$[\hat{a}_c(p, \lambda), \hat{a}_c^+(p', \lambda')] = 2p_0 \delta(\mathbf{p} - \mathbf{p}') \delta_{\lambda\lambda'}. \tag{9.4.16}$$

The transformation properties are given by

$$\hat{U}(a, \Lambda)\hat{V}_\mu(x)\hat{U}(a, \Lambda)^{-1} = \sum \Lambda_{\mu\nu} \hat{V}_\nu(\Lambda x + a), \tag{9.4.17}$$

$$\mathcal{P}\hat{V}_\mu(x)\mathcal{P}^{-1} = \eta_P \sum_\nu I_{s\mu\nu} \hat{V}_\nu(I_s x),$$
$$\mathcal{T}\hat{V}_\mu(x)\mathcal{T}^{-1} = \eta_T \sum_\nu I_{t\mu\nu} \hat{V}_\nu(I_t x), \tag{9.4.18}$$
$$\mathcal{C}\hat{V}_\mu(x)\mathcal{C}^{-1} = \eta_C \hat{V}_\mu^+(x).$$

The energy–momentum operator can be written as

$$\hat{P}_\mu = \sum_\lambda \int \frac{d^3 p}{2p_0} p_\mu \left\{ \hat{a}^+(p, \lambda)\hat{a}(p, \lambda) + \hat{a}_c^+(p, \lambda)\hat{a}_c(p, \lambda) \right\}. \tag{9.4.19}$$

## 9.5 Quantization of the Dirac Field. Weyl and Majorana Particles

For spin 1/2 particles we postulate the operators $\hat{\psi}(x)$, satisfying the Dirac equation:

$$(i\not{\partial} - m)\hat{\psi}(x) = 0. \tag{9.5.1}$$

Fourier expanding $\hat{\psi}(x)$ as we did for $\hat{\varphi}(x)$ we obtain an expression like (9.4.2b):

$$\hat{\psi}(x) = \frac{1}{(2\pi)^{3/2}} \int \frac{d^3p}{2p_0} \left\{ e^{-ip\cdot x}\hat{\beta}(p) + e^{ip\cdot x}\hat{\beta}(-p) \right\},$$

$$p_0 \equiv +\sqrt{m^2 + \mathbf{p}^2}.$$

The Dirac equation (9.5.1) implies that the $\hat{\beta}$ satisfy

$$(\not{p} - m)\hat{\beta}(p) = 0,$$

$$(\not{p} + m)\hat{\beta}_c(p) = 0,$$

where we have defined

$$\hat{\beta}_c(p) \equiv \hat{\beta}(-p).$$

We can then expand $\hat{\beta}$, $\hat{\beta}_c$ in terms of, respectively, $u(p, \lambda)$, $v(p, \lambda)$, where $\lambda$ is a spin quantum number. Denoting the coefficients by $\hat{b}(p, \lambda)$, $\hat{d}^+(p, \lambda)$ respectively, we obtain the expression

$$\hat{\psi}(x) = \frac{1}{(2\pi)^{3/2}} \sum_\lambda \int \frac{d^3p}{2p_0} \left\{ e^{-ip\cdot x}u(p, \lambda)\hat{b}(p, \lambda) \right.$$
$$\left. + e^{ip\cdot x}v(p, \lambda)\hat{d}^+(p, \lambda) \right\}. \tag{9.5.2}$$

We interpret the operator $\hat{b}(\hat{b}^+)$ as annihilating (creating) particles, and $\hat{d}(\hat{d}^+)$ as annihilating (creating) antiparticles. $\hat{d}^+$ would have been written as $\hat{b}_c$ if we had used the notation of the previous section; instead, we follow the standard one. In principle, it could happen that a spin $1/2$ is its own antiparticle (Majorana particle); this possibility does not appear to be realized in nature. It would be implemented in the formalism by simply identifying $\hat{b} = \hat{d}$.

Because we want spin $1/2$ particles to be *fermions*[4], i.e., we want the corresponding states to be antisymmetric, we will postulate *anticommutation relations*,

$$\{\hat{b}, \hat{b}'\} = \{\hat{d}, \hat{d}'\} = \{\hat{b}, \hat{d}\} = \{\hat{b}, \hat{d}'\} = \{\hat{b}, \hat{d}'^+\} = 0,$$

$$\{\hat{b}(p, \lambda), \hat{b}^+(p', \lambda')\} = 2\delta_{\lambda\lambda'}p_0\delta(\mathbf{p} - \mathbf{p}'), \tag{9.5.3}$$

$$\{\hat{d}(p, \lambda), \hat{d}^+(p', \lambda')\} = 2\delta_{\lambda\lambda'}p_0\delta(\mathbf{p} - \mathbf{p}');$$

---

[4] In fact, it can be proved with generality that in any relativistic quantum field theory bosons are necessarily integer spin particles, and fermionic particles must have half-integer spin. An elementary proof may be found in the book of Bjorken and Drell (1965); a rigorous one in Bogoliubov, Logunov and Todorov (1975).

of course, we assume that $\hat{b}$, $\hat{d}$ commute (anticommute, for fermions) with operators referring to other kinds of particle. Quantization with anticommutators guarantees the antisymmetry of the corresponding states, linear combinations of the basic ones

$$\hat{b}^+(p_1,\lambda_1)\ldots\hat{b}^+(p_N,\lambda_N)\hat{d}^+(p_1',\lambda_1')\ldots\hat{d}^+(p_{N'}',\lambda_{N'}')|0\rangle.$$

As for bosons, we assume that

$$b(p,\lambda)|0\rangle = d(p,\lambda)|0\rangle = 0,$$

$$\langle 0|0\rangle = 1.$$

The transformation properties of the field follow from those of the spinors $u$, $v$, and those of the operators $\hat{b}$, $\hat{d}^+$. For Lorentz transformations, for instance,

$$\hat{\psi}(x) \to \hat{U}(\Lambda)\hat{\psi}(x)\hat{U}(\Lambda)^{-1}$$

$$= \frac{1}{(2\pi)^{3/2}} \sum_\lambda \int \frac{d^3p}{2p_0} \left\{ e^{-ip\cdot x}u(p,\lambda)\hat{U}(\Lambda)\hat{b}(p,\lambda)\hat{U}(\Lambda)^{-1} \right.$$

$$\left. + e^{ip\cdot x}v(p,\lambda)\hat{U}(\Lambda)\hat{d}^+(p,\lambda)\hat{U}(\Lambda)^{-1} \right\}.$$

Using now (6.5.5), we obtain

$$\hat{U}(\Lambda)\hat{b}(p,\lambda)\hat{U}(\Lambda)^{-1} = \sum_{\lambda'} D^{(1/2)}_{\lambda\lambda'}(R(p,\Lambda))^*\hat{b}(\Lambda p,\lambda')$$

$$= \sum_{\lambda'} D_{\lambda'\lambda}(R^{-1}(p,\Lambda))\hat{b}(\Lambda p,\lambda'),$$

where we have taken into account that the $D^{(1/2)}_{\lambda\lambda'}$ are matrix elements of a unitary matrix, for which inverse and adjoint coincide. Changing variables to $p' = \Lambda p$, we get

$$\hat{U}(\Lambda)\hat{\psi}(x)\hat{U}(\Lambda)^{-1} = \frac{1}{(2\pi)^{3/2}}$$

$$\times \sum_{\lambda\lambda'} \int \frac{d^3p'}{2p_0'} \left\{ e^{-ip\cdot\Lambda x'}\hat{b}(p',\lambda')u(\Lambda^{-1}p',\lambda)D^{(1/2)}_{\lambda\lambda'}(L(\Lambda^{-1}p')^{-1}\Lambda^{-1}L(p')) \right.$$

$$\left. + \ldots \right\},$$

and we have not explicitly written the term in $v\hat{d}^+$.

We can now use the expression (6.6.5) for $u$ to obtain[5]

---

[5] The details of the calculation below may be found in Problem 9.3.

$$\sum_\lambda u_a(\Lambda^{-1}p', \lambda) D^{(1/2)}_{\lambda\lambda'}(L(\Lambda^{-1}p')^{-1}\Lambda^{-1}L(p'))$$

$$= \sum_\lambda D_{a\lambda}(L(\Lambda^{-1}p')) D^{(1/2)}_{\lambda\lambda'}(L(\Lambda^{-1}p')^{-1}\Lambda^{-1}L(p'))$$

$$= \sum_{a'} D_{aa'}(\Lambda^{-1}) u_{a'}(p', \lambda'),$$

and from $(Du)_a = \sum_b D_{ab}u_b$, we find that

$$\hat{U}(\Lambda)\hat{\psi}(x)\hat{U}(\Lambda)^{-1}$$

$$= \frac{1}{(2\pi)^{3/2}} \sum_{\lambda'} \int \frac{d^3p'}{2p'_0} \left\{ e^{-ip'\cdot\Lambda x} D(\Lambda^{-1})u(p', \lambda')\hat{b}(p', \lambda') + \ldots \right\}.$$

The term in $v\hat{d}^+$ can be likewise treated. Finally, we have, for an arbitrary Poincaré transformation,

$$\hat{U}(a, \Lambda)\hat{\psi}(x)\hat{U}(a, \Lambda)^{-1} = D(\Lambda^{-1})\hat{\psi}(\Lambda x + a). \tag{9.5.4}$$

The action of the transformations $\hat{P}$, $\hat{T}$ can be computed in a similar manner. We find that

$$\hat{P}\hat{\psi}(x)\hat{P}^{-1} = \eta_P \gamma_0 \hat{\psi}(I_s x),$$

$$\hat{T}\hat{\psi}(x)\hat{T}^{-1} = \eta_T C_T \hat{\psi}(I_s x). \tag{9.5.5}$$

The charge conjugation, $\hat{C}$, acts according to

$$\hat{C}\hat{\psi}(x)\hat{C}^{-1} = \eta_C C \hat{\psi}^+(x). \tag{9.5.6}$$

In the Pauli or Weyl representations,

$$C_T = i\gamma_2\gamma_3, \quad C = i\gamma_2.$$

$\hat{P}$, $\hat{C}$ are unitary, $\hat{T}$ is antiunitary.

**Exercise.** Check that (9.5.6) is equivalent to

$$\hat{C}\hat{b}(p, \lambda)\hat{C}^{-1} = \eta_C \hat{d}(p, \lambda), \quad \hat{C}\hat{d}\hat{C}^{-1} = \eta_C \hat{b} \;\blacksquare$$

A few extra words on notation are necessary to clarify the meaning of (9.5.6). We take $\hat{\psi}(x)$ to be a vertical matrix in Dirac space,

$$\hat{\psi}(x) = \begin{pmatrix} \hat{\psi}_1(x) \\ \hat{\psi}_2(x) \\ \hat{\psi}_3(x) \\ \hat{\psi}_4(x) \end{pmatrix}.$$

We *define* $\hat{\psi}^+(x)$ also to be a column matrix:

$$\hat{\psi}^+(x) = \begin{pmatrix} \hat{\psi}_1^+(x) \\ \hat{\psi}_2^+(x) \\ \hat{\psi}_3^+(x) \\ \hat{\psi}_4^+(x) \end{pmatrix}.$$

$\hat{\bar{\psi}}$, on the other hand, will be considered to be a row matrix; in Pauli's realization, where

$$\gamma_0 = \begin{pmatrix} 1 & 0 \\ 0 & -1 \end{pmatrix},$$

we have

$$\hat{\bar{\psi}} = (\hat{\psi}^+)^T \gamma_0 = (\hat{\psi}_1^+, \hat{\psi}_2^+, -\hat{\psi}_3^+, -\hat{\psi}_4^+).$$

Thus, $\hat{\bar{\psi}}\hat{\psi}$ is a scalar in Dirac space. In the Pauli realization,

$$\hat{\bar{\psi}}\hat{\psi} = \hat{\psi}_1^+\hat{\psi}_1 + \hat{\psi}_2^+\hat{\psi}_2 - \hat{\psi}_3^+\hat{\psi}_3 - \hat{\psi}_4^+\hat{\psi}_4.$$

The *current four-vector* operator is defined, for Dirac particles with charge $e$, by

$$\hat{J}_\mu(x) = e : \hat{\bar{\psi}}(x)\gamma_\mu\hat{\psi}(x) :; \tag{9.5.7}$$

the *four-momentum* is

$$\hat{P}_\mu = \sum_\lambda \int \frac{d^3p}{2p_0} p_\mu \left\{ \hat{b}^+(p,\lambda)\hat{b}(p,\lambda) + \hat{d}^+(p,\lambda)\hat{d}(p,\lambda) \right\}. \tag{9.5.8}$$

An alternative expression for the Hamiltonian, $\hat{P}_0 = \hat{H}_{0D}$ is

$$\begin{aligned} \hat{H}_{0D} &= \int d^3r : \hat{\bar{\psi}}(x)\gamma_0(-i\boldsymbol{\alpha}\boldsymbol{\nabla} + \beta m)\hat{\psi}(x) : \\ &= \int d^3r : \hat{\psi}^+(x)^T(-i\boldsymbol{\gamma}\boldsymbol{\nabla} + m)\hat{\psi}(x) : . \end{aligned} \tag{9.5.9}$$

Finally, we have, for the charge operator,

$$\begin{aligned} \hat{Q} &= \int d^3r\,\hat{J}_0(\mathbf{r},t) \\ &= \sum_\lambda \int \frac{d^3p}{2p_0} \left\{ e\hat{b}^+(p,\lambda)\hat{b}(p,\lambda) - e\hat{d}^+(p,\lambda)\hat{d}(p,\lambda) \right\}. \end{aligned} \tag{9.5.10}$$

This last expression shows that antiparticles have a charge opposite to that of particles. Equation (9.5.8) also implies that the energy is positive definite. Thus the field formalism (for the moment only, it is true, for free particles) manages to preserve the particle–antiparticle relations we guessed with the wave function formalism *and* to guarantee positive-definite Hamiltonians: we have got rid of the cumbersome sea, retaining the interesting feature of the existence of antiparticles.

We now finish this section by briefly considering the formalism for neutrinos, which are taken to be massless. We describe them by a field

$$\hat{\psi}_L(x) = \frac{1}{(2\pi)^{3/2}} \int \frac{d^3p}{2p_0} \left\{ e^{-ip\cdot x} u(p, \eta = -1)\hat{b}(p, \eta = -1) \right.$$
$$\left. + e^{ip\cdot x} v(p, \eta = +1)\hat{d}^+(p, \eta = +1) \right\}, \tag{9.5.11}$$

$$p_0 = |\mathbf{p}|,$$

in accordance with the experimental fact that neutrinos are left-handed, and antineutrinos right-handed (recall Sec. 3.9). Right-handed neutrinos apparently do not exist.

One can rewrite (9.5.11) as

$$\hat{\psi}_L(x) = \frac{1 - \gamma_5}{2} \frac{1}{(2\pi)^{3/2}} \sum_\eta \int \frac{d^3p}{2p_0} \left\{ e^{-ip\cdot x} u(p, \eta)\hat{b}(p, \eta) \right.$$
$$\left. + e^{ip\cdot x} v(p, \eta)\hat{d}^+(p, \eta) \right\} \tag{9.5.12}$$

$$\equiv \frac{1 - \gamma_5}{2} \hat{\psi}(x);$$

we have introduced nonexistent $\hat{b}(p, \eta = +1)$, $\hat{d}^+(p, \eta = 1)$ which causes no harm since the projector $(1 - \gamma_5)/2$ chops off their contributions (because e.g., $(1 - \gamma_5)u(p, +1) = 0$). If right-handed neutrinos were ever found, we would use (9.5.12) with $1 - \gamma_5$ replaced by $1 + \gamma_5$:

$$\hat{\psi}_R(x) = \frac{1 + \gamma_5}{2} \hat{\psi}(x). \tag{9.5.13}$$

The neutrinos we have described are called *Weyl* neutrinos. *Majorana* neutrinos would be particles for which $\hat{b} = \hat{d}$.

Let us return to Weyl neutrinos. $\hat{\mathcal{P}}$ is not defined for them; because of (9.5.5), and since $\gamma_0\gamma_5 = -\gamma_5\gamma_0$, $\hat{\mathcal{P}}$ would send left-handed neutrinos into nonexistent right-handed neutrinos. We can define $C\mathcal{P}$, however: in the Pauli or Weyl realizations

$$\hat{C}\hat{\mathcal{P}}\hat{\psi}_X(x)(C\mathcal{P})^{-1} = \eta_{CP} i\gamma_2\gamma_0\hat{\psi}_X^+(I_s x), \quad X = L \text{ or } R. \tag{9.5.14}$$

For $\hat{T}$,

$$\hat{T}\hat{\psi}_X(x)\hat{T}^{-1} = \eta_T i\gamma_2\gamma_3\hat{\psi}_X(I_t x). \tag{9.5.15}$$

**Exercise.** Verify that (9.4.14), (9.4.15) are consistent ∎

## 9.6 Covariant Quantization of the Electromagnetic Field

It can be shown (Strocchi, 1967) that one cannot at the same time have manifestly covariant quantization of the electromagnetic field and a positive-definite metric in the Hilbert space of states. In previous sections we have worked in a physical Hilbert space, but the quantization was not manifestly covariant. That formalism is the most convenient one for nonrelativistic or semirelativistic calculations. For fully relativistic situations (and even for some nonrelativistic ones) it is far more convenient to use a formalism in which the four-potential $A_\mu$ transforms in a manifestly covariant manner. For this, we will have to work in a space of states with indefinite metric (although, of course, physical states will correspond to the positive metric part of the space).

Let us start by constructing "Cartesian" polarization vectors associated with a photon. In the reference system in which the four-momentum of the photon is $\overline{k}$, $\overline{k}_0 = \overline{k}_3 = 1$, $\overline{k}_1 = \overline{k}_2 = 0$, we define

$$e_\mu(\overline{k}, j) = \delta_{\mu j}, \quad j = 1, 2, 3;$$

$$e_\mu(\overline{k}, 0) = \delta_{\mu 0}. \tag{9.6.1}$$

For $j = 0, 1$, these vectors are physical, in the sense that a photon characterized by them corresponds to a physical photon, with linear polarization given precisely by the three-vector $\mathbf{e}(\overline{k}, j)$, $j = 1, 2$. The other two $e_\mu(\overline{k}, 3, 0)$ are not physical; $\mathbf{e}(\overline{k}, 3)$ is parallel to $\mathbf{k}$ ("*longitudinal*") and $e(\overline{k}, 0)$ is invariant under ordinary rotations ("*scalar*").

The physical $e_\mu$ can be related to the helicity vectors $\epsilon$ defined in Sect. 5.2. Completing the last by a zero time component, i.e., $\epsilon_0(\overline{k}, \eta) = 0$, we have

$$\epsilon_\mu(\overline{k}, \eta) \equiv \epsilon_\mu^{(0)}(\eta) = \frac{1}{\sqrt{2}}(e_\mu(\overline{k}_1, 1) + \eta i e_\mu(\overline{k}, 2)). \tag{9.6.2}$$

Let us denote the $e_\mu$ collectively by $e_\mu(\overline{k}, \lambda)$, $\lambda = 0, 1, 2, 3$. We will extend their definition to an arbitrary reference system characterized by the four-momentum $k$, writing

$$e_\mu(k, \lambda) \equiv \sum_\nu H_{\mu\nu}(k) e_\nu(\overline{k}, \lambda) = H_{\mu\lambda}(k). \tag{9.6.3a}$$

Here $H$ is the Lorentz transformation

$$H(k) = R(\mathbf{z} \to \mathbf{k}) L(\overline{k} \to k^z); \quad H(k)\overline{k} = k. \tag{9.6.3b}$$

$L(\overline{k} \to k^z)$ is the pure boost that carries $\overline{k}$ over $k^z$, where $k^z = k_0\overline{k}$; and $R(\mathbf{z} \to \mathbf{k})$ is the rotation around the axis $\mathbf{z} \times \mathbf{k}$ that brings $OZ$ on $\mathbf{k}$. We have the following explicit expressions:

$$e_0(k, j = 1, 2) = 0; \quad \mathbf{e}(k, 1) = \frac{1}{\sqrt{2}}\{\epsilon(k, +1) + \epsilon(k, -1)\},$$

$$\tag{9.6.4a}$$

$$\mathbf{e}(k, 2) = \frac{1}{i\sqrt{2}}\{\epsilon(k, +1) - \epsilon(k, -1)\},$$

the $\epsilon(k,\eta)$ being given in (5.3.12b), and

$$e_0(k,3) = \frac{k_0^2 - 1}{2k_0}, \quad \mathbf{e}(k,3) = \frac{k_0^2 + 1}{2k_0^2}\mathbf{k},$$

$$e_0(k,0) = \frac{k_0^2 + 1}{2k_0}, \quad \mathbf{e}(k,0) = \frac{k_0^2 - 1}{2k_0^2}\mathbf{k}$$

$$(9.6.4b)$$

(the proof of (9.6.4b) may be found in Problem 5.3).

Let us now rewrite the electromagnetic field operator (four-potential) as[6]

$$\hat{A}_\mu^C(x) = \frac{1}{(2\pi)^{3/2}} \int \frac{d^3k}{2k_0} \sum_{j=1,2} \left\{ e^{-ik\cdot x} e_\mu(k,j)\hat{a}(k,j) \right.$$
$$\left. + e^{ik\cdot x} e_\mu(k,j)\hat{a}^+(k,j) \right\}.$$

$$(9.6.5)$$

In (9.6.5) we have defined the operators, associated with photons of well-defined linear polarization,

$$\hat{a}(k,1) = \frac{1}{\sqrt{2}}\{\hat{a}(k,+1) + \hat{a}(k,-1)\},$$

$$\hat{a}(k,2) = \frac{i}{\sqrt{2}}\{\hat{a}(k,+1) - \hat{a}(k,-1)\},$$

$$(9.6.6)$$

### 9.6.1 The Gupta–Bleuler Space

The idea now is to "complete" (9.6.5) to have a Minkowski product of $\hat{a}$'s and $e$'s. We thus write

$$\hat{A}_\mu(x) = \frac{1}{(2\pi)^{3/2}} \int \frac{d^3k}{2k_0} \sum_{\lambda=0}^{3} (-g_{\lambda\lambda}) \left\{ e^{-ik\cdot x} e_\mu(k,\lambda)\hat{a}(k,\lambda) \right.$$
$$\left. + e^{ik\cdot x} e_\mu(k,\lambda)\hat{a}^+(k,\lambda) \right\}, \quad k_0 = |k|.$$

$$(9.6.7)$$

We have introduced two new types of creation–destruction operator: the $\hat{a}(k,3)$ associated with $e_\mu(k,3)$, which annihilate nonexistent longitudinal photons, and $\hat{a}(k,0)$, associated with $e_\mu(k,0)$, which refers to scalar (also called "*timelike*", for obvious reasons) photons, which are also fictitious. We have too many states, so we will need supplementary conditions to get rid of them.

If (9.6.7) is to be covariant, we expect that the commutation relations will also be covariant. We then postulate

$$[\hat{a}_\mu(k), \hat{a}_\nu^+(k')] = -2g_{\mu\nu}k_0\delta(\mathbf{k} - \mathbf{k}'),$$

$$\hat{a}_\mu(k) \equiv -\sum g_{\lambda\lambda} e_\mu(k,\lambda)\hat{a}(k,\lambda),$$

$$[\hat{a}_\mu(k), \hat{a}_\nu(k')] = 0.$$

$$(9.6.8)$$

---

[6] Throughout this section we work consistently in natural Heaviside units.

Not only have we too many states, but (9.6.8) implies that some of them have a *negative* metric: in fact[7],

$$\parallel \hat{a}_0^+(k)|0\rangle \parallel^2 = -2g_{00}k_0\delta(\mathbf{0}) < 0. \tag{9.6.9}$$

As some other states have positive norm, states with *zero* norm also exist. We will have to get rid of them, too.

A possible solution was suggested by Fermi (1932), who proposed interpreting $\hat{a}(k,j)$, $j = 1,2,3$ as destruction, and $\hat{a}^+(k,j)$, $j = 1,2,3$ as creation, operators; but $\hat{a}(k,0)$ as a *creation*, and $\hat{a}^+(k,0)$ as a *destruction*, operator. The negative metric disappears, and one can single out a subspace of physical states on which the contributions of the nonphysical operators, $\hat{a}(k,3)$, $\hat{a}(k,0)$ cancel out.

Here we follow a method that is equivalent to Fermi's but maintains the interpretation of the $\hat{a}$ as destruction, and the $\hat{a}^+$ as creation operators. Of necessity, we then have to work in a space with indefinite metric, the *Gupta–Bleuler space*[8]. Working in terms of the $\hat{a}_\mu$ introduced in (9.6.8), we have

$$\hat{a}_\mu(k) \equiv - \sum g_{\lambda\lambda}e_\mu(k,\lambda)\hat{a}(k,\lambda),$$

$$\hat{a}(k,\lambda) = - \sum g_{\mu\mu}e_\mu(k,\lambda)\hat{a}_\mu(k).$$

We rewrite (9.6.7) as

$$\hat{A}_\mu(x) = \hat{A}_\mu^{(-)}(x) + [\hat{A}_\mu^{(-)}(x)]^+, \tag{9.6.10a}$$

$$\hat{A}_\mu^{(-)}(x) = \frac{1}{(2\pi)^{3/2}} \int \frac{d^3k}{2k_0} e^{-ik\cdot x}\hat{a}_\mu(k). \tag{9.6.10b}$$

The Gupta–Bleuler space $\mathcal{H}_{GB}$ is defined as the set of vectors of the form

$$|\Psi\rangle = \Psi_0|0\rangle + \sum_{n=1}^{\infty} \int \frac{d^3k_1}{2k_{10}} \cdots \frac{d^3k_n}{2k_{n0}} \sum g_{\mu_1\mu_1} \cdots g_{\mu_n\mu_n} \tag{9.6.11}$$

$$\times \Psi_{\mu_1\ldots\mu_n}(k_1,\ldots,k_n)\hat{a}_{\mu_1}^+(k_1)\ldots\hat{a}_{\mu_n}^+(k_n)|0\rangle;$$

$\Psi_0$ is a complex number and the vacuum, $\langle 0|0\rangle = 1$, is annihilated by all $\hat{a}$:

$$\hat{a}_\mu(k)|0\rangle = 0, \ \text{all} \ \mu, k. \tag{9.6.12}$$

If we were to fix the gauge by imposing conditions on the $\hat{A}_\mu$, we would violate manifest Lorentz covariance. So instead we leave $\hat{A}_\mu$ free and get rid of the offending states by imposing conditions on $\mathcal{H}_{GB}$. Because in the classical limit we want to recover the Lorentz condition,

$$\partial \cdot A_{\text{cl}}(x) = 0, \tag{9.6.13}$$

we start by selecting a subspace of $\mathcal{H}_{GB}$, which we call the space of *Lorentzian* vectors, $\mathcal{H}_L$, consisting of states $|\Psi^L\rangle$ such that

---

[7] Remember that the norm of a vector $|\Phi\rangle$ is defined by $\parallel \Phi \parallel = +(\langle\Phi|\Phi\rangle)^{1/2}$.

[8] Gupta (1950); Bleuler (1950).

$$\partial \cdot \hat{A}^{(-)}(x)|\Psi^L\rangle = 0, \text{ all } x, \tag{9.6.14a}$$

or, in momentum space,

$$\hat{L}(k)|\Psi^L\rangle = 0, \quad \hat{L}(k) \equiv \sum g_{\mu\mu}k_\mu \hat{a}_\mu(k). \tag{9.6.14b}$$

This is sufficient to guarantee that the expectation value on Lorentz vectors of $\partial \cdot \hat{A}$ vanishes, and thus to get (9.6.13) in the classical limit. The $|\Psi^L\rangle$ are then candidate physical states. However, this is not enough: indeed, $\mathcal{H}_L$ contains nonzero vectors with a zero norm.

We will next see this, and will show how to treat the problem of the zero-norm states. Although the arguments hold with generality, we will consider in detail mainly one-particle states,

$$|\Phi^L\rangle = \sum_\mu g_{\mu\mu} \int \frac{d^3k}{2k_0} \Phi_\mu^L(k)\hat{a}_\mu^+(k)|0\rangle. \tag{9.6.15}$$

The Lorentz condition (9.6.14b) translates into a subsidiary condition for the wave function $\Phi^L$:

$$k \cdot \Phi^L(k) = 0, \tag{9.6.16a}$$

or, more generally, for an $n$-particle wave function,

$$\sum_{\mu_i} g_{\mu_i\mu_i} k_{\mu_i} \Phi_{\mu_1\ldots\mu_n}^L(k_1,\ldots,k_n) = 0, \quad i = 1,\ldots,n. \tag{9.6.16b}$$

We will show that all Lorentzian vectors have *non-negative* norm, and specify the zero-norm ones. For this, we will use the fact that we can write any $\Phi_\mu$ as

$$\Phi_\mu(k) = k_\mu f(k) + \varphi_\mu(k), \quad \varphi_0(k) = 0, \tag{9.6.17a}$$

for which it is sufficient to choose

$$f(k) = \Phi_0(k)/k_0, \quad \boldsymbol{\varphi}(k) = \boldsymbol{\Phi}(k) - \mathbf{k}\Phi_0(k)/k_0. \tag{9.6.17b}$$

Let $|\Phi^{0L}\rangle$ be an arbitrary Lorentzian vector. Decomposing it as in (9.6.17), we see that the Lorentz condition implies that $k \cdot \varphi^{0L}(k) = 0$; using (9.6.8), and substituting (9.6.17), we get

$$0 = \langle \Phi^{0L}|\Phi^{0L}\rangle = \int \frac{d^3k'}{2k_0'} \int \frac{d^3k}{2k_0} \sum g_{\nu\nu} \sum g_{\mu\mu} \Phi_\nu^{0L}(k')^* \Phi_\mu^{0L}(k)$$

$$\times \langle 0|\hat{a}_\nu(k')\hat{a}_\mu^+(k)|0\rangle$$

$$\tag{9.6.18}$$

$$= -\int \frac{d^3k}{2k_0} \Phi^{0L}(k)^* \cdot \Phi^{0L}(k)$$

$$= \int \frac{d^3k}{2k_0} \varphi^{0L}(k)^* \varphi^{0L}(k).$$

This is clearly $\geq 0$. It will vanish if, and only if,

$$\varphi^{0L} \equiv 0.$$

We have thus identified zero-norm vectors in $\mathcal{H}_L$ as those we will call *pure gauge*,

$$\Phi_\mu^{0L}(k) = k_\mu f(k), \quad \text{for any } f.$$

More generally, we can characterize zero norm states with any number of particles as linear combinations of vectors of the form

$$\hat{L}^+(k)|\Phi^L\rangle, \quad \text{for any } |\Phi^L\rangle. \tag{9.6.19}$$

For example, we check that (9.6.19) is of zero norm. Because $k^2 = 0$, it follows that

$$[\hat{L}(k), \hat{L}^+(k')] = 0,$$

so that

$$\langle L^+(k)\Phi^L|L^+(k)\Phi^L\rangle = \langle \Phi^L|\hat{L}(k)\hat{L}^+(k)|\Phi^L\rangle$$

$$= \langle \Phi^L|\hat{L}^+(k)\hat{L}(k)|\Phi^L\rangle = 0.$$

Let $|\Phi^L\rangle$ now be an arbitrary vector in $\mathcal{H}_L$. We define its *gauge class* as the set of vectors $|\varphi\rangle$ such that

$$|\Phi^L\rangle = |\varphi^L\rangle + |f^{0L}\rangle, \tag{9.6.20}$$

where $|f^{0L}\rangle$ is any zero-norm Lorentzian vector. That is to say, $|\Phi^L\rangle$ is in the gauge class of $|\varphi^L\rangle$ if they differ by a pure gauge vector. With a calculation like that of (9.6.18) we see immediately that *the scalar product of two vectors* $\langle \Psi^L|\Phi^L\rangle$ *does not depend on the individual vectors, but on their gauge class:*

$$\langle \Psi^L|\Phi^L\rangle = \langle \psi^L|\varphi^L\rangle. \tag{9.6.21}$$

We can thus identify vectors that differ one from another in a vector of zero norm. Denoting this relation by the symbol $\sim$, we have

$$|\Phi^L\rangle \sim |\varphi^L\rangle \text{ if } |\Phi^L\rangle - |\varphi^L\rangle = |f^{0L}\rangle,$$

$$\langle f^{0L}|f^{0L}\rangle = 0. \tag{9.6.22}$$

**Exercise.** Verify that this is an equivalence relation, in the mathematical sense, and that it preserves linear combinations and scalar products ∎

Let us denote by $\mathcal{H}_C$ the space obtained from $\mathcal{H}_L$ with the identification (9.6.22). We will show (explicitly for one-particle states) that it is equivalent to the space of physical states that we introduced, for example, in the Coulomb gauge. This is not difficult. Given $\Phi_\mu^L(k)$, we identify it with $\varphi_\mu^L(k)$, given by

$$\Phi_\mu^L(k) - \varphi_\mu^L(k) = k_\mu f(k), \tag{9.6.23}$$

and we choose

$$f(k) = \Phi_0^L(k)/k_0.$$

Then, $\varphi_0^L(k)$ vanishes. Since the Lorentz condition implies that $k \cdot \varphi^L(k) = 0$, we get

$$\mathbf{k}\varphi^L = 0.$$

Thus, $\Phi_\mu^L(k)$ is equivalent to the wave function $\varphi_\mu^L(k)$,

$$\varphi_\mu^L(k) = \begin{pmatrix} 0 \\ \varphi^L(k) \end{pmatrix}, \quad \mathbf{k}\varphi^L(k) = 0,$$

i.e., a Coulomb gauge wave function. Moreover, for two states,

$$\langle \Psi^L | \Phi^L \rangle = \int \frac{d^3k}{2k_0} \psi^L(k)^* \varphi^L(k).$$

After having shown that the theory can be formulated in the Gupta–Bleuler space in a way to make it equivalent, for physical vectors, to the Coulomb gauge formulation, we will also show that *observables* have identical matrix elements. In particular, this will be the case for the four-momentum operator, $\hat{P}_0^{\mathrm{rad}} = \hat{H}_{\mathrm{rad}}$, and

$$\hat{H}_{\mathrm{rad}} = -\sum g_{\mu\mu}\frac{1}{2}\int d^3k\, \hat{a}_\mu^+(k)\hat{a}_\mu(k),$$

$$\hat{\mathbf{P}}_{\mathrm{rad}} = -\sum g_{\mu\mu}\int \frac{d^3k}{2k_0}\mathbf{k}\,\hat{a}_\mu^+(k)\hat{a}_\mu(k).$$

(9.6.24)

Consider, for example, the Hamiltonian. Its expectation value in Lorentzian vectors is

$$\langle \Phi^L | \hat{H}_{\mathrm{rad}} | \Phi^L \rangle = \frac{1}{2}\int d^3k \sum (-g_{\mu\mu})\langle \Phi^L | \hat{a}_\mu^+(k)\hat{a}_\mu(k) | \Phi^L \rangle. \qquad (9.6.25)$$

For a given $k$, choose the axis system with $OZ$ parallel to $\mathbf{k}$. The Lorentz condition then tells us that

$$0 = \hat{L}(k)|\Phi^L\rangle = k_0\{\hat{a}_0(k) - \hat{a}_3(k)\}|\Phi^L\rangle.$$

This equation is, of course, the key to the whole Gupta–Bleuler formalism: it guarantees that, as announced, the contributions of longitudinal and scalar photons cancel one another, for physical states. Then, and in the same reference system,

$$\langle \Phi^L | \{\hat{a}_0^+\hat{a}_0 - \hat{a}_3^+\hat{a}_3\} | \Phi^L \rangle = \langle \Phi^L | \{\hat{a}_0^+\hat{a}_0 - \hat{a}_3^+\hat{a}_0 | \Phi^L \rangle$$

$$= \langle \Phi^L | (\hat{a}_0 - \hat{a}_3)^+\hat{a}_0 | \Phi^L \rangle = \langle (\hat{a}_0 - \hat{a}_3)\Phi^L | \hat{a}_0 | \Phi^L \rangle = 0,$$

and therefore

$$-\sum g_{\mu\mu}\langle \Phi^L | \hat{a}_\mu^+(k)\hat{a}_\mu(k) | \Phi^L \rangle = \langle \Phi^L | \{\hat{a}_1^+(k)\hat{a}_1(k) + \hat{a}_2^+(k)\hat{a}_2(k)\} | \Phi^L \rangle.$$

Only the physical photons contribute to the right-hand side of (9.6.25) and thus $\langle \Phi^L | \hat{H}_{\mathrm{rad}} | \Phi^L \rangle$ coincides with what we calculated in the Coulomb gauge (and in particular is *positive definite*).

Let us next establish the connection between gauge classes for vectors, and gauge transformations for the field $\hat{A}_\mu$. Consider a matrix element

$$\langle \Phi^L | \hat{A}_\mu(x) | \Phi^L \rangle,$$

and change the gauge representative of $|\Phi^L\rangle$. The above matrix element is then to be compared with

$$\langle \Phi^L[f] | \hat{A}_\mu(x) | \Phi^L[f] \rangle, \tag{9.6.26a}$$

$$|\Phi^L[f]\rangle \equiv |\Phi^L\rangle + \sum_{n=1}^{\infty} \int \frac{d^3 k_1}{2k_{10}} \cdots \frac{d^3 k_n}{2k_{n0}} f_n(k_1, \ldots, k_n)$$
$$\times \hat{L}^+(k_1) \ldots \hat{L}^+(k_n) |\Phi^L\rangle. \tag{9.6.26b}$$

Calculating (9.2.26a), we find that

$$\langle \Phi^L[f] | \hat{A}_\mu(x) | \Phi^L[f] \rangle = \langle \Phi^L | (\hat{A}_\mu(x) - \partial_\mu f(x)) | \Phi^L \rangle, \tag{9.6.27}$$

$$f(x) = 2\mathrm{Re}\, \frac{i}{(2\pi)^{3/2}} \int \frac{d^3 k}{2k_0} e^{-ik \cdot x} f_1(k).$$

It follows that we can change the gauge for the expectation value of $\hat{A}_\mu$ by changing the representative of the vector in which the expectation value is taken.

The analysis we have carried out is relatively simple inasmuch as it refers to *free* photons. To be able to extend it to interactions it will be essential that these respect gauge invariance: otherwise the equivalence between the Coulomb gauge formalism and the Gupta–Bleuler one would break down and we would end up with a theory either not Lorentz invariant or with negative probabilities. Of course, the same argument shows that every *observable*, including the $S$ matrix, should be gauge invariant.

### 9.6.2 Covariant Transformation

Under Lorentz transformations we expect $\hat{A}_\mu(x)$ to behave covariantly,

$$\hat{U}(a, \Lambda) \hat{A}_\mu(x) \hat{U}(a, \Lambda)^{-1} = \sum_\nu (\Lambda^{-1})_{\mu\nu} \hat{A}_\nu(\Lambda x + a), \tag{9.6.28}$$

but the proof is not trivial for Lorentz transformations.

Let us start by considering the transformation properties of creation and annihilation operators in the Coulomb gauge; the transformation properties of the helicity states (6.4.9) imply that

$$\hat{U}(\Lambda) \hat{a}_C(k, \eta) \hat{U}(\Lambda)^{-1} = e^{i\eta\theta(p, \Lambda)} \hat{a}_C(k, \eta), \tag{9.6.29a}$$

where we have distinguished the operators by an index $C$ to emphasize that (9.6.29a) holds for physical photons, say in the Coulomb gauge. In terms of the "Cartesian" operators defined in (9.6.6), this gives

$$\hat{U}(\Lambda)\hat{a}_C(k,j)\hat{U}(\Lambda)^{-1} = \sum_{j'=1}^{2} R_{j'j}(\theta)\hat{a}_C(\Lambda k, j'),$$

$$(9.6.29b)$$

$$R = \begin{pmatrix} \cos\theta & -\sin\theta \\ \sin\theta & \cos\theta \end{pmatrix}.$$

Let us now turn to the covariant case. Taking into account (9.6.3a), $e_\mu(k,\lambda) = H_{\mu\lambda}$, we can write (9.6.7) as

$$\hat{A}_\mu(x) = \frac{1}{(2\pi)^{3/2}} \int \frac{d^3k}{2k_0} \sum_{\lambda\lambda'} \left\{ e^{-ik\cdot x} H_{\mu\lambda}(\overline{k}\rightarrow k)(-g_{\lambda\lambda'})\hat{a}(k,\lambda') \right.$$

$$\left. +\text{h.c.} \right\},$$

$$(9.6.30)$$

where "h.c." denotes the Hermitean conjugate of the preceeding term. We want to see that (9.6.28) is satisfied; to do so, we manipulate the right-hand side there, to get

$$\sum_{\nu}(\Lambda^{-1})_{\mu\nu}\hat{A}_\nu(\Lambda x)$$

$$= \frac{-1}{(2\pi)^{3/2}} \int \frac{d^3k}{2k_0} \sum \left\{ e^{-ik\cdot\Lambda x}(\Lambda^{-1}H(\overline{k}\rightarrow k)G)_{\mu\lambda'}\hat{a}(k,\lambda') + \text{h.c.} \right\}$$

$$= \frac{-1}{(2\pi)^{3/2}} \int \frac{d^3k'}{2k_0'} \sum \left\{ e^{-ik'\cdot x}(\Lambda^{-1}H(\overline{k}\rightarrow \Lambda k')G)_{\mu\lambda'}\hat{a}(\Lambda k',\lambda') + \text{h.c.} \right\},$$

where we have defined $\Lambda^{-1}k \equiv k'$. Changing the name of the integration variable, $k' \rightarrow k$, and after some simple manipulations, we find that

$$\sum_{\nu}(\Lambda^{-1})_{\mu\nu}\hat{A}_\nu(\Lambda x) = \frac{-1}{(2\pi)^{3/2}} \int \frac{d^3k}{2k_0} \sum \left\{ e^{-ik\cdot x} \right.$$

$$\times (H(\overline{k}\rightarrow k)GGH^{-1}(\overline{k}\rightarrow k)\Lambda^{-1}H(\overline{k}\rightarrow \Lambda k)G)_{\mu\lambda'}\hat{a}(\Lambda k,\lambda')$$

$$(9.6.31)$$

$$\left. +\text{h.c.} \right\}$$

$$= \frac{-1}{(2\pi)^{3/2}} \int \frac{d^3k}{2k_0} \sum \left\{ e^{-ik\cdot x} H_{\mu\lambda}(\overline{k}\rightarrow k)g_{\lambda\lambda'}\hat{a}_\Lambda(k,\lambda') + \text{h.c.} \right\},$$

and we defined

$$\hat{a}_\Lambda(k,\lambda) \equiv \sum_{\lambda'}(GH^{-1}(\overline{k}\rightarrow k)\Lambda^{-1}H(\overline{k}\rightarrow \Lambda k)G)_{\lambda\lambda'}\hat{a}(\Lambda k,\lambda').$$

Now, for any Lorentz transformation we have $G\Lambda G = \Lambda^{-1T}$, so that

$$\hat{a}_\Lambda(k,\lambda) = \sum_{\lambda'}\Gamma_{\lambda'\lambda}(k,\Lambda)\hat{a}(\Lambda k,\lambda'),$$

$$(9.6.32a)$$

where

$$\Gamma(k,\Lambda) = H^{-1}(\overline{k}\rightarrow \Lambda k)\Lambda H(\overline{k}\rightarrow k)$$

$$(9.6.32b)$$

coincides precisely with the transformation in the little group of $\overline{k}$ considered in Sect. 6.4 (see under (6.4.6)).

Calculating now the left-hand side in (9.6.28), we obtain

$$\hat{U}(\Lambda)\hat{A}(x)\hat{U}(\Lambda)^{-1} =$$

$$\frac{-1}{(2\pi)^{3/2}} \int \frac{d^3k}{2k_0} \sum \left\{ e^{-ik\cdot x} H_{\mu\lambda}(\overline{k}\to k) g_{\lambda\lambda'} \hat{U}(\Lambda)\hat{a}(k,\lambda')\hat{U}(\Lambda)^{-1} + \mathrm{h.c.} \right\};$$

comparing this with (9.6.31), and using (9.6.32), it follows that we will have covariant transformation properties if, and only if, we have

$$\hat{U}(\Lambda)\hat{a}(k,\lambda)\hat{U}(\Lambda)^{-1} = \sum_{\lambda'} \Gamma_{\lambda\lambda'}(k,\Lambda)\hat{a}(\Lambda k,\lambda'). \tag{9.6.33}$$

This is not yet what we had in (9.6.29b), but almost. A transformation $\Gamma$ in the little group of $\overline{k}$ can be written as in (6.4.1), i.e.,

$$\Gamma(k,\Lambda) = \Lambda_t R_z(\theta),$$

$$\hat{U}(\Lambda)\hat{a}(k,\lambda)\hat{U}(\Lambda)^{-1} = \sum R_{\lambda''\lambda}\hat{a}(\Lambda k,\lambda')\Lambda_{t\lambda'\lambda''}. \tag{9.6.34}$$

Equation (9.6.34) differs from (9.6.29b) in the transformation $\Lambda_t$. They are equivalent in the sense that, *on Lorentzian vectors*, the right-hand side of (9.6.34) differs from

$$\sum R_{\lambda''\lambda}\hat{a}(\Lambda k,\lambda''),$$

when applied to the same Lorentzian vector, in a vector of zero norm. This helps in identifying the gauge transformations with a representation of the $\Lambda_t$ (see also Problem 9.8).

### 9.6.3 The Metric Operator Method

The use of an indefinite metric for the Hilbert space of states may not look very appealing. One can sidestep it by considering a Hilbert space with positive definite metric, but by defining amplitudes $(\Psi|\Phi)$ in terms of matrix elements of an operator $\eta$, however, so that

$$(\Psi|\Phi) = \langle\Psi|\eta|\Phi\rangle$$

will be indefinite, for appropriately chosen $\eta$, even if $\langle\Psi|\Phi\rangle$ gave a positive-definite metric Hilbert space. The method is of course equivalent to the one we are following; the interested reader can find details in Schweber (1961) and work quoted there (p. 246).

## 9.7 Time-Ordered Product. Propagators

We can interpret field operators in an intuitive[9] manner as operators creating/annihilating a particle at a given point of space at a given time. This is made more suggestive if we remark that the amplitudes $\langle 0|\hat{F}(x)|\Psi\rangle$, where $\hat{F}$ is a generic field operator and $|\Psi\rangle$ a generic state, coincide with the corresponding wave functions. So,

$$\langle 0|\hat{\varphi}(x)|\Phi\rangle = \Phi(x),$$

$$\langle 0|\hat{\psi}(x)|\Psi\rangle = \Psi(x),$$

$$\langle 0|\hat{A}_\mu(x)|\Psi\rangle = \Psi_\mu(x),$$

with $\Phi$ the Klein–Gordon wave function (Chap. 2), $\Psi(x)$ the free Dirac wave function (Chap. 3), etc. .

This interpretation suggests that we consider objects called *propagators*, which represent the amplitude of probability for a particle to propagate from one spacetime point to another.

To define the propagators, we start with the notion of *time-ordered product* for field operators. For bosonic fields, $\hat{\varphi}_1$, $\hat{\varphi}_2$, we define

$$T\hat{\varphi}_1(r_1,t_1)\hat{\varphi}_2(r_2,t_2)$$

$$\equiv \theta(t_1 - t_2)\hat{\varphi}(r_1,t_1)\hat{\varphi}_2(r_2,t_2)$$

$$+\theta(t_2 - t_1)\hat{\varphi}_2(r_2,t_2)\hat{\varphi}_1(r_1,t_1);$$

for fermion fields $\hat{\psi}_1, \hat{\psi}_2$, we set

$$T\hat{\psi}_1(r_1,t_1)\hat{\psi}_2(r_2,t_2)$$

$$\equiv \theta(t_1 - t_2)\hat{\psi}(r_1,t_1)\hat{\psi}_2(r_2,t_2)$$

$$-\theta(t_2 - t_1)\hat{\psi}_2(r_2,t_2)\hat{\psi}_1(r_1,t_1);$$

that is to say, $T$ orders operators in the sense of increasing times (increasing from right to left) as if the operators commuted, for bosons, or anticommuted, for fermions. This last is the only difference with the definition given in Sect. 7.8. The $T$ product may be immediately defined to act on products of any number of operators, and is easily shown (Problem 9.5) to be relativistically invariant.

We now define the propagators. For a spin 0 particle, we call[10] the propagator $\Delta(x)$ and define it by

$$\Delta(x) \equiv \langle 0|T\hat{\varphi}(x)\hat{\varphi}^+(0)|0\rangle; \tag{9.7.1}$$

---

[9] Albeit not very rigorous. As emphasized in Sect. 9.2, it does not make much sense to try localize a particle at a point.

[10] The nomenclature for the various propagators is unfortunately not universal.

for a photon,

$$D_{\mu\nu}(x) \equiv \langle 0|T\hat{A}_\mu(x)\hat{A}_\nu(0)|0\rangle; \tag{9.7.2}$$

and for a spin 1/2 particle we set

$$S_{ab}(x) \equiv \langle 0|T\hat{\psi}_a(x)\hat{\bar{\psi}}_b(0)|0\rangle. \tag{9.7.3}$$

Natural units, and the Heaviside conventions, are used in this section.

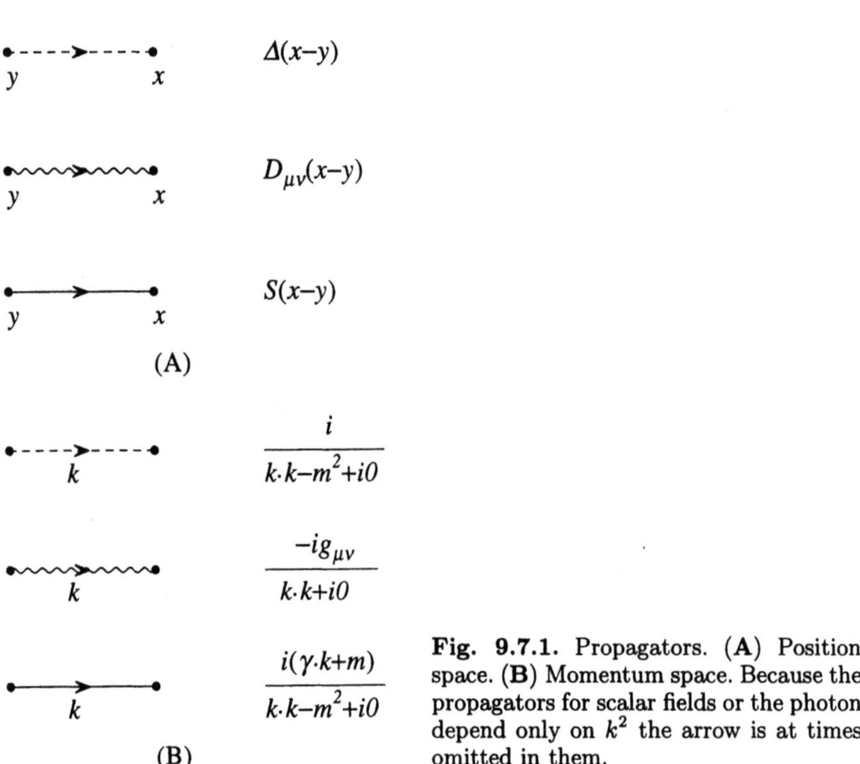

Fig. **9.7.1.** Propagators. (**A**) Position space. (**B**) Momentum space. Because the propagators for scalar fields or the photon depend only on $k^2$ the arrow is at times omitted in them.

We can imagine that the propagator is the quantum amplitude for a particle, created at **0** at time $t = 0$, to propagate to **x**, in the time $x_0$; or, conversely, that it represents an *antiparticle* created at $\mathbf{x}, x_0$ and travelling to $\mathbf{0}, t = 0$, where it is annihilated. Traditionally this propagation is denoted by the graphs of Fig. 9.7.1 (A), drawn for the more general propagation from $y$ to $x$.

**Exercise.** (A) Use the transformation properties of the fields to show that

$$\langle 0|T\hat{\varphi}(z)\hat{\varphi}^+(y)|0\rangle = \langle|T\hat{\varphi}(z-y)\hat{\varphi}^+(0)|0\rangle,$$

and corresponding formulas for the other propagators. (B) Check that the propagators act as Green functions for the equations of motion for the corresponding particles, so that

$(\partial \cdot \partial + m^2)\Delta(x) = i\delta_4(x)$, etc. ∎

It is not difficult to find the explicit expressions for the propagators; they involve Bessel functions, and shall not be given here[11]. We will instead present a Fourier representation. One has, in natural units and with $d^4k = dk_0 dk_1 dk_2 dk_3$, all $k_\mu$ being independent variables,

$$\Delta(x) = \int \frac{d^4k}{(2\pi)^4} e^{-ik\cdot x} \frac{i}{k\cdot k - m^2 + i0}, \tag{9.7.4}$$

$$D_{\mu\nu}(x) = \int \frac{d^4k}{(2\pi)^4} e^{-ik\cdot x} \frac{-ig_{\mu\nu}}{k\cdot k + i0}, \tag{9.7.5}$$

$$S_{ab}(x) = \int \frac{d^4k}{(2\pi)^4} e^{-ik\cdot x} \left( \frac{i}{\not{k} - m + i0} \right)_{ab}$$

$$= \int \frac{d^4k}{(2\pi)^4} e^{-ik\cdot x} \frac{i(\gamma\cdot k + m)_{ab}}{k\cdot k - m^2 + i0}; \tag{9.7.6}$$

the $i0$ in the denominators means, as usual, $i\epsilon$ with $\epsilon > 0$, $\epsilon \to 0$. The expression (9.7.5) for the electromagnetic field holds in the covariant quantization scheme of Sect. 9.6, and in Heaviside units. In Gaussian units, multiply the right-hand side by $4\pi$. In other gauges, the $-g_{\mu\nu}$ in the integral is replaced by an expression

$$-(g_{\mu\nu} + k_\mu f(k) + k_\nu f(k)),$$

where $f$ depends on the gauge[12]. The terms inside the integrals in the above equations, such as $i/(k\cdot k - m^2 + i0)$ for $\Delta$, are known as momentum space propagators. They are given in Fig. 9.7.1 (B).

The proof of (9.7.4)–(9.7.6) is fairly straightforward. As an example we supply the details of the first. Using the expression (9.4.3) for $\hat{\varphi}(x)$ and the definition of the $T$ product, we have, for $x_0 > 0$,

$$\Delta(x) = \langle 0|\hat{\varphi}(x)\hat{\varphi}^+(0)|0\rangle$$

$$= \frac{1}{(2\pi)^3} \int \frac{d^3p}{2p_0} \frac{d^3p'}{2p_0'} \langle 0|e^{-ip\cdot x}\hat{a}(p)\hat{a}^+(p')|0\rangle$$

$$= \frac{1}{(2\pi)^3} \int \frac{d^3p}{2p_0} e^{-ip\cdot x}, \quad x_0 > 0;$$

---

[11] They may be found in Bjorken and Drell (1965) or Bogoliubov and Shirkov (1959).

[12] In the Coulomb gauge

$$D_{\mu\nu}(x) = \int \frac{d^4k}{(2\pi)^4} e^{ik\cdot x} \frac{i}{k^2 + i0} \left( -g_{\mu\nu} + \frac{k_0(k_\nu\delta_{\mu 0} + k_\mu\delta_{\nu 0}) - k_\mu k_\nu - k^2\delta_{\mu 0}\delta_{\nu 0}}{k_0^2 - k^2} \right)$$

although the proof is not easy.

for $x_0 < 0$, on the other hand,

$$\Delta(x) = \langle 0|\hat{\varphi}^+(0)\hat{\varphi}(x)|0\rangle$$

$$= \frac{1}{(2\pi)^3} \int \frac{d^3p}{2p_0} e^{ip\cdot x}, \quad x_0 < 0.$$

Therefore

$$\Delta(x) = \frac{1}{(2\pi)^3} \int \frac{d^3p}{2p_0} \left\{ \theta(x_0)e^{-ip\cdot x} + \theta(-x_0)e^{ip\cdot x} \right\}. \tag{9.7.7}$$

To verify that this agrees with (9.7.4), we consider the integral there. Taking $x_0 > 0$, and separating the variable $dk_0$,

$$\Delta(x)|_{x_0>0} = i \int \frac{d^3k}{(2\pi)^4} e^{i\mathbf{k}\mathbf{r}} \int_{-\infty}^{+\infty} dk_0 e^{-ik_0 x_0} \frac{1}{k_0^2 - (\mathbf{k}^2 + m^2) + i0}.$$

The integral on $dk_0$ can be performed by closing the line $(-\infty, +\infty)$ with a half circle in the lower half plane, which vanishes (for $x_0 > 0$). The integral then equals $-2\pi i$ times the residues contained in that half plane, so that

$$\Delta(x)|_{x_0>0} = 2\pi \int \frac{d^3k}{(2\pi)^4} \frac{e^{i\mathbf{k}\mathbf{r}} e^{-ix_0\sqrt{\mathbf{k}^2+m^2}}}{2\sqrt{\mathbf{k}^2 + m^2}}.$$

This is the same as (9.7.7), up to a change of the name of the variable of integration. For $x_0 > 0$ the evaluation is similar, with one now closing the contour on the upper half plane.

The importance of propagators will be apparent in Chaps. 10 and 11. Here we will finish the section with a simple, but basic, result: *Wick's theorem*. For a time-ordered product of any two field operators, $\hat{f}_a(x)$, $\hat{f}_b(x)$ (referring to the same or different particles) the theorem asserts that

$$T\hat{f}_a(x)\hat{f}_b(y) = \langle 0|T\hat{f}_a(x)\hat{f}_b(y)|0\rangle + : \hat{f}_a(x)\hat{f}_b(y) : . \tag{9.7.8}$$

The verification of (9.7.8) is immediate. For a general $T$ product, let us call to $\langle 0|T\hat{f}_a\hat{f}_b|0\rangle$ the *pairing*, or *contraction*, of $\hat{f}_a$, $\hat{f}_b$, and denote it by

$$\underline{\hat{f}_a(x)\hat{f}_b(y)} \equiv \langle 0|T\hat{f}_a(x)\hat{f}_b(y)|0\rangle. \tag{9.7.9}$$

We also have

$$T\hat{f}_1(x_1)\ldots\hat{f}_n(x_n) = \sum_k \left( \underline{f_{i_1}(x_{i_1})f_{i'_1}(x_{i'_1})} \right) \ldots \left( \underline{\hat{f}_{i_k}(x_{i_k})\hat{f}_{i'_k}(x_{i'_k})} \right) \tag{9.7.10}$$

$$\times : \hat{f}_{i_k+1}(x_{i_k+1})\ldots\hat{f}_{i_n}(x_{i_n}) : .$$

The sum is extended to all possible partitions of the set $1,\ldots,n$. Equation (9.7.10) can be proved by induction; the details may be found in Schweber (1961), Itzykson and Zuber (1980), Bjorken and Drell (1965), etc.

# 9.8 Interactions in Quantum Field Theory. Lagrangian Formulation

### 9.8.1 Lagrangian Formalism for Fields

Fields, which for the moment we consider *classical*, are systems with an infinite number (actually, a continuum) of degrees of freedom. The most familiar field is the electromagnetic one, described by the field variables $\mathcal{E}(\mathbf{r}, t)$, $\mathcal{B}(\mathbf{r}, t)$ (or $A_\mu(x)$) with $\mathbf{r}, t$ varying over all space and time. For a system with a *finite* number of degrees of freedom, characterized thus by the set of variables $q(n, t)$, the dynamics may be obtained though the *Lagrangian* formalism. We have the *Lagrange function*, or *Lagrangian*,

$$L = L(q(n,t), \partial_t q(n,t)),$$

depending on the $q$ and their time derivatives. For free particles, we have the free Lagrangian,

$$L_0 = \sum_0 \mathcal{L}_0, \quad \mathcal{L}_0 = \frac{1}{2} m_n \left( \partial_t q(n,t) \right)^2. \tag{9.8.1}$$

$\mathcal{L}_0$ is called the *Lagrangian density*. When there is no danger of confusion, the name Lagrangian is also used for $\mathcal{L}_0$. The *action* is the time integral of $L$,

$$\mathcal{A} = \int dt\, L = \int dt \sum_n \mathcal{L}, \tag{9.8.2}$$

and $\mathcal{L}$ is the Lagrangian density including interactions. Demanding that $\mathcal{A}$ be stationary against variations of the $q$, we obtain the *Euler–Lagrange equations*:

$$\frac{\partial}{\partial t} \frac{\partial L}{\partial(\partial_t(q(n,t))} - \frac{\partial L}{\partial q(n,t)} = 0. \tag{9.8.3}$$

The *Hamiltonian* is obtained from $L$ through the relation

$$\begin{aligned}
H &= \sum_n \left\{ (\partial_t q(n,t)) \frac{\partial \mathcal{L}}{\partial(\partial_t q(n,t))} - \mathcal{L} \right\} \\
&\equiv \sum_n \mathcal{H},
\end{aligned} \tag{9.8.4}$$

and $\mathcal{H}$ is the *Hamiltonian density*,

$$\mathcal{H} = (\partial_t q(n,t)) \frac{\partial \mathcal{L}}{\partial(\partial_t q(n,t))} - \mathcal{L}. \tag{9.8.5}$$

Field theory may be obtained as a limiting case of this. We consider the fields $\varphi_a(\mathbf{r}, t)$, still classical. We replace the continuum $\mathbf{r}$ by a discrete set $\mathbf{r}_n$ spaced by the quantity $\delta$, and the whole of space by a finite volume $V$. For $\delta$, $V$ finite, the system has a finite number of degrees of freedom. Sums on

$n$, like those in (9.8.1) or (9.8.2), will become, in the limit $\delta \to 0$, integrals on $d^3r$. If the fields were confined in the finite volume $V$, we would integrate on $V$; for fields extended over all space, we let $V \to \infty$. Thus we have the Lagrangian

$$L = \int d^3r \, \mathcal{L}, \tag{9.8.6a}$$

and, because of relativistic invariance, we expect $\mathcal{L}$ to depend, for relativistic fields, not only on $\partial_t \varphi_a$, but also on space derivatives $\partial_i \varphi_a$. With covariant notation,

$$\mathcal{L} = \mathcal{L}(\varphi_a(x), \partial_\mu \varphi_a(x)). \tag{9.8.6b}$$

The action will be (we use units with $c = 1$)

$$\mathcal{A} = \int dt \int d^3r \, \mathcal{L} = \int d^4x \, \mathcal{L}. \tag{9.8.7}$$

Relativistic invariance will be obtained if $\mathcal{L}$ is a *Lorentz scalar*.

The equations of evolution will be the generalization of (9.8.3):

$$\sum_\mu \partial_\mu \frac{\partial \mathcal{L}}{\partial(\partial_\mu \varphi_a(x))} - \frac{\partial \mathcal{L}}{\partial \varphi_a(x)} = 0. \tag{9.8.8}$$

As a matter of fact, we will guess the form of the Lagrangians in field theory by requiring that (9.8.8) reproduce the field equations deduced in the previous sections.

The Hamiltonian is now

$$H = \int d^3r \, \mathcal{H},$$

$$\mathcal{H} = \sum_a (\partial_t \varphi_a(x)) \frac{\partial \mathcal{L}}{\partial(\partial_t \varphi_a(x))} - \mathcal{L}. \tag{9.8.9}$$

For any system, we may define the interaction Hamiltonian (Lagrangian) as $H_{\text{int}} = H - H_0$ ($L_{\text{int}} = L - L_0$), where $H_0(L_0)$ is the *free* Hamiltonian (Lagrangian). From (9.8.9) it easily follows that, if $L_{\text{int}}$ *does not involve the derivatives* $\partial_\mu \varphi_a$, then

$$L_{\text{int}} = -H_{\text{int}}. \tag{9.8.10}$$

The Lagrangian density is not unique. Because the integral to all four-space of a four-divergence vanishes[13], the Lagrangian densities

$$\mathcal{L}, \; \mathcal{L} + \partial \cdot f$$

for any $f_\mu = f_\mu(\varphi_a, \partial_\nu \varphi_a)$ are equivalent. This may be used to rewrite Lagrangians in particularly convenient forms.

---

[13] Provided the fields decrease at large $x$, which we assume.

For quantum fields the formulation only changes in that now the $\hat{\varphi}, \hat{\mathcal{L}}$, $\hat{\mathcal{H}}, \ldots$ are operators, to be denoted by carets. We will then list the *free* Lagrangian densities for scalar or Dirac fields[14]

$$\hat{\mathcal{L}}_0(\hat{\varphi}, \hat{\varphi}^+; \partial_\mu \hat{\varphi}, \partial_\mu \hat{\varphi}^+)$$

$$=: \sum g_{\mu\nu}(\partial_\mu \hat{\varphi}^+(x))(\partial_\nu \hat{\varphi}(x)) - m^2 \hat{\varphi}^+(x)\hat{\varphi}(x) : \qquad (9.8.11)$$

$$= - : \hat{\varphi}^+(x)(\partial \cdot \partial + m^2)\hat{\varphi}(x) : + \text{ four divergence};$$

$$\mathcal{L}_0(\hat{\psi}, \hat{\bar{\psi}}; \partial_\mu \hat{\psi}, \partial_\mu \hat{\bar{\psi}})$$

$$=: \hat{\bar{\psi}}(x)\left(\frac{i}{2}\gamma \cdot \overleftrightarrow{\partial} - m\right)\hat{\psi}(x) : \qquad (9.8.12)$$

$$=: \hat{\bar{\psi}}(x)(i\slashed{\partial} - m)\hat{\psi}(x) : + \text{ four divergence}.$$

We have Wick-ordered the products, which removes possible singularities and only alters the $\mathcal{L}_0$ in a constant. It is a simple exercise to check that the Euler–Lagrange equations applied to these $\mathcal{L}_0$ reproduce the known equations of motion,

$$(\partial \cdot \partial + m^2)\hat{\varphi}(x) = 0,$$

$$(i\slashed{\partial} - m)\hat{\psi}(x) = 0, \text{ etc.}$$

For the electromagnetic case the situation is complicated because of gauge invariance. A detailed discussion may be found in texts devoted specifically to field theory (e.g., Bogoliubov and Shirkov, 1959; Bjorken and Drell, 1965; Schweber, 1961). One gets the free Lagrangian, in the Heaviside system of units,

$$\mathcal{L}_{0\,\text{rad}}(\hat{A}_\mu, \partial_\nu \hat{A}_\mu) = -\frac{1}{4} : \sum g_{\mu\mu}g_{\nu\nu}\hat{F}_{\mu\nu}^2 : + \text{ g.t.},$$
$$(9.8.13)$$

$$F_{\mu\nu} = \partial_\mu \hat{A}_\nu - \partial_\nu \hat{A}_\mu,$$

and the gauge terms (g.t. in (9.8.13)) depend on the gauge in which we choose to quantize $\hat{A}$.


### 9.8.2 Interactions

So long as we do not introduce interactions, the field-theoretic formalism just presented is strictly equivalent to the wave function formalism. New physics, however, emerges when interactions are included. In particular, a very interesting feature of interactions introduced in the field-theoretic formalism is

---

[14] Note that, when a field is not Hermitean, $\hat{\varphi}$ and $\hat{\varphi}^+$, or $\hat{\psi}$ and $\hat{\bar{\psi}}$, should be considered independent variables.

that they take into account the possibility that particles may be created and annihilated, both real and virtually.

Among the various interactions, we may distinguish two kinds: *phenomenological* interactions that may, under certain circumstances, describe with a good approximation some processes but which can be reduced to other interactions; and *fundamental* interactions that cannot be so reduced, at least in the present state of our knowledge. With respect to the latter an empirical fact (the profound reason for which, if it exists, escapes our understanding) is that the *three* interactions[15] which appear to be fundamental are what are known as *gauge interactions*. This means the following: these interactions are mediated by spin 1 particles, the corresponding fields of which we denote generically by $\hat{W}_\mu(x)$; and the interactions may be introduced via the principle of minimum replacement. In particular, the interactions of matter with the $\hat{W}$ are obtained by replacing, in the Lagrangian of the elementary particles of matter (all the known ones are spin 1/2 particles), the derivative $i\partial_\mu$ by what is known as the *gauge covariant* derivative,

$$i\hbar\partial_\mu \rightarrow i\hbar\partial_\mu - \frac{g}{c}\hat{W}_\mu(x) \equiv i\hbar D_\mu^W. \tag{9.8.14}$$

When the matter particles have internal quantum numbers, so will the $W$, and (9.8.14) should include coefficients coupling these quantum numbers. $g$ is a constant associated with the strength of the interaction in question. In the electromagnetic case, the only one we will consider in detail, the constant $g$ is $e$, the electric charge of the matter particles, so that (9.8.14) is just the operator version of (3.4.1):

$$i\hbar\partial_\mu \rightarrow i\hbar D_\mu = i\hbar\partial_\mu - \frac{e}{c}\hat{A}_\mu(x). \tag{9.8.15}$$

If we have a set of fermions with charges $e_j$ and masses $m_j$, and represented by fields $\hat{\psi}_j$, then the free Lagrangian is the sum of the free Lagrangians for each of them. With $\hbar = c = 1$,

$$\hat{\mathcal{L}}_0 = \sum_j :\hat{\bar{\psi}}_j \left(\frac{i}{2}\overleftrightarrow{\partial} - m_j\right)\psi_j: +\hat{\mathcal{L}}_{0\text{ rad}} \tag{9.8.16a}$$

(recall (9.8.12)); $\hat{\mathcal{L}}_{0\text{ rad}}$ is the free Lagrangian of the electromagnetic field, given by (9.8.13). After the minimal replacement, we get the Lagrangian

$$\hat{\mathcal{L}} = \hat{\mathcal{L}}_0 + \hat{\mathcal{L}}_{\text{int}},$$

$$\hat{\mathcal{L}}_{\text{int}} = -\sum_j e_j :\hat{\bar{\psi}}_j(x)\gamma \cdot \hat{A}(x)\hat{\psi}_j(x): . \tag{9.8.16b}$$

The corresponding interaction Hamiltonian is thus

---

[15] These are the strong, weak and electromagnetic interactions. A fourth fundamental interaction, viz., the gravitational one, will not be considered here as there exists at present no established quantum theory of it.

$$\hat{H}_{\text{int}} = -\int d^3r\,\hat{\mathcal{L}}_{\text{int}}$$

$$= \sum_j e_j \int d^3r : \hat{\bar{\psi}}_j(x)\gamma \cdot \hat{A}(x)\hat{\psi}_j(x) : .$$

(9.8.17)

For scalar particles of charge $e$, (9.8.11) gives[16]

$$\hat{L}_{\text{int}} = \sum_{\mu\nu} g_{\mu\nu} \int d^3r : ie(\partial_\mu\hat{\varphi}^+(x))\hat{A}_\nu(x)\hat{\varphi}(x)$$

$$-ie\hat{\varphi}^+(x)\hat{A}_\mu(x)\partial_\nu\hat{\varphi}(x) + e^2\hat{\varphi}^+(x)\hat{\varphi}(x)\hat{A}_\mu(x)\hat{A}_\nu(x) : .$$

(9.8.18)

Two comments may be made with respect to (9.8.17), (9.8.18). First of all, because $\hat{H}_{\text{int}}$ (say) contains $\hat{A}$ linearly, it changes the number of photons; likewise, and as we will see, the number of fermions is also not conserved. Secondly, it is not *a priori* obvious that (9.8.17), (9.8.18) be equivalent to the wave function Schrödinger-type equations with the familiar Coulomb potentials. That this is so we will prove in subsequent chapters, where we will furthermore evaluate the semirelativistic corrections that the form of the interaction, (9.8.17), say, implies.

For phenomenological interactions the requisites (apart, of course, from that of agreement with experiment) are that $\hat{\mathcal{L}}_{\text{int}}$ be Hermitean and a Minkowski scalar. For interactions of particles of spin $1/2$ a useful guide is the classification of Sect. 3.8, generalized to bilinears in quantum fields. For example, if we denote by $\hat{\psi}_N$ fields corresponding to neutrons or protons, and $\hat{\varphi}_\pi$ is a pseudoscalar field, we may consider the interaction

$$\hat{\mathcal{L}}_{\text{int},Y} = ig_Y : \hat{\bar{\psi}}_N\gamma_5\hat{\psi}_N\hat{\varphi}_\pi :$$

(9.8.19)

(the *Yukawa interaction*). If $\hat{\varphi}_s$ were scalar, we would have

$$\hat{\mathcal{L}}_{\text{int},s} = g_s : \hat{\bar{\psi}}_N\hat{\psi}_N\hat{\varphi}_s : .$$

(9.8.20)

The $i$ in (9.8.19) is introduced so that $\hat{\mathcal{L}}_{\text{int},Y}$ (and $\hat{\mathcal{L}}_{\text{int},s}$) is Hermitean, for real $g_Y, g_s$. Both (9.8.19) and (9.8.20) also preserve parity. If we had the combination of interactions

$$\mathcal{L}_{\text{int}} =: g_1\hat{\bar{\psi}}\hat{\psi}\hat{H} + ig_2\hat{\bar{\psi}}\gamma_5\hat{\psi}\hat{H} :,$$

with $\hat{\psi}$ spin $1/2$ fields and $\hat{H}$ a spinless one, the interaction would violate parity.

---

[16] No elementary scalar particles are known; but the $\pi$, $K$ may, to a certain approximation, be considered elementary. Note that the Lagrangian (9.8.18) contains derivatives, so here $H_{\text{int}} \neq -L_{\text{int}}$.

## 9.9 Gauge Invariance in Quantum Electrodynamics

The theory of quantum fields for electromagnetic interactions is usually called *quantum electrodynamics*, or *QED* for short. In this section we will discuss briefly the basic question of gauge invariance in QED.

We define the gauge transformations on fields as suggested by the wave function formalism,(3.4.1), (3.4.2). Thus, we set

$$\hat{A}_\mu(x) \to \hat{A}_\mu(x) - \partial_\mu \hat{f}(x),$$

$$\hat{\psi}(x) \to e^{ie\hat{f}(x)/c\hbar}\hat{\psi}(x). \tag{9.9.1}$$

Here $e$ is the charge associated with the field $\hat{\psi}$; $\hat{f}$ is in general an operator that we will assume to commute with $\hat{A}, \partial_\mu \hat{f}$. We simplify the discussion by considering a single fermion, and will furthermore take units with $\hbar = c = 1$. Under (9.9.1) the covariant gauge derivative behaves as follows:

$$iD_\mu\hat{\psi}(x) \equiv \left(i\partial_\mu - e\hat{A}_\mu(x)\right)\hat{\psi}(x) \to e^{ie\hat{f}(x)}iD_\mu\hat{\psi}(x). \tag{9.9.2}$$

Therefore the *full* Lagrangian, given by (9.8.16), and which may be written (up to a four-divergence) as

$$\hat{\mathcal{L}} = \hat{\mathcal{L}}_{0\,\text{rad}} + : \hat{\bar{\psi}}i\gamma \cdot D\psi : -m : \overline{\psi}\psi :, \tag{9.9.3}$$

is gauge invariant.

For the simple case $\hat{f}(x) = \lambda$, $\lambda$ being a $c$ number independent of $x$, we can apply Noether's theorem to the Lagrangian to conclude that one has the *continuity equation*,

$$\partial \cdot \hat{J}(x) = 0, \ \hat{J}_\mu(x) =: \hat{\bar{\psi}}(x)\gamma_\mu\hat{\psi}(x) : . \tag{9.9.4}$$

From this, one deduces the conservation of the electromagnetic charge. One defines the charge operator $\hat{Q}$ as

$$\hat{Q} = \int d^3r \hat{J}_0(x). \tag{9.9.5}$$

Then, (9.9.4) gives

$$\partial_t\hat{Q} = \int d^3r \partial_0\hat{J}_0 = -\int d^3r \nabla\hat{\mathbf{J}} = 0, \tag{9.9.6}$$

and the last expression holds because we integrate a divergence to the whole space.

A current that satisfies (9.9.4) is called a *conserved current*.

The formulation of QED depends on the gauge one chooses (the "g.t." terms in (9.8.13)), although gauge invariance guarantees that the *physical* consequences are gauge-independent. For example, the expression for the propagator is altered when changing the gauge. The structure (9.9.1) indicates that the general structure of the propagator will be

$$D_{\mu\nu} = \int \frac{d^4k}{(2\pi)^4} e^{-ik\cdot x} i \frac{-g_{\mu\nu} + k_\mu a_\nu(k) + k_\nu a_\mu(k) + k_\mu k_\nu b(k)}{k\cdot k + i0}. \qquad (9.9.7)$$

The functions $a, b$ will depend on the gauge. The proof that physical results are independent of $a, b$ will be given, in a few simple examples, in Chap. 11. More general discussions, as well as the study of the Coulomb gauge formulation (which is a somewhat peculiar case), may be found in the treatise of Bjorken and Drell (1965).

## Problems

**P.9.1.** *Number operator.* Check that the operator

$$\hat{N}_a \equiv \sum_\lambda \int \frac{d^3p}{2p_0} \hat{a}^+(p, \lambda)\hat{a}(p, \lambda)$$

"counts" particles of type $a$, that is to say,

$$\hat{N}_a\{\hat{a}^+(p_1, \lambda_1)\ldots\hat{a}^+(p_n, \lambda_n)|0\rangle\}$$

$$= n\{\hat{a}^+(p_1, \lambda_1)\ldots\hat{a}^+(p_n, \lambda_n)|0\rangle\}.$$

This is true for both bosons and fermions.

**P.9.2.** *Zitterbewegung.* If, for a Dirac particle, we insisted on having as position operator $\hat{X}$, with

$$\hat{X}_i\psi(\mathbf{r}, t) = r_i\Psi(\mathbf{r}, t),$$

then the velocity operator, for a free particle, would be

$$\hat{\mathbf{v}} = \frac{i}{\hbar}[\hat{H}_0, \hat{\mathbf{X}}] = c\boldsymbol{\alpha}.$$

The eigenvalues of $\hat{\mathbf{v}}$ are $\pm c$. Moreover, $\hat{\mathbf{v}}$ does not commute with $\hat{H}_0$: the particle, although free, would follow a twisted path (Zitterbewegung, German for shaky movement), each stretch being covered at the speed of light. Note that this holds for particles with mass *different* from zero.

Prove all of this. For details, see Bjorken and Drell (1964).

**P.9.3.** Give a detailed proof of the transformation properties of the Dirac field, (9.5.4).

*Solution.* In the Weyl representation, and recalling Sect. 6.6, Problem 6.5 and Appendix A.3, we have

$$u(p, \tau) = \begin{pmatrix} u^R(p, \tau) \\ \tilde{u}^L(p, \tau) \end{pmatrix},$$

$\tau = 1, 2$ and

$$u_\alpha^R(p,\tau) = D_{\alpha\tau}^{(1/2)}(L(p)) = ((p\cdot\tilde\sigma)^{1/2})_{\alpha\tau},$$

$$\tilde u_\alpha^L(p,\tau) = \sum_\beta (p\cdot\sigma)_{\alpha\beta} u_\beta^R(p,\tau) = ((p\cdot\sigma)^{1/2})_{\alpha\tau};$$

we take $m = 1$ to lighten the notation. We then have

$$(\mathrm{I}) \equiv \sum_\tau u_a^R(\Lambda^{-1}p',\tau)D_{\tau\tau'}^{(1/2)}(L^{-1}(\Lambda^{-1}p')\Lambda^{-1}L(p'))$$

$$= \sum_\tau D_{\alpha\tau}^{(1/2)}(L(\Lambda^{-1}p'))D_{\tau\tau'}^{(1/2)}(L^{-1}(\Lambda^{-1}p')\Lambda^{-1}L(p'))$$

$$= \sum_{\alpha'} D_{\alpha\alpha'}^{(1/2)}(\Lambda^{-1})D_{\alpha'\tau'}^{(1/2)}(L(p'))$$

$$= \sum_{\alpha'} D_{\alpha\alpha'}^{(1/2)}(\Lambda^{-1})u_{\alpha'}^R(p',\tau') = \sum_{\alpha'}(D^{(1/2)}(\Lambda^{-1}))_{\alpha\alpha'}u_{\alpha'}^R(p',\tau').$$

On the other hand,

$$(\mathrm{II}) \equiv \sum_\tau \tilde u_\alpha^L(\Lambda^{-1}p',\tau)D_{\tau\tau'}^{(1/2)}(L^{-1}(\Lambda^{-1}p')\Lambda^{-1}(L(p'))$$

$$= \sum (\Lambda^{-1}p'\cdot\sigma)_{\alpha\beta}D_{\beta\beta'}^{(1/2)}(\Lambda^{-1})(p'\cdot\tilde\sigma)_{\beta'\alpha'}\tilde u_{\alpha'}^L(p',\tau').$$

Defining $\Lambda p = p'$, and recalling (Appendix A.3) that

$$\Lambda p\cdot\tilde\sigma = A(\Lambda)p\cdot\tilde\sigma A(\Lambda)^+, \quad D_{\alpha\beta}^{1/2}(\Lambda) = A_{\alpha\beta}(\Lambda),$$

we see that (II) becomes

$$(\mathrm{II}) = ((p\cdot\sigma)A^{-1}(\Lambda)A(\Lambda)p\cdot\tilde\sigma A(\Lambda)^+\tilde u^L(p',\tau'))_\alpha$$

$$= \sum_{\alpha'}(A(\Lambda^{-1})^{-1+})_{\alpha\alpha'}\tilde u_{\alpha'}^L(p',\tau')$$

$$= \sum_{\alpha'}(\tilde D^{(1/2)}(\Lambda^{-1}))_{\alpha\alpha'}\tilde u_{\alpha'}^L(p',\tau').$$

Combining (I), (II) with (A.3.7), (A.3.8) of Appendix A.3, we find that

$$\sum_\tau u(\Lambda^{-1}p',\tau)D_{\tau\tau'}^{(1/2)}(L^{-1}(\Lambda^{-1}p')\Lambda^{-1}L(p')) = D(\Lambda^{-1})u(p',\tau'),$$

which proves the desired transformation formula for the piece $u\hat b$ of the Dirac field. For the piece $v\hat d^+$ use the fact that, in the Weyl representation,

$$v = i\gamma_2 u^* = \begin{pmatrix} -i\sigma_2\tilde u^{L*} \\ i\sigma_2 u^{R*} \end{pmatrix},$$

and then the calculation proceeds as in the previous case.

**P.9.4.** Prove the following commutation relations.

(i) If $(x - y)^2 < 0$, then

$$[\hat{\varphi}_1(x), \hat{\varphi}_2(y)] = 0,$$

$$[\hat{A}_\mu(x), \hat{A}_\nu(y)] = 0,$$

$$\{\hat{\psi}_{1a}(x), \hat{\psi}_{2b}(y)\} = 0.$$

Here $\hat{\varphi}_1, \hat{\varphi}_2$ are any two (equal or different) scalar fields, including the possibility that $\hat{\varphi}_2 = \hat{\varphi}_1^+$; likewise $\hat{\psi}_1, \hat{\psi}_2$ are fermion fields.

(ii)

$$[i\partial_t \hat{\varphi}(\mathbf{r}, t), \hat{\varphi}^+(\mathbf{r}', t)] = \delta(\mathbf{r} - \mathbf{r}'),$$

$$\{\hat{\psi}_a(\mathbf{r}, t), \hat{\psi}_b^+(\mathbf{r}', t)\} = \delta_{ab}\delta(\mathbf{r} - \mathbf{r}').$$

The connection between (i) and *microscopic causality*, and that between (ii) and the *canonical formulation* of field theory may be seen in, for example, Bjorken and Drell (1965); Bogoliubov and Shirkov (1959); Bogoliubov, Logunov and Todorov (1975).

**P.9.5.** Show that the $T$ product of field operators is relativistically invariant.

*Solution.* We consider the scalar field case; others are similar. We have

$$T\hat{\varphi}_1(x)\hat{\varphi}_2(y) = \theta(x_0 - y_0)\hat{\varphi}_1(x)\hat{\varphi}_2(y) + \theta(y_0 - x_0)\hat{\varphi}_2(y)\hat{\varphi}_1(x).$$

If $(x - y)^2 \geq 0$, the sign of $x_0 - y_0$ is Lorentz invariant, and so then is the definition of the $T$ product. For $(x - y)^2 < 0$, the sign of $x_0 - y_0$ is not invariant: but, as shown in Problem 9.4, in that case $\hat{\varphi}_1, \hat{\varphi}_2$ commute, so their order is irrelevant.

**P.9.6.** Check that the Newton–Wigner operator $\mathbf{Q}_{NW}$ can be obtained by a Foldy–Wouthysen transformation for $\mathbf{X}$ (as given in Problem 9.2).

**P.9.7.** Obtain propagators for a particle with spin 1/2 and (a) of Majorana type; (b) left-handed; (c) right-handed.

In the two-component formalism, (b), for example, can be written as

$$i/(\tilde{\sigma} \cdot p + i0) = i\sigma \cdot p/(p^2 + i0).$$

**P.9.8.** Let $v$ be any vector orthogonal to the lightlike vector $k$. Let $\Gamma$ be in the little group of $k$, and write $\Gamma = \Lambda_{kt} R_{\mathbf{k}}$, where $R_{\mathbf{k}}$ is a rotation around $\mathbf{k}$. Prove that $\Lambda_{kt}v$ differs from $v$ only in a vector proportional to $k$:

$$\Lambda_{kt}v = v + f(\Gamma)k.$$

*Solution.* Choose $OZ$ along $\mathbf{k}$. If we set

$$v = \alpha_1 n^{(1)} + \alpha_2 n^{(2)} + \alpha k,$$

$$n_\mu^{(j)} = \delta_{j\mu}, \text{ and recall (6.4.1), we can write}$$

$$\Lambda_{kt} v = \alpha_1 n^{(1)} + \alpha_2 n^{(2)} + (\xi\alpha_1 + \eta\alpha_2 + \alpha)k$$

$$= v + (\xi\alpha_1 + \eta\alpha_2)k.$$

**P.9.9.** Show that

$$[\hat{a}_\mu(k), \hat{a}_\nu^+(k')] = -2k_0\delta(\mathbf{k} - \mathbf{k}')g_{\mu\nu}$$

implies that

$$[\hat{a}(k, \lambda), \hat{a}^+(k', \lambda')] = -2k_0\delta(\mathbf{k} - \mathbf{k}')g_{\lambda\lambda'}.$$

**P.9.10.** *Equivalence between the wave function and field formulations for free particles.* Consider an $n$-particle state for spinless particles,

$$|\varPhi\rangle = \int \frac{d^3p_1}{2p_{10}} \dots \frac{d^3p_n}{2p_{n0}} \varPhi(p_1, \dots, p_n; t)\hat{a}^+(p_1)\dots\hat{a}^+(p_n)|0\rangle.$$

Show that the action of the field-theoretic Hamiltonian (9.4.6) on $|\varPhi\rangle$ is equivalent to

$$\varPhi \to \left[(\mathbf{p}_1^2 + m^2)^{1/2} + \dots + (\mathbf{p}_n^2 + m^2)^{1/2}\right]\varPhi(\mathbf{p}_1, \dots, \mathbf{p}_n; t),$$

that is to say, to the action of the KGS Hamiltonian (in $p$ space) of Sect. 2.2.

Obtain the equivalent equations in the Dirac case.

**P.9.11.** Let the spin 1/2 fermions represented by the fields $\psi_j$ be massless. Show that if $V$ is a vector and $\phi$ a scalar field, the interactions, for $j$ equal or different to $j'$,

$$\sum g_{\mu\nu}\overline{\psi}_j\gamma_\mu\frac{1}{2}(1 \pm \gamma_5)\psi_{j'}V_\nu, \quad \overline{\psi}_j\frac{1}{2}(1 \pm \gamma_5)\psi_{j'}\phi,$$

only involve right-handed (for +), or, left-handed (for −) fermions, for the first; and left with right, for the second and both choices of signs ±. The first interaction exists in nature; the second is postulated in standard theories, but has not been verified yet.

**P.9.12.** Let $V_{\mu\nu} = \partial_\mu V_\nu - \partial_\nu V_\mu$ for a massless vector field $V$. Let $\psi_j$ be spin 1/2 fields (massive or not). Show that the phenomenological interactions (*Veltman's interactions*)

$$\sum g_{\mu\mu}g_{\nu\nu}\overline{\psi}_j(\sigma_{\mu\nu} \pm \tilde{\sigma}_{\mu\nu})\psi_{j'}V_{\mu\nu},$$

$$\tilde{\sigma}_{\mu\nu} = \frac{i}{2}\sum g_{\alpha\alpha}g_{\beta\beta}\epsilon_{\mu\nu\alpha\beta}\sigma_{\alpha\beta}$$

only involve right-handed/left-handed vector particles.

**P.9.13.** Check that, acting on antisymmetric tensors, $T_{\mu\nu}$, the operation

$$T_{\mu\nu} \to T_{\mu\nu}^\pm \equiv \frac{1}{2}\left(T_{\mu\nu} \pm \frac{i}{2}\sum g_{\alpha\alpha}g_{\beta\beta}\epsilon_{\mu\nu\alpha\beta}T_{\alpha\beta}\right),$$

is a projection. Use this to verify that the functions

$$F^{\pm}_{\mu\nu}(x),$$

for the electromagnetic field, represent photons with definite helicity.

**P.9.14.** Show that the parity of a fermion and its antifermion are opposite. Show that the parity of a boson and its antiparticle are equal. Show that the *charge* partity $\eta_C$ of a fermion is equal to that of its antifermion.

*Solution.* Consider a spin $1/2$ fermion with field $\psi$; the general case may be reduced to this using the Bargmann–Wigner formalism. From (9.5.5), we have (we assume $\eta_P = \pm 1$)

$$\hat{\mathcal{P}}\hat{\psi}(x)\hat{\mathcal{P}}^{-1} = \eta_P \gamma_0 \hat{\psi}(I_s x),$$

$$\hat{\mathcal{P}}\hat{\psi}^+(x)\hat{\mathcal{P}}^{-1} = \eta_P \gamma_0 \hat{\psi}^+(I_s x).$$

Apply this to the vacuum. From the first of the above,

$$\hat{\mathcal{P}} \int \frac{d^3 p}{2p_0} \sum_\lambda v(p,\lambda) e^{ip\cdot x} \hat{d}^+(p,\lambda)|0\rangle$$

$$= \eta_P \int \frac{d^3 p}{2p_0} \sum_\lambda \gamma_0 v(p,\lambda) e^{ip_0 x_0 + i\mathbf{p}\mathbf{r}} \hat{d}^+(p,\lambda)|0\rangle,$$

and hence

$$\hat{\mathcal{P}} \sum_\lambda v(p,\lambda)|p,\lambda\rangle_c = \eta_P \sum_\lambda \gamma_0 v(I_s p,\lambda)|I_s p,\lambda\rangle_c.$$

Because $v = i\gamma_2 u^*$, this equation becomes, after simple algebra,

$$\hat{\mathcal{P}} i\gamma_2 \sum_\lambda u^*(p,\lambda)|p,\lambda\rangle_c = -i\gamma_2 \eta_P \sum_\lambda \gamma_0 u^*(I_s p,\lambda)|I_s p,\lambda\rangle_c.$$

From the second equation (for $\hat{\psi}^+$) above, on the other hand,

$$\hat{\mathcal{P}} \sum_\lambda u^*(p,\lambda)|p,\lambda\rangle = \eta_P \sum_\lambda \gamma_0 u^*(I_s p,\lambda)|I_s p,\lambda\rangle.$$

Choosing to simplify $OZ \parallel \mathbf{p}$, and $\lambda$ the spin component along $\mathbf{p}$, we get from the two last equations, respectively,

$$\hat{\mathcal{P}}|p,\lambda\rangle_c = -\eta_P|I_s p,\lambda\rangle_c,$$

$$\hat{\mathcal{P}}|p,\lambda\rangle = +\eta_P|I_s p,\lambda\rangle,$$

as was to be proved.

The calculation is similar for $\mathcal{C}$.

**P.9.15.** Use the fact that the electromagnetic interaction is

$$e \sum g_{\mu\nu} \hat{\bar{\psi}} \gamma_\mu \hat{\psi} \hat{A}_\mu,$$

and the Yukawa interaction

$$ig \hat{\bar{\psi}}_N \gamma_5 \hat{\psi}_N \hat{\phi}_\pi,$$

to deduce that the intrinsic parities $\eta_P$ of the photon and pion are both $(-1)$. Show that the charge-conjugation parities are also $(-1)$.

**P.9.16.** For particles of spin 1, 0, with fields, respectively, $\hat{V}_\mu, \hat{\phi}_S$, and interaction with fermion fields proportional to

$$\sum g_{\mu\nu} \hat{\bar{\psi}} \gamma_\mu \gamma_5 \hat{\psi} \hat{V}_\mu, \quad \hat{\bar{\psi}} \hat{\psi} \hat{\phi}_S,$$

prove that the intrinsic parities are $(+1)$.

**P.9.17.** Use the previous results to check that interactions

$$\sum_\mu \hat{\bar{\psi}} \gamma_\mu (v + a\gamma_5) \hat{\psi} \hat{W}_\mu, \quad \hat{\bar{\psi}} (s + p\gamma_5) \hat{\psi} \hat{\phi},$$

*cannot* preserve parity, if $v, a, s, p$ are all different from zero, no matter what intrinsic parities one chooses for the $\hat{\psi}, \hat{W}, \hat{\phi}$.

# 10. Interactions in Quantum Field Theory. Nonrelativistic Limit. Reduction to Equivalent Potential

## 10.1 Potentials Equivalent to Field-Theoretic Interactions. General Method

It is impossible to solve interactions such as those introduced in Sect. 9.8 exactly. To study them we therefore have to resort to approximation methods, notably to expansions in powers of the parameter characterizing the strength of the interaction, $e$ for the electromagnetic interactions, $g_Y, g_s \ldots$ in other cases. This method, which will be described in some detail, gives excellent results for scattering problems in electromagnetic interactions of elementary particles. Here the expansion parameter is effectively the fine structure constant, $\alpha \simeq 1/137$, which is small: hence, in all but exceptional situations, one or at most two terms in the perturbation expansion give very accurate results.

This method of perturbations is useless for solving directly bound state problems, which are manifestly nonanalytic in the coupling constant: but we can still use it indirectly, at least in the strict nonrelativistic and semirelativistic limits. In fact, we will show that, in the NR limit, the field-theoretic interactions of Sect. 9.8 are equivalent to potential interactions. The strict nonrelativistic limit will then reproduce standard potentials, like the Coulomb one; the $O(1/c^2)$ corrections will give relativistic corrections to them.

Before plunging into the calculations, let us recapitulate some relations and set up notations. First, recall that the $S$ matrix can be written as (Sect. 7.8, (7.8.9))

$$\hat{S} = \sum \left( \frac{-i}{\hbar} \right)^n \frac{1}{n!} \int dt_1 \ldots dt_n T(\hat{H}_{\text{int}}(t_1) \ldots \hat{H}_{\text{int}}(t_n)); \tag{10.1.1}$$

throughout the rest of this chapter we will distinguish operators by putting carets over them. In (10.1.1) $\hat{H}_{\text{int}}$ is the interaction Hamiltonian in the interaction representation,

$$\hat{H}_{\text{int}} = e^{i\hat{H}_0 t/\hbar} H_{\text{int}}^{\text{Schrödinger}} e^{-i\hat{H}_0 t/\hbar}.$$

*For collision states,* and because these are free, $\hat{H}_0$ can be taken to be the total Hamiltonian; and $\hat{H}_{\text{int}}$ in (10.1.1) can thus be considered to be the *Heisenberg* picture interaction Hamiltonian. Moreover, and because we can evaluate the terms in (10.1.1) as derivatives of $\hat{S}$ with respect to the interaction, calculated at zero interaction, it follows that the field operators that enter into the expression for $\hat{H}_{\text{int}}$, e.g. the $\hat{\psi}, \hat{A}$ in (9.8.17), can be considered *free* fields. Thus, in this case we can also write

$$\hat{S} = \sum \left(\frac{i}{n}\right)^n \int d^4x_1 \ldots d^4x_n T(\hat{\mathcal{L}}_{\text{int}}(x_1) \ldots \hat{\mathcal{L}}_{\text{int}}(x_n)), \tag{10.1.2}$$

with

$$\hat{\mathcal{L}}_{\text{int}}(x) = -\sum e_j : \bar{\hat{\psi}}_j(x)\gamma \cdot \hat{A}(x)\hat{\psi}_j(x),$$

where $\hat{\psi}, \hat{A}$ are free fields.

We define the invariant relativistic transition amplitude as in (7.3.5b):

$$\langle f|S|i\rangle = \langle f|i\rangle + i\delta(\mathbf{p}_i - \mathbf{p}_f)\delta(E_f - E_i)F(i \to f).$$

For elastic scattering, and with all the interactions (9.8.17) to (9.8.19), the lowest nonzero term in the contribution of (10.1.1) to $F$ is the second-order one. Thus, in these cases[1], the Born approximation will be

$$\langle f|\hat{S}|i\rangle_B = i\delta(\mathbf{p}_f - \mathbf{p}_i)\delta(E_f - E_i)F_B(i \to f)$$

$$= \left(\frac{i}{\hbar}\right)^2 \frac{1}{2!} \int dt_1 dt_2 \langle f|T\hat{H}_{\text{int}}(t_1)\hat{H}_{\text{int}}(t_2)|i\rangle. \tag{10.1.3}$$

We take $|i\rangle \neq |f\rangle$, and normalize states relativistically:

$$\langle p, \lambda|p', \lambda'\rangle = 2cp_0\delta(\mathbf{p} - \mathbf{p}')\delta_{\lambda\lambda'}, \quad p_0 = \sqrt{m^2c^2 + \mathbf{p}^2}.$$

For nonrelativistic scattering we define the transition amplitude by

$$_{NR}\langle f|\hat{S}|i\rangle_{NR} = i\delta(\mathbf{p}_f - \mathbf{p}_i)\delta(E_f - E_i)T_{NR}(i \to f),$$

with states normalized to

$$_{NR}\langle \mathbf{p}, \lambda|p'\lambda'\rangle_{NR} = \delta_{\lambda\lambda'}\delta(\mathbf{p} - \mathbf{p}'),$$

so that

$$|p, \lambda\rangle = \sqrt{2p_0c}|p, \lambda\rangle_{NR}. \tag{10.1.4}$$

For two-particle scattering with states

$$|i\rangle = |p_1, \lambda_1; p_2, \lambda_2\rangle,$$

$$|f\rangle = |p_1', \lambda_1'; p_2', \lambda_2'\rangle, \tag{10.1.5}$$

(10.1.4) tells us that

$$T_{NR}(i \to f) = \frac{1}{4c^2\sqrt{p_{10}p_{20}p_{10}'p_{20}'}}F(i \to f). \tag{10.1.6}$$

---

[1] The extension to the cases where the first nonvanishing term is linear in $\hat{H}_{\text{int}}$ is left to the reader.

In potential theory the Born approximation gives

$$_{NR}\langle f|\hat{S}|i\rangle_{NR} = -2\pi i \delta(E_f - E_i)$$

$$\times \int d^3 r_1 d^3 r_2 \underset{\sim}{\psi}^{(\mathbf{p}\,'_1,\lambda'_1)}(\mathbf{r}_1)^+ \underset{\sim}{\psi}^{(\mathbf{p}\,'_2,\lambda'_2)}(\mathbf{r}_2)^+ \underset{\sim}{V}(\mathbf{r}) \qquad (10.1.7a)$$

$$\times \underset{\sim}{\psi}^{(\mathbf{p}_1,\lambda_1)}(\mathbf{r}_1) \underset{\sim}{\psi}^{(\mathbf{p}_2,\lambda_2)},$$

where $\mathbf{r} = \mathbf{r}_1 - \mathbf{r}_2$ and the $\underset{\sim}{\psi}^{(\mathbf{p}\lambda)}$ are the nonrelativistic wave functions for momentum $\mathbf{p}$ and spin $\lambda$. For spin 1/2 particles, the only case we will consider explicitly, they are the two-component vertical matrices

$$\underset{\sim}{\psi}^{(\mathbf{p},\lambda)}(\mathbf{r}) = \frac{1}{(2\pi\hbar)^{3/2}} e^{i\mathbf{p}\mathbf{r}/\hbar} \underset{\sim}{\chi}(\lambda),$$

and the Pauli spinors are normalized to

$$\underset{\sim}{\chi}(\lambda)^+ \underset{\sim}{\chi}(\lambda') = \delta_{\lambda\lambda'}.$$

If we integrate the centre of mass coordinate, (10.1.7a) can also be written as

$$_{NR}\langle f|\hat{S}|i\rangle_{NR}|_{\mathrm{Born}} = i\delta(\mathbf{p}_f - \mathbf{p}_i)\delta(E_f - E_i)T^B_{NR}(i \to f),$$

$$\qquad (10.1.7b)$$

$$T^B_{NR}(i \to f) = \frac{-2\pi}{(2\pi\hbar)^3} \int d^3 k\, e^{i\mathbf{r}(\mathbf{p}_1 - \mathbf{p}'_1)/\hbar}$$

$$\times \sum_{aa'bb'} \chi^*_{1a'}(\lambda'_1)\chi^*_{2b'}(\lambda'_2)V_{a'a;b'b}(\mathbf{r})\chi_{1a}(\lambda_1)\chi_{2b}(\lambda_2). \qquad (10.1.7c)$$

Note that we have taken into account the possibility that $\underset{\sim}{V}$ is a matrix in spin space, i.e., that the potential is spin dependent. In the strictly nonrelativistic limit one can easily show that the most general form of $\underset{\sim}{V}$ respecting parity and time-reversal invariance is

$$\underset{\sim}{V}(r) = V_0(r) + V_s(r)\underset{\sim}{\sigma}_1\underset{\sim}{\sigma}_2 + V_t(r)\underset{\sim}{S}_{12}, \qquad (10.1.8)$$

where the Pauli matrices[2] $\sigma_\alpha$ act on spinor $\chi_\alpha$, $\alpha = 1, 2$. Even the first relativistic corrections will introduce other types of term (velocity dependent).

The essence of the method should now be clear. We evaluate $F_B$ using field theory, say (10.1.3). Equation (10.1.6) then gives $T^B_{NR}$; inverting the Fourier transform (10.1.7c), we find the equivalent potential. The method is known as the *reduction* to an *equivalent potential*.

One may think that this does not imply a nonrelativistic approximation. However, $T^B_{NR}$, as given by (10.1.6), need not a priori be of the form (10.1.7c); it will certainly be so in the strict nonrelativistic limit, because then we know

---

[2] From now on we will omit the tilde under matrix objects where the matrix character should be clear from the context. Carets over operators will, however, be kept.

that interactions can be represented by potentials. Thus, the nonrelativistic approximation will be essential.

In general we expand $F(i \rightarrow f)$ in powers of $1/c$ (for parity-conserving interactions only even powers appear),

$$F_B(i \rightarrow f) = F_B^{(0)}(i \rightarrow f) + \frac{1}{c^2}F_B^{(2)}(i \rightarrow f) + \dots, \qquad (10.1.9)$$

and from this we obtain an effective $T_n^B$ that coincides with $F_B$ at any given order $n$. It may then happen that (10.1.7) can be inverted and we deduce a potential equivalent to the original interaction, $\hat{H}_{\text{int}}$, to order $n$ in $1/c^2$:

$$\hat{V} = \hat{V}^{(0)} + \frac{1}{c^2}\hat{V}^{(2)} + \dots + \left(\frac{1}{c^2}\right)^n \hat{V}^{(2n)}. \qquad (10.1.10)$$

$\hat{V}^{(0)}$ will coincide with the purely nonrelativistic potential.

Once we have an equivalent potential, say $\hat{V}_n$ at a given order, we also expand the free Hamiltonian, in powers of $1/c^2$, and obtain a Schrödinger equation,

$$\hat{H}_n \Psi = E\Psi, \qquad (10.1.11a)$$

$$\hat{H}_n = H_{NR} + \hat{V}^{(0)} + \hat{H}_n', \qquad (10.1.11b)$$

$$\hat{H}_n' = \frac{1}{c^2}\hat{V}^{(2)} + \dots + \left(\frac{1}{c^2}\right)^n \hat{V}^{(2n)} + \frac{1}{c^2}\hat{H}_{\text{kin}}^{(2)} + \dots + \left(\frac{1}{c^2}\right)^n \hat{H}_{\text{kin}}^{(2n)}, \quad (10.1.11c)$$

where $\hat{H}_{\text{kin}}^{(2n)}$ are the relativistic corrections to the kinetic energy. Equation (10.1.11a) is then dealt with by solving the standard nonrelativistic Schrödinger equation

$$(\hat{H}_{NR} + \hat{V}^{(0)})\Psi_{NR} = E_{NR}\Psi_{NR},$$

and treating $\hat{H}_n'$ as a perturbation.

In the present text we will only evaluate the first relativistic corrections to nonrelativistic potentials; this is sufficient for most applications.


## 10.2 Equivalent Potential for Two Particles in Electromagnetic Interaction

### 10.2.1 Elastic Collision of Two Charged Particles in the Born Approximation

Let us consider two particles with spin 1/2 and charges $e_1, e_2$ in electromagnetic interaction. We start with the case in which both are particles; later we will consider particle–antiparticle interactions. Moreover we will also assume them to be distinguishable.

The interaction Hamiltonian can be written as in (9.8.17):

$$\hat{H}_{\text{int}}(x_0) = \int d^3r : e_1\hat{\bar{\psi}}_1(x)\hat{A}\!\!\!/(x)\hat{\psi}_1(x) + e_2\hat{\bar{\psi}}_2(x)\hat{A}\!\!\!/(x)\hat{\psi}_2(x) : \qquad (10.2.1)$$

(for the moment we will work in natural units, $\hbar = c = 1$). The Born approximation is of second order in $\hat{H}_{\text{int}}$. If we evaluate it using (10.1.3), it is easy to see that only the crossed terms of the product $\hat{H}_{\text{int}}\hat{H}_{\text{int}}$ contribute; and both contribute equally, which cancels the 2! in the denominator. Then,

$$\langle f|\hat{S}|i\rangle_{\text{Born}} = i^2 e_1 e_2 \sum_{\mu\nu} g_{\mu\mu}g_{\nu\nu} \int d^4x_1 d^4x_2 \langle p'_1, \lambda'_1; p'_2\lambda'_2|$$

$$\times T(: \hat{\bar{\psi}}_1(x_1)\gamma_\mu\hat{\psi}_1(x_1) :: \hat{\bar{\psi}}_2(x_2)\gamma_\nu\hat{\psi}(x_2) : \hat{A}_\mu(x_1)\hat{A}_\nu(x_2)) \qquad (10.2.2)$$

$$\times |p_1, \lambda_1; p_2, \lambda_2\rangle.$$

Because the states $|i\rangle$, $|f\rangle$ do not contain photons, they act like the vacuum for $\hat{A}_\mu$, $\hat{A}_\nu$. We can thus replace

$$T(\hat{A}_\mu(x_1)\hat{A}_\nu(x_2)) \to \langle 0|T\hat{A}_\mu(x_1)\hat{A}_\nu(x_2)|0\rangle$$

$$= D_{\mu\nu}(x_1 - x_2),$$

where $D_{\mu\nu}$ is the photon propagator, (9.7.2).

Let us write $|i\rangle$, $|f\rangle$ as

$$|i\rangle = |p_1, \lambda_1; p_2, \lambda_2\rangle = \hat{a}_1^+(p_1, \lambda_1)\hat{a}_2^+(p_2, \lambda_2)|0\rangle,$$

$$\langle f| = \langle p'_1, \lambda'_1; p'_2, \lambda'_2| = \langle 0|\hat{a}_2(p'_2, \lambda'_2)\hat{a}_1(p'_1, \lambda'_1),$$

where the $\hat{a}$ represent $\hat{b}$ or $\hat{d}$ according to whether we have two particles, or two antiparticles[3]. Substituting (9.5.2), and after some manipulations, we get the very transparent expression

$$\langle f|\hat{S}|i\rangle = i^2 e_1 e_2 \sum_{\mu\nu} g_{\mu\mu}g_{\nu\nu}$$

$$\times \int d^4x_1 d^4x_2 \overline{\Psi}^{(p'_1\lambda'_1)}(x_1)\gamma_\mu\Psi^{(p_1\lambda_1)}(x_1)D_{\mu\nu}(x_1 - x_2) \qquad (10.2.3)$$

$$\times \overline{\Psi}^{(p'_2\lambda'_2)}(x_2)\gamma_\nu\Psi^{(p_2\lambda_2)}(x_2),$$

where the $\Psi^{(p\lambda)}$ are the relativistic wave functions corresponding to momentum $\mathbf{p}$, and third component of spin $\lambda$:

---

[3] The *order* in which we place the creation–annihilation operators in states $\langle f|$, $|i\rangle$ is in principle arbitrary. We *fix* it, however, to get a precise $(+)$ sign for the scalar product,

$$\langle f|i\rangle = +2p_{10}2p_{20}\delta_{\lambda_1\lambda'_1}\delta_{\lambda_2\lambda'_2}\delta(\mathbf{p}_1 - \mathbf{p}'_1)\delta(\mathbf{p}_2 - \mathbf{p}'_2).$$

$$\Psi^{(p\lambda)}(x) = \frac{1}{(2\pi)^{3/2}} e^{-ip\cdot x} u(p,\lambda),$$

$$\overline{\Psi}^{(p\lambda)}(x) = \frac{1}{(2\pi)^{3/2}} e^{ip\cdot x} \overline{u}(p,\lambda).$$

$$(10.2.4)$$

(We shall not delve any further into the derivation of these formulas because a detailed calculation of a process similar to this will be presented in Sect. 10.4.1). We can represent the process by diagrams, similar to those of Sects. 8.3, 8.4, known as *Feynman diagrams*, drawn in Fig. 10.2.1A: particles 1 and 2 (continuous lines) interact by emitting a photon in $x_2 = (t_2, \mathbf{r}_2)$ that propagates until it is absorbed at $x_1 = (t_1, \mathbf{r}_1)$ (or vice versa). The photon, which is called *virtual*[4] because it does not appear in either the initial or the final state, is represented by a wavy line.

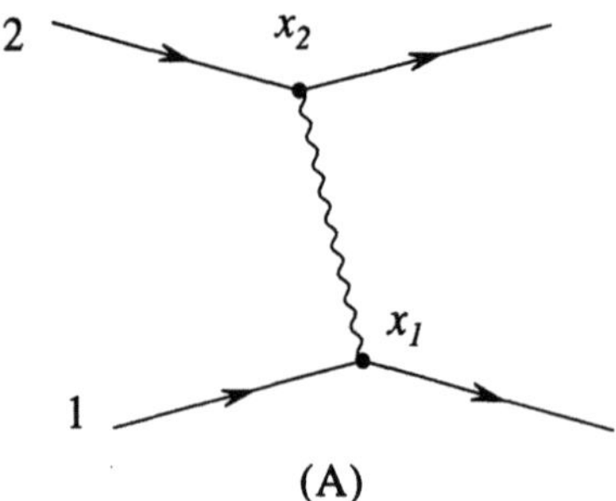

(A)

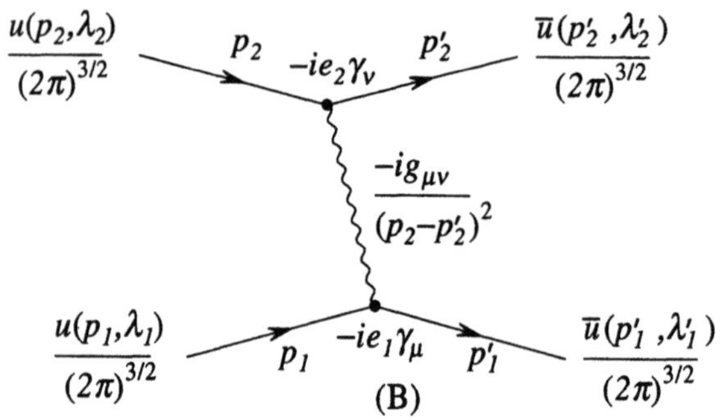

(B)

Fig. **10.2.1.** Interaction between two charged particles in the Born approximation. (**A**) Position space. (**B**) Momentum space.

The $d^4x_1 d^4x_2$ integrals may be performed by introducing the representation (9.7.5) for the photon propagator. The result, which we express *in the Heaviside system of units*, is

---

[4] This will be general; all particles that do not appear in initial or final states will be called *virtual*.

$$\langle f|\hat{S}|i\rangle_{\mathrm{Born}} = (2\pi)^4 \delta_4(p_f - p_i)(-1)^2 \sum_{\mu\nu} \frac{\bar{u}(p_1',\lambda_1')}{(2\pi)^{3/2}} ie_1\gamma_\mu \frac{u(p_1,\lambda_1)}{(2\pi)^{3/2}}$$

$$\times \frac{-ig_{\mu\nu}}{(p_2 - p_2')^2} \frac{\bar{u}(p_2',\lambda_2')}{(2\pi)^{3/2}} ie_2\gamma_\nu \frac{u(p_2,\lambda_2)}{(2\pi)^{3/2}}; \tag{10.2.5}$$

$$\delta_4(p_f - p_i) \equiv \delta(p_{10} + p_{20} - p_{10}' - p_{20}')\delta(\mathbf{p}_1 + \mathbf{p}_2 - \mathbf{p}_1' - \mathbf{p}_2').$$

If we draw a diagram like that of Fig. 10.2.1 (B), we can put every term in (10.2.5), except the $(2\pi)^4\delta_4(p_f - p_i)$, associated with overall energy–momentum conservation, in correspondence with an element of the diagram, as shown in Fig. 10.2.1 (B). We will follow this development in next chapter; here we will return to (10.2.3).

We now want to go to the NR limit (eventually with relativistic corrections). To do so, it is convenient to introduce[5] the mean time in which the process takes place, $t = (t_1 + t_2)/2$, and the time retardation, $\tau = t_1 - t_2$, related to the finite speed with which the interaction, carried by the virtual photon, propagates. Now substituting (10.2.4) into (10.2.3) and replacing

$$d^3r_1 d^3r_2 \to d^3r d^3R, \ \mathbf{r} \equiv \mathbf{r}_1 - \mathbf{r}_2, \ \mathbf{R} \equiv \frac{1}{2}(\mathbf{r}_1 + \mathbf{r}_2),$$

we obtain the expression

$$\langle f|\hat{S}|i\rangle_{\mathrm{Born}} = \frac{i^2 e_1 e_2}{(2\pi)^6}$$

$$\times \sum_{\mu\nu} g_{\mu\mu} g_{\nu\nu} \bar{u}(p_1',\lambda_1')\gamma_\mu u(p_1,\lambda_1)\mathcal{M}_{\mu\nu}\bar{u}(p_2',\lambda_2')\gamma_\nu u(p_2,\lambda_2), \tag{10.2.6a}$$

where

$$\mathcal{M}_{\mu\nu} = \int d^3r \int d^3R\, e^{i\mathbf{R}(\mathbf{p}_1 + \mathbf{p}_2 - \mathbf{p}_1' - \mathbf{p}_2')} e^{-i\mathbf{r}(\mathbf{p}_2 - \mathbf{p}_2' - (\mathbf{p}_1 - \mathbf{p}_1'))/2}$$

$$\times \int dt \int d\tau e^{-it(p_{10}+p_{20}-p_{10}'-p_{20}')} e^{-i\tau(p_{10}-p_{10}'-(p_{20}-p_{20}'))/2} D_{\mu\nu}(\mathbf{r},t) \tag{10.2.6b}$$

$$= (2\pi)^4 \delta_4(p_f - p_i) \int d^3r\, e^{i\mathbf{r}(\mathbf{p}_1 - \mathbf{p}_1')} U_{\mu\nu},$$

and finally,

$$U_{\mu\nu} = \int d\tau e^{-i\tau(p_{10}-p_{10}'-(p_{20}-p_{20}'))/2} D_{\mu\nu}(\mathbf{r},\tau). \tag{10.2.7a}$$

On comparing this with (10.1.3), we obtain the expression for $F_B$, in the Born approximation but still *without* nonrelativistic approximations:

---

[5] Note that we will use both notations $(x_0, x_i)$ and $(t, \mathbf{r})$ interchangeably. So $t_1 = x_{10}$, $t_2 = x_{20}$ $(c = 1)$.

$$F_B(i \to f) = \frac{ie_1 e_2}{(2\pi)^2} \sum_{\mu\nu} \overline{u}(p_1', \lambda_1')\gamma_\mu u(p_1, \lambda_1) g_{\mu\mu} g_{\nu\nu}$$

$$\times \left[ \int d^3 r e^{i\mathbf{r}(\mathbf{P_1} - \mathbf{P_1'})} U_{\mu\nu} \right] \overline{u}(p_2', \lambda_2')\gamma_\nu u(p_2, \lambda_2). \tag{10.2.7b}$$

Note that, because of energy–momentum conservation,

$$\mathbf{p}_1 - \mathbf{p}_1' = -(\mathbf{p}_2 - \mathbf{p}_2'), \quad p_{10} - p_{10}' = -(p_{20} - p_{20}'). \tag{10.2.8}$$

Expression (10.2.7) looks a lot like (10.1.7), with Pauli spinors replaced by Dirac spinors, as would be expected of a relativistic expression. However, they differ in a fundamental feature: as (10.2.7a) shows, $U_{\mu\nu}$ depends not only on $\mathbf{r}$, but on the energies $p_{10}, p_{10}', \ldots$ as well, and hence, and through them, on the momenta of the particles. To be able to link this up with the potential formalism we will have to consider the NR limit, as stated at the end of Sect. 10.1.

### 10.2.2 Nonrelativistic Limit

We then expand the expression (10.2.7) in powers of $1/c$. Then introducing $c$ explicitly (but keeping $\hbar = 1$), and *returning to the Gauss system of units*, we see that the term in square brackets in (10.2.7b) is

$$\int d^3 r e^{i\mathbf{r}(\mathbf{P_1} - \mathbf{P_1'})} U_{\mu\nu},$$

$$U_{\mu\nu} = c \int_{-\infty}^{+\infty} d\tau e^{i\tau(E_1(p_1) - E_1(p_1'))} D_{\mu\nu}(\mathbf{r}, c\tau), \tag{10.2.9a}$$

$$D_{\mu\nu}(\mathbf{r}, c\tau) = 4\pi \frac{-ig_{\mu\nu}}{(2\pi)^4} \int d^4 k \frac{e^{i\mathbf{kr} - ick_0 r}}{k_0^2 - \mathbf{k}^2 + i0},$$

or, if we integrate the space part $d^3 k$ of $d^4 k$,

$$D_{\mu\nu}(\mathbf{r}, c\tau) = \frac{4\pi}{r} \frac{ig_{\mu\nu}}{2(2\pi)^2} \int_{-\infty}^{+\infty} dk_0 e^{ick_0\tau + ir|k_0|};$$

substituting into $U_{\mu\nu}$, we get

$$U_{\mu\nu} = \frac{4\pi}{r} \left\{ \frac{+ig_{\mu\nu}}{2(2\pi)^2} c \int dk_0 e^{ir|k_0|} \right.$$

$$\left. \times \int d\tau e^{-i\tau(E_1(p_1) - E_1(p_1')) - ick_0\tau} \right\}$$

$$= \frac{4\pi}{r} \left\{ \frac{ig_{\mu\nu}}{4\pi} c \int dk_0 e^{ir|k_0|} \delta\Big(ck_0 + (E_1(p_1) - E_1(p_1'))\Big) \right\}$$

$$= \frac{1}{r} ig_{\mu\nu} \exp \frac{ir|E_1(p_1') - E_1(p_1)|}{c}.$$

Finally,

$$U_{\mu\nu} = \frac{ig_{\mu\nu}}{r}\exp\frac{ir|E_1(p_1') - E_1(p_1)|}{c},$$

(10.2.9b)

$$E_a(p) = \sqrt{m_a^2 c^4 + \mathbf{p}^2 c^2}, \ a = 1, 2.$$

Expression (10.2.9b) is still exact, in the sense that we have not made nonrelativistic approximations. Although we still have a way to go, (10.2.9b) already shows how the Coulomb potential appears in the NR limit: here we can approximate the exponential in (10.2.9b) by unity, so that $U_{\mu\nu} \simeq ig_{\mu\nu}(1/r)$: *the potential is connected to the three-dimensional Fourier transform of the propagator*, in the strict nonrelativistic limit.

For a detailed calculation, which we will carry out including terms of order $1/c^2$, we expand (10.2.9b):

$$U_{\mu\nu} \simeq \frac{ig_{\mu\nu}}{r}\left\{1 + ir\frac{|E_1(p_1) - E_1(p_1')|}{c}\right.$$

$$\left. + \frac{r^2}{2!c^2}(E_1(p_1) - E_1(p_1'))(E_2(p_2) - E_2(p_2')) + \ldots\right\},$$

where we have used (10.2.8) to write the quadratic term in a form symmetric in particles 1, 2. Here and in the rest of this section, the symbol "$\simeq$" will mean "up to corrections of order $1/c^4$".

When this expression for $U_{\mu\nu}$ is substituted into (10.2.7b) it is easy to check that the contribution of the term of $U_{\mu\nu}$ linear in $1/c$ vanishes. We then get

$$F_B(i \to f) \simeq \frac{-e_1 e_2}{(2\pi)^2}\sum_\mu g_{\mu\mu}\bar{u}(p_1', \lambda')\gamma_\mu u(p_1, \lambda_1)\int d^3 r\, e^{i\mathbf{r}(\mathbf{p_1}-\mathbf{p_1'})}$$

$$\times\left\{\frac{1}{r} + \frac{r}{2c^2}(E_1(p_1) - E_1(p_1'))(E_2(p_2) - E_2(p_2'))\right\}$$

(10.2.10)

$$\times\bar{u}(p_2', \lambda_2')\gamma_\mu u(p_2, \lambda_2).$$

Before we continue it is necessary to choose an explicit realization for the $\gamma$ matrices. We of course take Pauli's, so that

$$\gamma_0\gamma_i = \alpha_i, \ \boldsymbol{\alpha} = \begin{pmatrix} 0 & \boldsymbol{\sigma} \\ \boldsymbol{\sigma} & 0 \end{pmatrix},$$

and for the spinors (cf. (3.5.11)) one has

$$u(p, \lambda) \equiv \begin{pmatrix} u_b \\ u_s \end{pmatrix} = \sqrt{mc^2 + E(p)}\begin{pmatrix} \chi(\lambda) \\ \dfrac{c}{mc^2 + E(p)}\mathbf{p}\boldsymbol{\sigma}\chi(\lambda) \end{pmatrix}; \quad (10.2.11)$$

the "small" $u_s$ components are of order $1/c$ with respect to the "large" $u_b$ ones. Because of this, it follows that

$$\bar{u}\gamma u = u^+ \gamma_0 \gamma u = u^+ \boldsymbol{\alpha} u = u_b^+ \boldsymbol{\sigma} u_s + u_s^+ \boldsymbol{\sigma} u_b$$

is of order $1/c$ with respect to

$$\bar{u}\gamma_0 u = u^+ u = u_b^+ u_b + u_s^+ u_s.$$

To the precision to which we are working, this tells us that only the term $\bar{u}\gamma_0 u \ldots \bar{u}\gamma_0 u$ among the $\bar{u}\gamma_\mu u \ldots \bar{u}\gamma_\mu u$ in (10.2.10) will receive a non-negligible contribution from the term in $1/c^2$ in the curly bracket there. Hence,

$$F_B(i \to f) \simeq \frac{-e_1 e_2}{(2\pi)^2} \left\{ u^+(p_1', \lambda_1') u(p_1, \lambda_1) \int d^3 r e^{i\mathbf{r}(\mathbf{p_1} - \mathbf{p_1'})} \right.$$

$$\times \left[ \frac{1}{r} + \frac{r}{2c^2} (E_1(p_1) - E_1(p_1'))(E_2(p_2) - E_2(p_2')) \right]$$

$$\times u^+(p_2', \lambda_2') u(p_2, \lambda_2)$$

$$\left. + u^+(p_1', \lambda_1') \boldsymbol{\alpha} u(p_1, \lambda_1) u^+(p_2', \lambda_2') \boldsymbol{\alpha} u(p_2, \lambda_2) \int \frac{d^3 r}{r} e^{i\mathbf{r}(\mathbf{p_1} - \mathbf{p_1'})} \right\}. \tag{10.2.12}$$

The Dirac equation for spinors $u, u^+$ implies that

$$E(p) u(p, \lambda) = (c\boldsymbol{\alpha}\mathbf{p} + mc^2 \beta) u(p, \lambda),$$

$$u^+(p, \lambda) E(p) = u^+(p, \lambda)(c\boldsymbol{\alpha}\mathbf{p} + mc^2 \beta).$$

Using this we see that the terms with $mc^2 \beta$ cancel one against the other and we are left with

$$F_B(i \to f) \simeq \frac{-e_1 e_2}{(2\pi)^2} \int d^3 r e^{i\mathbf{r}(\mathbf{p_1} - \mathbf{p_1'})}$$

$$\times \left\{ u^+(p_1', \lambda_1') u(p_1, \lambda_1) \frac{1}{r} u^+(p_2', \lambda_2') u(p_2, \lambda_2) \right.$$

$$+ u^+(p_1', \lambda_1')(c\boldsymbol{\alpha}\mathbf{p_1} - c\boldsymbol{\alpha}\mathbf{p_1'}) u(p_1, \lambda_1) \frac{r}{2c^2} \tag{10.2.13}$$

$$\times u^+(p_2', \lambda_2')(c\boldsymbol{\alpha}\mathbf{p_2} - c\boldsymbol{\alpha}\mathbf{p_2'}) u(p_2, \lambda_2)$$

$$\left. + u^+(p_1', \lambda_1') \boldsymbol{\alpha} u(p_1, \lambda_1) \frac{1}{r} u^+(p_2', \lambda_2') \boldsymbol{\alpha} u(p_2, \lambda_2) \right\}.$$

From (10.2.13) we may obtain $T_{NR}$ with the help of (10.1.7):

$$T_{NR}(i \to f) \simeq \frac{-e_1 e_2}{(2\pi)^2} \int d^3 r e^{i\mathbf{r}(\mathbf{p}_1 - \mathbf{p}_1')}$$

$$\times \left\{ u_{NR}^+(p_1', \lambda_1') u_{NR}(p_1, \lambda_1) \right.$$

$$+\frac{1}{r} u_{NR}^+(p_2', \lambda_2') u_{NR}(p_2, \lambda_2)$$

$$+u_{NR}^+(p_1', \lambda_1')(\mathbf{p}_1 - \mathbf{p}_1')\boldsymbol{\alpha} u_{NR}(p_1, \lambda_1)$$

$$\times \frac{r}{2} u_{NR}^+(p_2', \lambda_2')(\mathbf{p}_2 - \mathbf{p}_2')\boldsymbol{\alpha} u_{NR}(p_2, \lambda_2)$$

$$\left. +u_{NR}^+(p_1', \lambda_1')\boldsymbol{\alpha} u_{NR}(p_1, \lambda_1)\frac{1}{r} u_{NR}^+(p_2', \lambda_2')\boldsymbol{\alpha} u_{NR}(p_2, \lambda_2) \right\}$$

with $u_{NR} \equiv u/\sqrt{2E}$, or, to the order to which we are working,

$$u_{NR}(p, \lambda) \simeq \begin{pmatrix} (1 - \mathbf{p}^2/8m^2c^2)\chi(\lambda) \\ (\mathbf{p}\boldsymbol{\sigma}/2mc)\chi(\lambda) \end{pmatrix}. \tag{10.2.14}$$

Substituting this, we find that

$$u_{NR}^+(p', \lambda')\boldsymbol{\alpha} u_{NR}(p, \lambda)$$

$$= \frac{1}{2mc} \left\{ (\mathbf{p} + \mathbf{p}')\chi^+(\lambda')\chi(\lambda) + i\chi^+(\lambda')(\mathbf{p} - \mathbf{p}') \times \boldsymbol{\sigma}\chi(\lambda) \right\}$$

$$+O(1/c^3),$$

$$\tag{10.2.15}$$

$$u_{NR}^+(p', \lambda')u_{NR}(p, \lambda) = \chi^+(\lambda')\chi(\lambda)$$

$$+\frac{1}{4m^2c^2} \left\{ -\frac{\mathbf{p}^2 + \mathbf{p}'^2}{2}\chi^+(\lambda')\chi(\lambda) + \chi^+(\lambda')(\boldsymbol{\sigma}\mathbf{p}')(\boldsymbol{\sigma}\mathbf{p})\chi(\lambda) \right\}$$

$$+0(1/c^4),$$

and therefore

$$T_{NR}^B(i \to f) = \frac{-e_1 e_2}{(2\pi)^2} \int d^3 r e^{i\mathbf{r}(\mathbf{p}_1 - \mathbf{p}_1')}$$

$$\tag{10.2.16a}$$

$$\times \left\{ \chi^+(\lambda_1')\chi(\lambda_1)\frac{1}{r}\chi^+(\lambda_2')\chi(\lambda_2) + \frac{1}{4c^2}U_1 \right\},$$

$$U_1 = \frac{1}{rm_2^2}\chi^+(\lambda_1')\chi(\lambda_1)\left[-\frac{\mathbf{p}_2^2+\mathbf{p}_2'^2}{2}\chi^+(\lambda_2')\chi(\lambda_2)\right.$$

$$\left.+\chi^+(\lambda_2')(\boldsymbol{\sigma}\mathbf{p}_2')(\boldsymbol{\sigma}\mathbf{p}_2)\chi(\lambda_2)\right]$$

$$+\frac{1}{rm_1^2}\left[-\frac{\mathbf{p}_1^2+\mathbf{p}_1'^2}{2}\chi^+(\lambda_1')\chi(\lambda_1)\right.$$

$$\left.+\chi^+(\lambda_1')(\boldsymbol{\sigma}\mathbf{p}_1')(\boldsymbol{\sigma}\mathbf{p}_1)\chi(\lambda_1)\right]\chi^+(\lambda_2')\chi(\lambda_2) \qquad (10.2.16\text{b})$$

$$+\frac{r}{2m_1m_2}(\mathbf{p}_1^2-\mathbf{p}_1'^2)\chi^+(\lambda_1')\chi(\lambda_1)(\mathbf{p}_2^2-\mathbf{p}_2'^2)\chi^+(\lambda_2')\chi(\lambda_2)$$

$$+\frac{1}{rm_1m_2}\left[(\mathbf{p}_1+\mathbf{p}_1')\chi^+(\lambda_1')\chi(\lambda_1)+i\chi^+(\lambda_1')(\mathbf{p}_1-\mathbf{p}_1')\times\boldsymbol{\sigma}\chi(\lambda_1)\right]$$

$$\cdot\left[(\mathbf{p}_2+\mathbf{p}_2')\chi^+(\lambda_2')\chi(\lambda_2)+i\chi^+(\lambda_2')(\mathbf{p}_2-\mathbf{p}_2')\times\boldsymbol{\sigma}\chi(\lambda_2)\right].$$

In the strictly nonrelativistic limit we can neglect the term $(1/rc^2)U_1$ in (10.2.16). Then comparing the rest with (10.1.7c) we see that the equivalent potential $V^{(0)}$ is

$$V^{(0)} = e_1e_2/r,$$

as was to be desired: up to relativistic corrections, two charged particles interact via a Coulomb potential.

### 10.2.3 Relativistic Corrections. The Breit Term

To evaluate the relativistic corrections it is convenient to return momentarily to $\langle f|\hat{S}|i\rangle$. From (10.2.16) it follows that we can write

$$_{NR}\langle f|\hat{S}|i\rangle_{NR}|_{\text{Born}} = \frac{-2\pi i}{(2\pi)^6}\delta(E_f - E_i)$$

$$\times \int d^3r_1 d^3r_2 e^{-i\mathbf{p}_1'\mathbf{r}_1-i\mathbf{p}_2'\mathbf{r}_2}W e^{i\mathbf{p}_1\mathbf{r}_1+i\mathbf{p}_2\mathbf{r}_2} \qquad (10.2.17\text{a})$$

with

$$W = e_1e_2\left\{\chi^+(\lambda_1')\chi(\lambda_1)\frac{1}{r}\chi^+(\lambda_2')\chi(\lambda_2) + \frac{1}{4c^2}U_1\right\}. \qquad (10.2.17\text{b})$$

To check this it is sufficient to integrate the centre of mass coordinate and use the definition of $T$ in terms of $\hat{S}$.

The quantity $W$ in (10.2.17b), with $U_1$ given in (10.2.16b), contains the momenta $\mathbf{p}_1,\ldots,\mathbf{p}_2'$. We can eliminate them by using the following trick. Inside the integral (10.2.17a) we can replace

$$\mathbf{p}_1 \to -i\boldsymbol{\nabla}_1, \ \mathbf{p}_2 \to -i\boldsymbol{\nabla}_2, \qquad (10.2.18\text{a})$$

with the gradients located at the extreme right in $W$; and, also,

$$\mathbf{p}_1' \to -i\boldsymbol{\nabla}_1, \quad \mathbf{p}_2' \to -i\boldsymbol{\nabla}_2, \tag{10.2.18b}$$

but now the gradients are to be located at the extreme left in $W$. This can be easily verified by partial integration, or by noticing that we can write

$$\int d^3r_1 d^3r_2 e^{-i\mathbf{p}_1'\mathbf{r}_1 - i\mathbf{p}_2'\mathbf{r}_2} W e^{i\mathbf{p}_1\mathbf{r}_1 + i\mathbf{p}_2\mathbf{r}_2} \sim \langle \varphi_{p_1'} \varphi_{p_2'} | \hat{W} | \varphi_{p_1'} \varphi_{p_2} \rangle,$$

where the $|\varphi_{\mathbf{p}}\rangle$ represent states with momentum $\mathbf{p}$,

$$\mathbf{p}|\varphi_{\mathbf{p}}\rangle = \hat{\mathbf{P}}|\varphi_{\mathbf{p}}\rangle \to -i\boldsymbol{\nabla}\varphi_{\mathbf{p}}(\mathbf{r}),$$

so if $\hat{W}$ contains a term like $\mathbf{p}_2'\hat{O}\mathbf{p}_1$, for example, we can replace it by $\hat{\mathbf{P}}_2'\hat{O}\hat{\mathbf{P}}_1$.

Carrying out the substitutions (10.2.18), we obtain, for the first term in the formula (10.2.16b) for $U_1$, the expression

$$\frac{1}{rm_2^2}\chi^+(\lambda_1')\chi(\lambda_1)\frac{-(\mathbf{p}_2^2 + \mathbf{p}_2'^2)}{2}\chi^+(\lambda_2')\chi(\lambda_2) \to$$

$$\to \frac{-1}{2m_2^2}\chi^+(\lambda_1')\chi(\lambda_1)\chi^+(\lambda_2')\chi(\lambda_2)\left\{\frac{1}{r}(i\boldsymbol{\nabla}_2)^2 + (i\boldsymbol{\nabla}_2)^2\frac{1}{r}\right\}$$

$$= \chi^+(\lambda_1')\chi(\lambda_1)\chi^+(\lambda_2')\chi(\lambda_2)$$

$$\times \left\{\frac{1}{m_2^2 r}\triangle_2 + \frac{1}{m_2^2 r^3}\mathbf{r}\,\boldsymbol{\nabla}_2 - \frac{2\pi}{m_2^2}\delta(\mathbf{r})\right\}, \quad \mathbf{r} = \mathbf{r}_1 - \mathbf{r}_2,$$

where we have used the identity

$$\triangle_2\frac{1}{r} = \frac{1}{r}\triangle_2 + \frac{2}{r^3}\mathbf{r}\,\boldsymbol{\nabla}_2 - 4\pi\delta(\mathbf{r}).$$

The rest of the terms in (10.2.16b) may be evaluated in the same manner. A computational trick useful for treating singularities like the $\delta(\mathbf{r})$ we have just found is the following. We perform the straightforward calculation for $r \neq 0$; no singularities will be encountered. The result is then fixed up to terms of the form[6] (constant) $\times\delta(\mathbf{r})$. The coefficient of this $\delta$ can be obtained by integrating with a function $f(r)$ that is spherically symmetric and zero except in a small neighbourhood of $r = 0$. The result will be the (constant) $\times f(0)$ which allows identification of the coefficient of $\delta(\mathbf{r})$. The final result is due to Breit, for the evaluation of the retardation effect, and Bethe and Fermi, for the explicit evaluation of the hyperfine interactions containing the operator $S_{12}$; see below. We then have, including the first relativistic corrections to the kinetic part of the Hamiltonian,

$$\hat{V}_{\text{eff}} = e_1 e_2 \hat{V}_{12} - \frac{\hbar^4}{8m_1^3 c^2}\triangle_1^2 - \frac{\hbar^4}{8m_2^3 c^2}\triangle_2^2, \tag{10.2.19a}$$

---

[6] For our specific case. In general, other terms involving derivatives of the $\delta$ function will also be present. See Sect. 10.6 for an example.

$$\hat{V}_{12} = \frac{1}{r} + \frac{\hbar^2}{2c^2}\left\{ \frac{i}{2r^3}\left( \frac{1}{m_1^2}\sigma_1(\mathbf{r}\times\boldsymbol{\nabla}_1) - \frac{1}{m_2^2}\sigma_2(\mathbf{r}\times\boldsymbol{\nabla}_2)\right) \right.$$

$$-\pi\left(\frac{1}{m_1^2} + \frac{1}{m_2^2}\right)\delta(\mathbf{r})$$

$$+\frac{1}{m_1 m_2 r}\left( \boldsymbol{\nabla}_1\boldsymbol{\nabla}_2 + \sum_{ij}\frac{r_i r_j}{r^2}\nabla_{1i}\nabla_{2j}\right)$$

$$-\frac{i}{m_1 m_2 r^3}(\sigma_1(\mathbf{r}\times\boldsymbol{\nabla}_2) - \sigma_2(\mathbf{r}\times\boldsymbol{\nabla}_1))$$

$$-\frac{1}{2m_1 m_2 r^3}S_{12}$$

$$\left. -\frac{4\pi}{3m_1 m_2}\sigma_1\sigma_2\delta(\mathbf{r}) \right\} + 0(1/c^4),$$

$$(10.2.19b)$$

with

$$S_{12} = \frac{3}{r^2}(\mathbf{r}\sigma_1)(\mathbf{r}\sigma_2) - \sigma_1\sigma_2, \quad (\boldsymbol{\nabla}_a)_j = \frac{\partial}{\partial r_{aj}}, \quad \Delta_a = \sum_j \frac{\partial^2}{\partial r_{aj}^2},$$

and where $\sigma_a$ acts on spinor $\chi(\lambda_a)$.

Before we continue, a few more words have to be said about the singularities in (10.2.19b). $\delta$-function singularities have already been dealt with. The terms above that are singular, such as those containing $1/r^3$, vanish when averaged over angular directions. This averaging *is* justified because, recalling how (10.2.19) were obtained, one should consider everything but the $\delta$ as defined for $r \neq 0$ and nonsingular.

The corresponding Hamiltonian may then be written as

$$\hat{H} = \frac{-\hbar^2}{2m_1}\Delta_1 + \frac{-\hbar^2}{2m_2}\Delta_2 + \hat{V}_{\text{eff}}. \qquad (10.2.19c)$$

If $m_1 = m \ll m_2$, we can neglect $m$ as compared to $m_2$, and the Hamiltonian simplifies to

$$\hat{H} \simeq \frac{-\hbar^2}{2m}\Delta - \frac{\hbar^4}{8m^3 c^2}\Delta^2$$

$$+ e_1 e_2\left\{ \frac{1}{r} + \frac{i\hbar^2}{4m^2 c^2 r^3}\sigma(\mathbf{r}\times\boldsymbol{\nabla}) - \frac{\pi\hbar^2}{2m^2 c^2}\delta(\mathbf{r})\right\},$$

$$(10.2.20)$$

identical to what we got for a particle in a potential (produced by the very heavy particle) using the Foldy–Wouthuysen method in (4.5.11).

It is possible to generate effective potentials that produce the same effect as the field-theoretic interaction, when they are substituted in equations of the Dirac type:

$$(-\hbar c\boldsymbol{\alpha}_1\boldsymbol{\nabla}_1 + m_1c^2\beta_1 - i\hbar c\boldsymbol{\alpha}_2\boldsymbol{\nabla}_2 + m_2c^2\beta_2 + V_B)\psi(\mathbf{r}_1,\mathbf{r}_2)$$
$$\equiv \hat{H}_B\psi(\mathbf{r}_1,\mathbf{r}_2); \tag{10.2.21}$$

in the case we are considering, and to order $1/c^2$,

$$V_B = e_1e_2\left\{\frac{1}{r} - \frac{1}{2r}\left(\boldsymbol{\alpha}_1\boldsymbol{\alpha}_2 + \frac{(\mathbf{r}\boldsymbol{\alpha}_1)(\mathbf{r}\boldsymbol{\alpha}_2)}{r^2}\right)\right\} + 0(1/c^4) \tag{10.2.22}$$

(called the *Breit potential*).

**Exercise.** (A) Using the Foldy–Wouthuysen method show that (10.2.22) is equivalent to (10.2.20) up to order $1/c^4$. (B) Identify the Breit correction to the Coulomb potential in (10.2.22) as being due to the retardation of the interaction, that is to say, originated by the term $1/c^2$ in the expansion of $U_{\mu\nu}$ (see, for example, (10.2.10)) ∎

# 10.3 Hydrogenlike Atoms: Hyperfine Structure. System with Two Electrons: the Helium Atom

### 10.3.1 Hydrogenlike Atoms

The results of the previous section are valid for interactions of elementary particles. We can extend them to the case of an electron interacting with a nucleus, which we take to be of spin 1/2, by introducing a phenomenological interaction between the nucleus and the electromagnetic field, of a kind suggested by (3.4.11). We let $\hat{\Psi}$ be the field of the nucleus, $Ze$ its charge and $m_N$ its mass; then we write

$$\hat{\mathcal{L}}_{\text{int, nucleus}} =: \hat{\bar{\Psi}}(x)\left\{Ze\,\hat{A}\,(x) - \frac{Ze\delta}{4m_N}\sigma\cdot\hat{F}(x)\right\}\hat{\Psi}(x) :, \tag{10.3.1}$$

where $\delta$ is given in terms of the magnetic moment of the nucleus by

$$\mu_N = \frac{Ze\hbar}{2mc}(1+\delta).$$

In this situation it is convenient to expand not in powers of $1/c$, but in terms of $m_e/m_N$. The first term will give a purely static potential. Indeed, consider (10.2.9) taking particle 2 to be the nucleus. The argument of the exponential in (10.2.9b) is $(E_2(\mathbf{p}_2) - E_2(\mathbf{p}'_2))/c$, which is of order $1/m_N$:

$$\frac{E_2(\mathbf{p}_2) - E_2(\mathbf{p}'_2)}{c} \simeq \frac{\mathbf{p}_2^2 - \mathbf{p}'^2_2}{2m_Nc}.$$

In hydrogenlike atoms we know that the average value of the momentum is $m_ev_e$, with $v_e$ the average velocity, $v_e^2 = (Z\alpha)^2c^2$; and the average value of $r$ is $a_B = 1/(m_eZ\alpha c)$. Then,

$$e^{-ir(E_2(\mathbf{p}_2)-E_2(\mathbf{p}_2'))/c} = 1 + 0\left(\frac{m_e^2 v_e^2 a_B}{m_N c}\right)$$

$$= 1 + 0\left(\frac{m_e}{m_N} Z\alpha\right).$$

It follows that, with this precision $O(m_e Z\alpha/m_N)$, we can replace the field-theoretic description by one in terms of a purely Coulombic potential. (Actually this is not strictly true. So-called *radiative corrections*, independent of $m_N$ or $Z$, but of higher order in $\alpha$, exist, owing to the virtual interaction of the electron with the radiation field. They can be found in field-theoretic texts such as Akhiezer and Berestetskii (1963).)

A more precise evaluation is obtained by keeping the first-order terms in $m_e/m_N$, as given in (10.2.19), to which one has to add the extra piece due to the interaction,

$$(-Z\delta/4m_N)\sigma \cdot \hat{F},$$

in (10.3.1). Among these corrections a particularly important one is that producing the *hyperfine structure*, coming from the last term in (10.2.19). For $Z = 1$, the corresponding potential is

$$\hat{V}_H = \frac{2\pi\hbar^3}{3m_e m_N c}(1+\delta)\alpha\boldsymbol{\sigma}_e\boldsymbol{\sigma}_N\delta(\mathbf{r}),\ \alpha \simeq 1/137. \tag{10.3.2}$$

This shifts the levels in the hydrogen atoms, shifts depending on the total spin of the electron–proton system; note that, if $s$ is this spin ($s = 0, 1$),

$$\boldsymbol{\sigma}_e\boldsymbol{\sigma}_N \to 2s(s+1) - 3.$$

### 10.3.2 System With Two Electrons. The Helium Atom

The computation of the scattering amplitude for two electrons is practically identical to that performed in Sect. 10.2. We list the differences.

(i)  Because the particles are now identical, a properly normalized state would now be

$$|p_1, \lambda_1; p_2, \lambda_2\rangle = \frac{1}{\sqrt{2!}}\hat{b}^+(p_1, \lambda_1)\hat{b}^+(p_2, \lambda_2)|0\rangle,$$

and the same for $|f\rangle$. It is, however, more convenient (and this is what we will do here) to use the convention

$$|p_1, \lambda_1; p_2, \lambda_2\rangle = \hat{b}^+(p_1, \lambda_1)\hat{b}^+(p_2, \lambda_2)|0\rangle,$$

*for identical particles also.* We thus have

$$\langle p_1', \lambda_1'; p_2', \lambda_2' | p_1, \lambda_1; p_2, \lambda_2 \rangle$$

$$= 2p_{10}\delta(\mathbf{p}_1 - \mathbf{p}_1')\delta_{\lambda_1\lambda_1'} \, 2p_{20}\delta(\mathbf{p}_2 - \mathbf{p}_2')\delta_{\lambda_2\lambda_2'}$$

$$\pm 2p_{10}\delta(\mathbf{p}_1 - \mathbf{p}_2')\delta_{\lambda_1\lambda_2'} \, 2p_{20}\delta(\mathbf{p}_2 - \mathbf{p}_1')\delta_{\lambda_2\lambda_1'} \, ,$$

$$|p_1, \lambda_1; p_2, \lambda_2\rangle \equiv \hat{c}^+(p_1, \lambda_1)\hat{c}^+(p_2, \lambda_2)|0\rangle,$$

$$\langle p_1', \lambda_1'; p_2', \lambda_2'| \equiv \langle 0|\hat{c}(p_2', \lambda_2')\hat{c}(p_1', \lambda_1'); \hat{c} = \hat{a}, \hat{b}, \hat{d},$$

$(+/-)$ for bosons/fermions. This implies that, for completeness sums, we will have to divide by $n!$, where $n$ is the number of identical particles, to avoid redundant combinations. For example, in the two-particle sector we will have

$$1 = \frac{1}{2!}\int \frac{d^3p}{2p_{10}}\frac{d^3p_2}{2p_{20}} \sum_{\lambda_1\lambda_2} |p_1, \lambda_1; p_2, \lambda_2\rangle\langle p_1, \lambda_1; p_2, \lambda_2|.$$

The equations of Sect. 10.1 are still valid with the present convention; had we used

$$|p_1, \lambda_1; p_2, \lambda_2\rangle = \frac{1}{\sqrt{2!}}\hat{b}^+(p_1, \lambda_1)\hat{b}^+(p_2, \lambda_2)|0\rangle,$$

we would have had to introduce extra factors $1/\sqrt{2!}$.

(ii)  Of course, $e_1 = e_2 = -|e|$.

(iii)  The amplitude now contains *two* terms, owing to the identity of the particles. One is exactly like the right-hand side of (10.2.5) with $e_1 = e_2 = -|e|$. The other is, if we let $e = -|e|$,

$$(-1)(2\pi)^4\delta_4(p_f - p_i)\sum_{\mu\nu} \frac{\overline{u}(p_2', \lambda_2')}{(2\pi)^{3/2}} ie\gamma_\mu \frac{u(p_1, \lambda_1)}{(2\pi)^{3/2}}$$

$$\times \frac{-ig_{\mu\nu}}{(p_2 - p_1')^2}\frac{\overline{u}(p_1', \lambda_1')}{(2\pi)^{3/2}} ie\gamma_\nu \frac{u(p_2, \lambda_2)}{(2\pi)^{3/2}},$$
(10.3.3)

and it differs from (10.2.5) only in the exchange of particles 1 and 2 in the final state (we could have exchanged them in the initial state with identical result) and a global $(-1)$ sign. We can associate a term like (10.2.5) with diagram I in Fig. 10.3.1, exactly like that of Fig. 10.2.1 (B), and the new term (10.3.3) with diagram II in Fig. 10.3.1. The corresponding potential can still be written as that in (10.2.19) with $e_1 = e_2 = e$, $m_1 = m_2 = m_e$. It is not necessary to introduce a term due to the exchange of 1 and 2: it is sufficient to require that $\hat{V}_{eff}$ be applied only to wave functions $\Psi_{\lambda_1\lambda_2}(\mathbf{r}_1, \mathbf{r}_2; t)$ antisymmetric under the exchange $1 \leftrightarrow 2$.

As an application, let us consider the relativistic corrections to the Hamiltonian of a heliumlike atom, i.e., one with two electrons and a nucleus without spin, and of change $Ze$. These corrections will be of two types: those due to the interactions between the electrons, like $\hat{V}_{\text{eff}}$ in (10.2.19); and those due

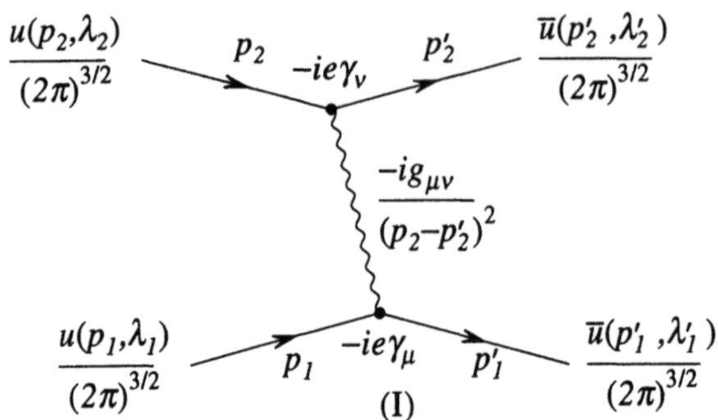

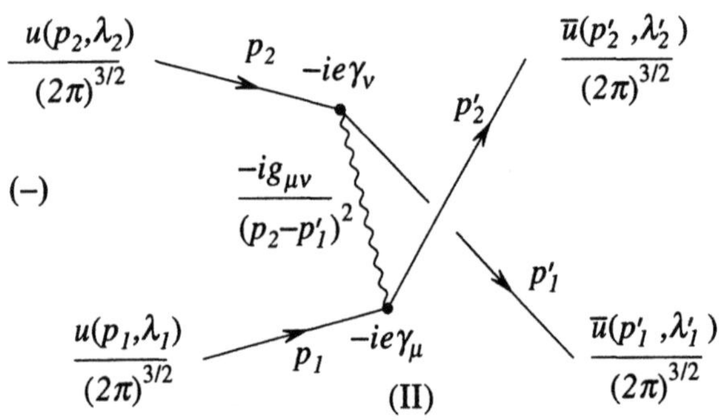

Fig. 10.3.1. Electron–electron collision.

to the potential seen by the electons in the field of the nucleus (considered infinitely heavy), that is to say, corrections like those in (10.2.20), for each electron. Summing up, we have

$$\hat{H} = \hat{H}_0 + \frac{e^2}{r} - \frac{Ze^2}{r_1} - \frac{Ze^2}{r_2} + \hat{V}^{(2)},$$

(10.3.4a)

$$\hat{H}_0 = \frac{-\hbar^2}{2m}\Delta_1 + \frac{-\hbar^2}{2m}\Delta_2, \ \hat{V}^{(2)} = \frac{-\hbar^4}{8m^3c^2}(\Delta_1^2 + \Delta_2^2) + \hat{V}_{12} + \hat{V}';$$

$\hat{V}_{12}$ is given by (10.2.19) with $e_1 = e_2 = -e$, $m_1 = m_2 = m_e = m$, and

$$\hat{V}' = \frac{-ie^2Z\hbar^2}{4m^2c^2}\left\{\frac{1}{r_1^3}\sigma_1(\mathbf{r}_1 \times \boldsymbol{\nabla}_1) + \frac{1}{r_2^3}\sigma_2(\mathbf{r}_2 \times \boldsymbol{\nabla}_2)\right\}$$

(10.3.4b)

$$+\pi\frac{e^2Z\hbar^2}{2m^2c^2}\left\{\delta(\mathbf{r}_1) + \delta(\mathbf{r}_2)\right\}.$$

For details on the numerical evaluation, and more information, see the monograph of Das (1973) or the text of Bethe and Salpeter (1957).

## 10.4 Electron–Positron Collisions: Effective Potential. Positronium

### 10.4.1 Scattering Amplitude in the Born Approximation

The field-theoretic interaction Hamiltonian is now

$$\hat{H}_{\text{int}} = e \int d^3r : \hat{\bar{\psi}}(x)\gamma \cdot \hat{A}(x)\psi(x) :; \tag{10.4.1}$$

the electron–positron states will be defined as

$$|i\rangle = \hat{b}^+(p_1,\lambda_1)\hat{d}^+(p_2,\lambda_2)|0\rangle,$$

$$\langle f| = \langle 0|\hat{d}(p_2',\lambda_2')\hat{b}(p_1',\lambda_1'),$$

i.e., with the same normalization conventions (including sign) as in the case of two different fermions, Sect. 10.2.1. We have

$$\langle f|\hat{S}|i\rangle_{\text{Born}} = \frac{1}{2!}i^2 e^2 \int d^4x_1 d^4x_2 \sum_{\mu\nu} g_{\mu\mu}g_{\nu\nu}\langle 0|\hat{d}(p_2',\lambda_2')\hat{b}(p_1',\lambda_1')$$

$$\times T(: \hat{\bar{\psi}}(x_1)\gamma_\mu\hat{\psi}(x_1) : \hat{A}_\mu(x_1)\hat{A}_\nu(x_2) : \hat{\bar{\psi}}(x_2)\gamma_\nu\hat{\psi}(x_2) :) \tag{10.4.2a}$$

$$\times \hat{b}^+(p_1,\lambda_1)\hat{d}^+(p_2,\lambda_2)|0\rangle.$$

Writing the $T$ product explicitly, we get

$$\int d^4x_1 d^4x_2 \ldots T(: \hat{\bar{\psi}}(x_1)\gamma_\mu\hat{\psi}(x_1) : \hat{A}_\mu(x_1)\hat{A}_\nu(x_2) : \hat{\bar{\psi}}(x_2)\gamma_\nu\hat{\psi}(x_2) :) \ldots$$

$$= \int d^4x_1 d^4x_2 \ldots (\theta(x_{10} - x_{20})\hat{A}_\mu(x_1)\hat{A}_\nu(x_2)$$

$$\times : \hat{\bar{\psi}}(x_1)\gamma_\mu\hat{\psi}(x_1) :: \hat{\bar{\psi}}(x_2)\gamma_\nu\hat{\psi}(x_2) :$$

$$+ \theta(x_{20} - x_{10})\hat{A}_\nu(x_2)\hat{A}_\mu(x_1) : \hat{\bar{\psi}}(x_2)\gamma_\nu\hat{\psi}(x_2) :: \hat{\bar{\psi}}(x_1)\gamma_\mu\hat{\psi}(x_1) :) \ldots.$$

We can exchange, in the second term, the factors

$$: \hat{\bar{\psi}}(x_2)\gamma_\nu\hat{\psi}(x_2) :, \quad : \hat{\bar{\psi}}(x_1)\gamma_\mu\hat{\psi}(x_1) :,$$

since the commutator gives zero when substituted into (10.4.2a). We may thus write

$$\int d^4x_1 d^4x_2 \ldots (\theta(x_{10} - x_{20})\hat{A}_\mu(x_1)\hat{A}_\nu(x_2) + \theta(x_{20} - x_{10})\hat{A}_\nu(x_2)\hat{A}_\mu(x_1))$$

$$\times : \hat{\bar{\psi}}(x_1)\gamma_\mu\hat{\psi}(x_1) :: \hat{\bar{\psi}}(x_2)\gamma_\nu\hat{\psi}(x_2) : \ldots =$$

$$\int d^4x_1 d^4x_2 \ldots (T\hat{A}_\mu(x_1)\hat{A}_\nu(x_2)) : \hat{\bar{\psi}}(x_1)\gamma_\mu\hat{\psi}(x_1) :: \hat{\bar{\psi}}(x_2)\gamma_\nu\hat{\psi}(x_2) : \ldots.$$

Since there are no photons in the initial or final state, these states act as the vacuum for the field $\hat{A}$: we may then use (9.7.8) and replace

$$T\hat{A}_\mu(x_1)\hat{A}_\nu(x_2) \to D_{\mu\nu}(x_1 - x_2),$$

and we obtain the expression, to be compared with (10.2.2),

$$\langle f|\hat{S}|i\rangle_{\text{Born}} = \frac{1}{2!}i^2 e^2 \int d^4x_1 d^4x_2 \sum_{\mu\nu} g_{\mu\nu} D_{\mu\nu}(x_1 - x_2)$$

$$\times \langle 0|\hat{d}(p_2',\lambda_2')\hat{b}(p_1',\lambda_1') : \bar{\hat{\psi}}(x_1)\gamma_\mu\hat{\psi}(x_1) : \tag{10.4.2b}$$

$$\times : \bar{\hat{\psi}}(x_2)\gamma_\nu\hat{\psi}(x_2) : \hat{b}^+(p_1,\lambda_1)\hat{d}^+(p_2,\lambda_2)|0\rangle.$$

Expression (10.4.2) can be split into two pieces, which we will call $E$ (*exchange*) and $A$ (*annihilation*) terms,

$$\langle f|S|i\rangle_{\text{Born}} = \langle f|\hat{S}|i\rangle_E + \langle f|\hat{S}|i\rangle_A; \tag{10.4.3}$$

they are represented graphically in Fig. 10.4.1, and we will now evaluate them in full detail.

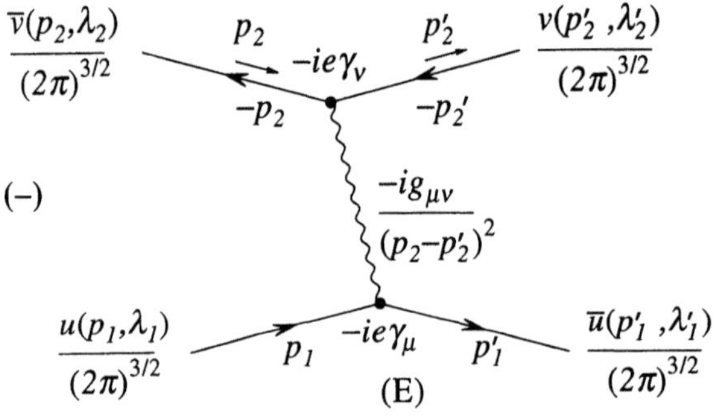


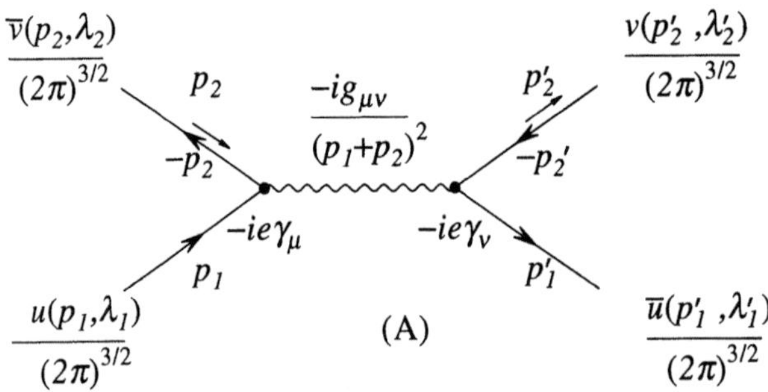


**Fig. 10.4.1.** $e^+e^-$ collision. (**E**) Exchange channel. (**A**) Annihilation channel.

We first write $\hat{\psi}(x)$ as

$$\hat{\psi}(x) = \hat{\psi}^{(-)}(x) + \hat{\psi}^{(+)}(x),$$

$$\hat{\psi}^{(-)}(x) \equiv \frac{1}{(2\pi)^{3/2}} \sum_\lambda \int \frac{d^3p}{2p_0} e^{-ip\cdot x} u(p,\lambda)\hat{b}(p,\lambda),$$

$$\hat{\psi}^{(+)}(x) \equiv \frac{1}{(2\pi)^{3/2}} \sum_\lambda \int \frac{d^3p}{2p_0} e^{+ip\cdot x} v(p,\lambda)\hat{d}^+(p,\lambda),$$

$$p_0 = +\sqrt{m^2 + p^2},$$

(10.4.4)

(cf. (9.5.2)). Then we substitute into (10.4.2) and obtain

$$\langle f|\hat{S}|i\rangle_{\text{B.}} = \frac{i^2 e^2}{2!} \int d^4x_1 d^4x_2 \sum g_{\mu\nu} D_{\mu\nu}(x_1 - x_2)(\mathcal{M}_{1\mu\nu} - \mathcal{M}_{2\mu\nu}), \quad (10.4.5a)$$

$$\mathcal{M}_{1\mu\nu} = \langle f| : \hat{\bar{\psi}}(x_1)\gamma_\mu\hat{\psi}(x_1) :: \hat{\bar{\psi}}(x_2)\gamma_\nu\hat{\psi}^{(-)}(x_2) : |i\rangle, \quad (10.4.5b)$$

$$\mathcal{M}_{2\mu\nu} = \langle f| : \hat{\bar{\psi}}(x_1)\gamma_\mu\hat{\psi}(x_1) :: \hat{\bar{\psi}}(x_2)\gamma_\nu\hat{\psi}^{(+)}(x_2) : |i\rangle. \quad (10.4.5c)$$

Substituting for $\hat{\psi}^{(-)}(x_2)|i\rangle$ its explicit expression we find for the $\mathcal{M}_1$ term that

$$\mathcal{M}_1 = \cdots \frac{1}{(2\pi)^{3/2}} \sum_\lambda \int \frac{d^3p}{2p_0} e^{-ip\cdot x_2} u(p,\lambda)\hat{b}(p,\lambda)\hat{b}^+(p_1,\lambda_1)\hat{d}^+(p_2,\lambda_2)|0\rangle.$$

We now write

$$\hat{b}(p,\lambda)\hat{b}^+(p_1,\lambda_1) = \left\{\hat{b},\hat{b}^+\right\} - \hat{b}^+\hat{b} = 2p_{10}\delta(\mathbf{p}_1 - \mathbf{p})\delta_{\lambda\lambda_1} - \hat{b}^+(p_1,\lambda_1)\hat{b}(p,\lambda).$$

The term $-\hat{b}^+\hat{b}$ here gives a vanishing contribution; only the anticommutator will thus survive, so we have

$$\mathcal{M}_{1\mu\nu} = \langle f| : \hat{\bar{\psi}}(x_1)\gamma_\mu\hat{\psi}(x_1) : \hat{\bar{\psi}}(x_2)\gamma_\nu \frac{u(p_1,\lambda_1)}{(2\pi)^{3/2}} e^{-ip_1\cdot x_2}\hat{d}^+(p_2,\lambda_2)|0\rangle.$$

Next we will use a decomposition similar to (10.4.4) for $\hat{\bar{\psi}}$:

$$\hat{\bar{\psi}}(x) = \hat{\bar{\psi}}^{(-)}(x) + \hat{\bar{\psi}}^{(+)}(x),$$

$$\hat{\bar{\psi}}^{(-)}(x) \equiv \frac{1}{(2\pi)^{3/2}} \sum_\lambda \int \frac{d^3p}{2p_0} e^{ip\cdot x}\bar{u}(p,\lambda)\hat{b}^+(p,\lambda),$$

$$\hat{\bar{\psi}}^{(+)}(x) \equiv \frac{1}{(2\pi)^{3/2}} \sum_\lambda \int \frac{d^3p}{2p_0} e^{-ip\cdot x}\bar{v}(p,\lambda)\hat{d}(p,\lambda);$$

(10.4.6)

substituting this, we find that

$$\mathcal{M}_{1\mu\nu} = \mathcal{M}^E_{1\mu\nu} + \mathcal{M}^A_{1\mu\nu},$$

where

$$\mathcal{M}^A_{1\mu\nu} = \langle f| : \hat{\bar{\psi}}(x_1)\gamma_\mu\hat{\psi}(x_1) : \hat{\bar{\psi}}^{\,(+)}(x_2)\gamma_\nu \frac{u(p_1,\lambda_1)}{(2\pi)^{3/2}}$$

$$\times e^{-ip_1\cdot x_2}\hat{d}^+(p_2,\lambda_2)|0\rangle, \tag{10.4.7a}$$

$$\mathcal{M}^E_{1\mu\nu} = \langle f| : \hat{\bar{\psi}}(x_1)\gamma_\mu\hat{\psi}(x_1) : \hat{\bar{\psi}}^{\,(-)}(x_2)\gamma_\nu \frac{u(p_1,\lambda_1)}{(2\pi)^{3/2}}$$

$$\times e^{-ip_1\cdot x_2}\hat{d}^+(p_2,\lambda_2)|0\rangle. \tag{10.4.7b}$$

Just as before, it so happens that, in the term

$$\ldots \hat{\bar{\psi}}^{\,(+)}(x_2)\ldots\hat{d}^+(p_2,\lambda_2)|0\rangle,$$

only the anticommutator of $\hat{\bar{\psi}}^{\,(+)}$ and $\hat{d}^+$ will give a nonzero contribution to (10.4.7a). Thus,

$$\mathcal{M}^A_{1\mu\nu} = \langle 0|\hat{d}(p_2',\lambda_2')\hat{b}(p_1',\lambda_1') : \hat{\bar{\psi}}(x_1)\gamma_\mu\hat{\psi}(x_1) : |0\rangle$$

$$\times \frac{\bar{v}(p_2,\lambda_2)}{(2\pi)^{3/2}}e^{-ip_2\cdot x_2}\gamma_\nu \frac{u(p_1,\lambda_1)}{(2\pi)^{3/2}}e^{-ip_1\cdot x_2}.$$

Substituting next $\hat{\bar{\psi}}(x_1), \hat{\psi}(x_1)$ by expressions like (10.4.6), (10.4.4), we see that only the term

$$(: \hat{\bar{\psi}}^{\,(-)}(x_1)\gamma_\mu\hat{\psi}^{(+)}(x_1) :)$$

gives a nonvanishing contribution; and, in this term, only the anticommutators with the annihilators will survive. Finally, using $\langle 0|0\rangle = 1$, we obtain

$$\mathcal{M}^A_{1\mu\nu} = +\frac{\bar{u}(p_1',\lambda_1')}{(2\pi)^{3/2}}e^{ip_1'\cdot x_1}\gamma_\mu \frac{v(p_2',\lambda_2')}{(2\pi)^{3/2}}e^{ip_2'\cdot x_1}$$

$$\times \frac{\bar{v}(p_2,\lambda_2)}{(2\pi)^{3/2}}e^{-ip_2\cdot x_2}\gamma_\nu \frac{u(p_1,\lambda_1)}{(2\pi)^{3/2}}e^{-ip_1\cdot x_2}.$$

The term $\mathcal{M}^E_1$ can be treated in a similar manner to $\mathcal{M}^A_1$. We start by commuting $\hat{\bar{\psi}}^{\,(-)}(x_2)$ and $\hat{\bar{\psi}}\gamma_\mu\hat{\psi}(x_1)$, which is possible because the commutator gives a vanishing contribution, so that

$$\mathcal{M}^E_{1\mu\nu} = \langle 0|\hat{d}(p_2',\lambda_2')\hat{b}(p_1',\lambda_1')\frac{1}{(2\pi)^{3/2}}\sum_\lambda \int \frac{d^3p}{2p_0}e^{ip\cdot x_2}\bar{u}(p,\lambda)$$

$$\times \hat{b}^+(p,\lambda) : \hat{\bar{\psi}}(x_1)\gamma_\mu\hat{\psi}(x_1) : \gamma_\nu \frac{u(p_1,\lambda_1)}{(2\pi)^{3/2}}e^{-ip_1\cdot x_2}\hat{d}^+(p_2,\lambda_2)|0\rangle.$$

From the term $\langle 0|\hat{d}\hat{b}\ldots\hat{b}^+$ only the anticommutator $\{\hat{b},\hat{b}^+\}$ survives. Hence,

$$\mathcal{M}^E_{1\mu\nu} = \langle 0|\hat{d}(p'_2,\lambda'_2) : \hat{\overline{\psi}}(x_1)\gamma_\mu\hat{\psi}(x_1) : \hat{d}^+(p_2,\lambda_2)|0\rangle$$

$$\times \frac{\overline{u}(p'_1,\lambda'_1)}{(2\pi)^{3/2}} e^{ip'_1\cdot x_2}\gamma_\nu \frac{u(p_1,\lambda_1)}{(2\pi)^{3/2}} e^{-ip_1\cdot x_2}. \qquad (10.4.8)$$

Next, we note that in the product

$$: \hat{\overline{\psi}}(x_1)\gamma_\mu\hat{\psi}(x_1) :$$

the only nonzero contribution is that of the piece

$$: \hat{\overline{\psi}}^{\,(+)}(x_1)\gamma_\mu\hat{\psi}^{(+)}(x_1) :,$$

and, of this, only anticommmutators with the $\hat{d}$, $\hat{d}^+$ do not give zero. Taking into account a change of sign when writting

$$: \hat{\overline{\psi}}^{\,(+)}(x_1)\gamma_\mu\hat{\psi}^{(+)}(x_1) := \sum_{\lambda\lambda'}\int \frac{d^3p}{2p_0}\int\frac{d^3p'}{2p'_0}\dots : \hat{d}(p,\lambda)\dots\hat{d}^+(p',\lambda') : \dots$$

$$= -\sum\int\frac{d^3p}{2p_0}\int\frac{d^3p'}{2p'_0}\dots\hat{d}^+(p',\lambda')\dots\hat{d}(p,\lambda)\dots+\dots,$$

we get the final result

$$\mathcal{M}^E_{1\mu\nu} = -\frac{\overline{v}(p_2,\lambda_2)}{(2\pi)^{3/2}}e^{-ip'_2\cdot x_1}\gamma_\mu\frac{v(p'_2,\lambda'_2)}{(2\pi)^{3/2}}e^{ip'_2\cdot x_1}$$

$$\times\frac{\overline{u}(p'_1,\lambda'_1)}{(2\pi)^{3/2}}e^{ip'_1\cdot x_2}\gamma_\nu\frac{u(p_1,\lambda_1)}{(2\pi)^{3/2}}e^{-ip_1\cdot x_2}. \qquad (10.4.9)$$

The evaluation of $\mathcal{M}_2$ follows the same pattern. We get two terms, one like $\mathcal{M}^E_1$ and another like $\mathcal{M}^A_1$, replacing $x_1$ by $x_2$ and $\mu$ by $\nu$. Since $D_{\mu\nu}(x_1-x_2)$ is not altered by this substitution, the new terms will merely duplicate the contribution of $\mathcal{M}^E_1$, $\mathcal{M}^A_1$: this cancels the $1/2!$ in the expression, (10.4.5) say, for $\langle f|\hat{S}|i\rangle$.

Substituting then (10.4.8) and (10.4.9) into (10.4.5), we get

$$\langle f|\hat{S}|i\rangle_E = -i^2e^2\sum_{\mu\nu}g_{\mu\mu}g_{\nu\nu}\int d^4x_1 d^4x_2 \overline{\Psi}^{(p'_1\lambda'_1)}(x_1)\gamma_\mu\Psi^{(p_1\lambda_1)}(x_1)$$

$$\times D_{\mu\nu}(x_1-x_2)\overline{\Psi}_c^{(p_2\lambda_2)}(x_2)\gamma_\nu\Psi_c^{(p'_2\lambda'_2)}(x_2), \qquad (10.4.10a)$$

with $\Psi^{(p\lambda)}$ as defined in (10.2.4), and

$$\Psi_c^{(p\lambda)}(x) \equiv (2\pi)^{-3/2}e^{ip\cdot x}v(p,\lambda),$$

$$\overline{\Psi}_c^{(p\lambda)}(x) \equiv (2\pi)^{-3/2}e^{-ip\cdot x}\overline{v}(p,\lambda), \qquad (10.4.10b)$$

on the one hand, and

$$\langle f|\hat{S}|i\rangle_A = i^2 e^2 \sum_{\mu\nu} g_{\mu\mu} g_{\nu\nu} \int d^4 x_1 d^4 x_2 \overline{\Psi}^{(p_1'\lambda_1')}(x_1)\gamma_\mu \Psi_c^{(p_2'\lambda_2')}(x_1)$$

$$\times D_{\mu\nu}(x_1 - x_2)\overline{\Psi}_c^{(p_2\lambda_2)}(x_2)\gamma_\nu \Psi^{(p_1\lambda_1)}(x_2),$$

$$(10.4.11)$$

on the other.

Expression (10.4.10) is similar to (10.2.3), found for particle–particle scattering in Sect. 10.2. Replacing the photon propagator by its Fourier representation and integrating in $d^4 x_1$, $d^4 x_2$ exactly as in Sect. 10.2, we find the fully relativistic expression for $\langle f|\hat{S}|i\rangle_E$. *In the Heaviside system of units*, and with $\hbar = c = 1$,

$$\langle f|\hat{S}|i\rangle_E = -(2\pi)^4 \delta_4(p_f - p_i) \sum_{\mu\nu} \frac{\overline{u}(p_1',\lambda_1')}{(2\pi)^{3/2}} ie\gamma_\mu \frac{u(p_1,\lambda_1)}{(2\pi)^{3/2}}$$

$$\times \frac{-ig_{\mu\nu}}{(p_2 - p_2')^2} \frac{\overline{v}(p_2,\lambda_2)}{(2\pi)^{3/2}} ie\gamma_\nu \frac{v(p_2',\lambda_2')}{(2\pi)^{3/2}},$$

$$(10.4.12)$$

in manifest correspondence with the diagram of Fig. 10.4.1 (E). (In this diagram we have introduced the convention, which we will systematically follow, of denoting antiparticles by lines with the arrow *against* the direction of time.)

Let us return to (10.4.10). Using the identity

$$\overline{v}(p_2,\lambda_2)\gamma_\nu v(p_2',\lambda_2') = \overline{u}(p_2',\lambda_2')\gamma_\nu u(p_2,\lambda_2),$$

we see that we can write $\langle f|\hat{S}|i\rangle_E$ in a form exactly like (10.2.5) *except* for the replacement of the product $e_1 e_2$ by $-e^2$: just a relative $(-)$ sign between the case of particle–particle and particle–antiparticle, to be expected because of the opposite signs of the particle and antiparticle charges. Because of this, the *exchange* piece will give us a reduced potential for particle–antiparticle interactions equal to that for two particles, with a change of sign. We write this in the Gauss system of units, and with $\hbar, c$ explicit as

$$\hat{V}_{E\text{eff}} = -e^2 \hat{V}_e - \frac{\hbar^4}{8m_e^2 c^2}\Delta_1^2 - \frac{\hbar^4}{8m_e^2 c^2}\Delta_2^2,$$

$$(10.4.13)$$

with $\hat{V}_e$ identical to $\hat{V}_{12}$ in (10.2.19b) and $m_1 = m_2 = m_e$.

### 10.4.2 Annihilation Channel

Let us now turn to the annihilation piece (10.4.11). Substituting the explicit expressions for $\Psi^{(p\lambda)}$, $\Psi_c^{(p\lambda)}$ and the Fourier expansion of $D_{\mu\nu}$ we can integrate $d^4 x_1 d^4 x_2$. We find that, in the Heaviside system of units,

$$\langle f|\hat{S}|i\rangle_A = (2\pi)^4 \delta_4(p_f - p_i) \sum_{\mu\nu} \frac{\overline{u}(p_1',\lambda_1')}{(2\pi)^{3/2}} ie\gamma_\mu \frac{v(p_2',\lambda_2')}{(2\pi)^{3/2}}$$

$$\times \frac{-ig_{\mu\nu}}{(p_1 + p_2)^2} \frac{\overline{v}(p_2,\lambda_2)}{(2\pi)^{3/2}} ie\gamma_\nu \frac{u(p_1,\lambda_1)}{(2\pi)^{3/2}}.$$

$$(10.4.14)$$

We can put this result in correspondence with diagram A in Fig. 10.4.1. The reason for the name[7] "annihilation channel" is clear in this diagram: we can say that the initial particle–antiparticle pair $e^+e^-$ annihilates into a (virtual) photon which later materializes into the final $e^+e^-$.

Let us now return to (10.4.11). Because the exchange piece $\langle f|\hat{S}|i\rangle_E$ already contains, in the strict nonrelativistic limit, the Coulomb potential $-e^2/r$, it is to be expected that (10.4.11) only yield corrections of order $1/c^2$. This is indeed the case. Substituting (10.2.4), (10.4.10b) into (10.4.11) and writing

$$d^4x_1 = dx_{10}d^3r_1, \ d^4x_2 = dx_{20}d^3r_2,$$

we get

$$\langle f|\hat{S}|i\rangle_A = \frac{i^2e^2}{(2\pi)^6} \int dx_{10}dx_{20} \int d^3r_1 d^3r_2$$

$$\times \sum_{\mu\nu} g_{\mu\nu} e^{ip'_{10}x_{10} - i\mathbf{p}'_1\mathbf{r}_1} \overline{u}(p'_1, \lambda'_1)\gamma_\mu v(p'_2, \lambda'_2)$$

$$\times e^{ip'_{20}x_{10} - i\mathbf{p}'_2\mathbf{r}_1} D_{\mu\nu}(x_1 - x_2) e^{-ip_{20}x_{20} + i\mathbf{p}_2\mathbf{r}_2}$$

$$\times \overline{v}(p_2, \lambda_2)\gamma_\nu u(p_1, \lambda_1) e^{-ip_{10}x_{20} + i\mathbf{p}_1\mathbf{r}_2}.$$

Changing integration variables,

$$d^3r_1 d^3r_2 \to d^3R d^3r, \ dx_{10}dx_{20} \to dt d\tau,$$

$$\mathbf{R} = \frac{\mathbf{r}_1 + \mathbf{r}_2}{2}, \ \mathbf{r} = \mathbf{r}_1 - \mathbf{r}_2; \ t = \frac{x_{10} + x_{20}}{2}, \ \tau = x_{10} - x_{20},$$

and integrating in $dt, d^3R$, we find that

$$\langle f|\hat{S}|i\rangle_A = -\frac{e^2}{(2\pi)^2}\delta(E(p_1) + E(p_2) - E(p'_1) - E(p'_2))\delta(\mathbf{p}_f - \mathbf{p}_i)$$

$$\times \sum_{\mu\nu} g_{\mu\nu}\overline{u}(p'_1, \lambda'_1)\gamma_\mu v(p'_2, \lambda'_2)\overline{v}(p_2, \lambda_2)\gamma_\nu u(p_1, \lambda_1) \qquad (10.4.15)$$

$$\times c \int d^3r \int d\tau e^{i(E(p_1)+E(p_2))\tau} e^{-i(\mathbf{p}_1+\mathbf{p}_2)\mathbf{r}} D_{\mu\nu}(r, c\tau),$$

where we have reintroduced $c$ explicitly. We can then treat $D_{\mu\nu}$ as we did to obtain (10.2.9b). Now, however, instead of the difference of energies $E(p_1) - E(p'_1)$ we have the *sum* $E(p_1) + E(p_2)$. In Gaussian units, then,

$$c \int d\tau e^{i(E(p_1)+E(p_2))r} D_{\mu\nu}(\mathbf{r}, c\tau)$$

$$\qquad (10.4.16)$$

$$= \frac{ig_{\mu\nu}}{r}\exp ir\frac{E(p_1) + E(p_2)}{c},$$

---

[7] In some texts, and for historical reasons, the name exchange diagram is used for the annihilation diagram, and direct diagram is used for the exchange diagram.

to be compared with (10.2.9b). In the limit $c \to \infty$,

$$E(p) \simeq mc^2 + \mathbf{p}^2/2m$$

and

$$\exp ir \frac{E(p_1) + E(p_2)}{c} \simeq e^{2imcr} + O(1/c),$$

so that, from (10.4.15), (10.4.16), and up to higher orders in $1/c$,

$$\langle f|\hat{S}|i\rangle_A \simeq -\frac{ie^2}{(2\pi)^2}\delta(E_f - E_i)\delta(\mathbf{p}_f - \mathbf{p}_i)$$

$$\times \sum_\mu g_{\mu\mu}\bar{u}(p_1', \lambda_1')\gamma_\mu v(p_2', \lambda_2')\bar{v}(p_2, \lambda_2)\gamma_\mu u(p_1, \lambda_1)J,$$

(10.4.17a)

where

$$J = \int \frac{d^3r}{r} e^{2imcr} e^{-i(\mathbf{p}_1 + \mathbf{p}_2)\mathbf{r}}.$$

(10.4.17b)

This last integral can be evaluated using a convergence factor,

$$J = \lim_{\epsilon \to 0} \int \frac{d^3r}{r} e^{2imcr} e^{i\mathbf{k}\mathbf{r}} e^{-\epsilon r}$$

$$= \frac{4\pi}{\mathbf{k}^2 - 4m^2c^2} = \frac{-\pi}{m^2c^2} + 0\left(\frac{1}{c^4}\right).$$

(10.4.17c)

As we promised, the annihilation channel contributes to the order $1/c^2$. The result (10.4.17c) for $J$ shows that we can obtain the same value by replacing (10.4.17b) by[8]

$$J = \int d^3r \; e^{ir(\mathbf{p}_1 - \mathbf{p}_1')} \frac{-\pi}{m^2c^2}\delta(\mathbf{r}) + 0(1/c^4),$$

(10.4.17d)

so that we have

$$\langle f|S|i\rangle_A \simeq -i\frac{e^2}{(2\pi)^2}\delta(E_f - E_i)\delta(\mathbf{p}_f - \mathbf{p}_i)$$

$$\times \sum_\mu g_{\mu\mu}\bar{u}(p_1', \lambda_1')\gamma_\mu v(p_2', \lambda_2')\bar{v}(p_2, \lambda_2)\gamma_\mu u(p_1, \lambda_1)$$

(10.4.18)

$$\times \int d^3r \; e^{ir(\mathbf{p}_1 - \mathbf{p}_1')} \frac{-\pi}{m^2c^2}\delta(r).$$

Since this expression is already of order $1/c^2$, it follows that we can neglect, when evaluating it, all relativistic corrections. Then using (10.1.6) and (10.1.3) we thus write

---

[8] Equation (10.4.17d) seems to imply that the annihilation potential is a contact interaction. Actually, a more accurate evaluation shows that the potential is extended in a region of size $\sim 1/mc$, falling to zero outside it. To the order of $1/c$ to which we are working this is not distinguishable from a purely pointlike potential.

$$T^A_{NR} \simeq \frac{1}{(2mc^2)^{4/2}} \frac{e^2}{(2\pi)^2} \int d^3r \; e^{ir(\mathbf{p_1}-\mathbf{p'_1})} \frac{\pi}{m^2c^2} \delta(\mathbf{r})M, \qquad (10.4.19a)$$

where

$$M = \sum_\mu g_{\mu\mu} \bar{u}(p'_1, \lambda'_1)\gamma_\mu v(p'_2, \lambda'_2)\bar{v}(p_2, \lambda_2)\gamma_\mu u(p_1, \lambda_1).$$

In the nonrelativistic limit (10.2.11) implies that

$$u(p, \lambda) \simeq (2mc^2)^{1/2} \begin{pmatrix} \chi(\lambda) \\ 0 \end{pmatrix},$$

$$v(p, \lambda) = i\gamma_2 u^*(p, \lambda) \simeq (2mc^2)^{1/2} \begin{pmatrix} 0 \\ -i\sigma_2\chi^*(\lambda) \end{pmatrix},$$

so that $u^+v \sim v^+u \sim 0$ and therefore

$$\begin{aligned}
M &= u^+(p'_1, \lambda'_1)v(p'_2, \lambda'_2)v^+(p_2, \lambda_2)u(p_1, \lambda_1) \\
&\quad - u^+(p'_1, \lambda'_1)\boldsymbol{\alpha}v(p'_2, \lambda'_2)v^+(p_2, \lambda_2)\boldsymbol{\alpha}u(p_1, \lambda_1) \\
&\simeq (2mc^2)^{4/2} \left\{ i\chi^+(\lambda'_1)\boldsymbol{\sigma}\sigma_2\chi^*(\lambda'_2) \right. \\
&\quad \left. \times i\chi^T(\lambda_2)\sigma_2\boldsymbol{\sigma}\chi(\lambda_1) \right\} \\
&= -(2mc^2)^{4/2} \sum_{aa',bb'} \chi^*_a(\lambda'_1)\chi^*_b(\lambda'_2) \\
&\quad \times (\boldsymbol{\sigma}\sigma_2)_{ab}(\sigma_2\boldsymbol{\sigma})_{a'b'}\chi_{b'}(\lambda_1)\chi_{a'}(\lambda_2).
\end{aligned}$$

With use of the identity

$$(\boldsymbol{\sigma}\sigma_2)_{ab}(\sigma_2\boldsymbol{\sigma})_{a'b'} = \frac{3}{2}\delta_{ab'}\delta_{ba'} + \frac{1}{2}\boldsymbol{\sigma}_{ab'}\boldsymbol{\sigma}_{ba'},$$

this becomes

$$\begin{aligned}
M \simeq -\frac{1}{2}(2mc^2)^{4/2} &\left\{ 3\chi^+(\lambda'_1)\chi(\lambda_1)\chi^+(\lambda'_2)\chi(\lambda_2) \right. \\
&\left. + \chi^+(\lambda'_1)\boldsymbol{\sigma}\chi(\lambda_1)\chi^+(\lambda'_2)\boldsymbol{\sigma}\chi(\lambda_2) \right\}.
\end{aligned} \qquad (10.4.19b)$$

Substituting into (10.4.19a), and comparing with (10.1.7c), we obtain the potential generated by the annihilation channel (found first by Pirenne), $\hat{V}_{A\text{ eff}}$:

$$\hat{V}_{A\text{ eff}} = \frac{\pi\hbar^2 e^2}{2m_e^2 c^2}(3 + \boldsymbol{\sigma}_1\boldsymbol{\sigma}_2)\delta(\mathbf{r}) + 0(1/c^4), \qquad (10.4.20)$$

where we have written $\hbar$ explicitly. The full electron–positron potential is $\hat{V}_{E\text{ eff}} + \hat{V}_{A\text{ eff}}$, with $\hat{V}_{E\text{ eff}}$ given by (10.4.13).

### 10.4.3 Positronium

The bound state of a positron and an electron is called positronium. The effective Hamiltonian of the system, in the centre of mass frame, and including terms of order $1/c^2$, can be written, from (10.4.13) and (10.4.20), as

$$\hat{H}_{\text{eff}} = \hat{H}_{NR}^{(0)} + \hat{H}^{(1)}, \tag{10.4.21a}$$

where $\hat{H}_{NR}^{(0)}$ is the purely nonrelativistic Hamiltonian; in Gauss units,

$$\hat{H}_{NR}^{(0)} = \frac{-\hbar^2}{m_e}\Delta - \frac{e^2}{r}. \tag{10.4.21b}$$

The difference with the familiar Hamiltonian for the hydrogen atom lies in that, because of the equality of $e^+, e^-$ masses, the reduced mass for positronium is $m = m_e/2$. The relativistic corrections $\hat{H}^{(1)}$ may conveniently be split into five pieces:

$$\hat{H}^{(1)} = \hat{V}_{\text{kin}} + \hat{V}_{\text{orb}} + \hat{V}_{SL} + \hat{V}_{\text{mag}} + \hat{V}_A, \tag{10.4.21c}$$

where $\hat{V}_{\text{kin}}$ is the relativistic correction to the kinetic energies,

$$\hat{V}_{\text{kin}} = -\frac{\hbar^4}{4m_e^3 c^2}\Delta^2, \tag{10.4.21d}$$

$\hat{V}_{\text{orb}}$ containing the (orbital) spin-independent terms,

$$\hat{V}_{\text{orb}} = \frac{e^2\hbar^2}{m_e^2 c^2}\left\{\pi\delta(\mathbf{r}) + \frac{1}{2r}\left(\Delta + \frac{1}{r^2}\sum_{ij} r_i r_j \nabla_i \nabla_j\right)\right\}, \tag{10.4.21e}$$

and $\hat{V}_{SL}$ is the spin–orbit coupling term,

$$\hat{V}_{SL} = \frac{e^2\hbar^2}{m_e^2 c^2}\frac{3}{4\hbar r}\hat{\mathbf{L}}(\boldsymbol{\sigma}_1 + \boldsymbol{\sigma}_2) = \frac{3e^2}{2m_e^2 c^2 r^3}\hat{\mathbf{L}}\hat{\mathbf{S}}. \tag{10.4.21f}$$

Note that

$$\hat{\mathbf{S}} = (\hbar/2)(\boldsymbol{\sigma}_1 + \boldsymbol{\sigma}_2)$$

is the operator for the total spin of the system. Then, $\hat{V}_{\text{mag}}$ describes the hyperfine interaction between the magnetic moments of electron and positron:

$$\hat{V}_{\text{mag}} = \frac{e^2\hbar^2}{4m_e^2 c^2}\left\{\frac{3(\mathbf{r}\boldsymbol{\sigma}_1)(\mathbf{r}\boldsymbol{\sigma}_2)}{r^5} - \frac{\boldsymbol{\sigma}_1\boldsymbol{\sigma}_2}{r^3} + \frac{8\pi}{3}\boldsymbol{\sigma}_1\boldsymbol{\sigma}_2\delta(\mathbf{r})\right\} =$$

$$\frac{e^2}{4m_e^2 c^2}\left\{\frac{6}{r^3}\sum_{ij}\left(\frac{r_i r_j}{r^2} - \frac{1}{3}\delta_{ij}\right)\hat{S}_i\hat{S}_j + \frac{8\pi}{3}(2\hat{S}^2 - 3\hbar^2)\delta(\mathbf{r})\right\}. \tag{10.4.21g}$$

Finally, $\hat{V}_A$ is the potential due to the annihilation channel, (10.4.20):

$$\hat{V}_A = \frac{\pi\hbar^2 e^2}{2m_e^2 c^2}(3 + \boldsymbol{\sigma}_1\boldsymbol{\sigma}_2)\delta(\mathbf{r})$$

$$= \frac{\pi e^2}{m_e^2 c^2}\hat{S}^2\delta(\mathbf{r}).$$

(10.4.21h)

It is useful to realize that $\hat{H}^{(1)}$ depends only on $\hat{\mathbf{S}}$, and not on the individual spins. Because of this, $\hat{\mathbf{S}}^2$ commutes with $\hat{H}$ and we can classify the positronium states as *orthopositronium* (total spin $s = 1$) or *parapositronium* (with $s = 0$).

To evaluate the energy levels we exactly solve the equation

$$\hat{H}_{NR}^{(0)}\psi_E^{(0)} = E^{(0)}\psi_E^{(0)}, \tag{10.4.22}$$

and treat $\hat{H}^{(1)}$ as a first-order perturbation, which is consistent with the neglect of $O(1/c^4)$ corrections. Thus, the energies will be given by

$$E \simeq E^{(0)} + \langle\psi_E^{(0)}|\hat{H}^{(1)}|\psi_E^{(0)}\rangle / \langle\psi_E^{(0)}|\psi_E^{(0)}\rangle.$$

A detailed calculation can be found in the text of Akhiezer and Berestetskii (1963). We will merely exemplify what one gets by evaluating the difference in energies between the fundamental levels of ortho- and parapositronium. The corresponding wave function is, for $s = 0, 1$,

$$\psi_s = \chi^{(s)}\psi_{10}^{(0)}(r), \ \ \psi_{10}^{(0)}(r) = \frac{1}{\sqrt{4\pi}}\,\frac{2}{a^{3/2}}e^{-r/a}, \ \ a = \frac{2\hbar^2}{m_e e^2},$$

with $\chi^{(s)}$ the spin wave function (which need not be written down explicitly). The energy splitting is thus $\Delta E_0 = $ (energy ortho $-$ energy para), with

$$\Delta E_0 = \langle\psi_{s=1}|\hat{H}^{(1)}|\psi_{s=1}\rangle - \langle\psi_{s=0}|\hat{H}^{(1)}|\psi_{s=0}\rangle$$

$$= \frac{\pi e^2}{m_e^2 c^2}\left\{\left\langle\psi_1\left|\left(\frac{4}{3}\hat{S}^2 - 2\hbar^2 + \hat{S}^2\right)\delta(\mathbf{r})\right|\psi_1\right\rangle\right.$$

$$\left. - \left\langle\psi_0\left|\left(\frac{4}{3}\hat{S}^2 - 2\hbar^2 + \hat{S}^2\right)\delta(\mathbf{r})\right|\psi_0\right\rangle\right\}$$

$$= \frac{14}{3}\,\frac{\pi e^2\hbar^2}{m_e^2 c^2}|\psi_{10}^{(0)}(0)|^2.$$

Numerically,

$$\Delta E_0 = \frac{7}{12}\alpha^4 m_e c^2 \simeq 8.45 \times 10^{-4} \ \text{eV},$$

to be compared with the experimental value[9]

---

[9] The slight disagreement, $\Delta E_0 - \Delta E_0(\exp) \simeq 0.04\times 10^{-4}$ eV disappears when one takes into account radiative corrections, as shown in a calculation by Karplus and Klein (1952). The details may be found there or in the text of Itzykson and Zuber (1980).

$$\Delta E_0(\exp) \simeq 8.412 \times 10^{-4} \text{ eV}.$$

As we see, orthopositronium is slightly heavier than parapositronium, into which it can decay emitting a photon. Both states can annihilate directly into photons,

$$\text{ortho} \to 3\gamma, \quad \text{para} \to 2\gamma,$$

with mean lives of $1.386 \times 10^{-7}$ s and $1.25 \times 10^{-10}$ s, respectively. Detailed calculations of all these decays may be found in Akhiezer and Berestetskii (1963) or Jauch and Rohrlich (1959), as well as in Sect. 11.5 in this text.

## 10.5 Scalar and Pseudoscalar Interactions. The Yukawa Potential

At distances of the order of 1 fm ($= 10^{-13}$ cm) or more the interactions between nucleons (neutrons and protons) can be approximated by effective field-theoretic interactions. We introduce fields $\hat{\psi}_N$ for the nucleons and also $\hat{\varphi}_\pi$ , $\hat{\varphi}_s$ for pions and scalar particles (effective: these scalar particles do not really exist). The masses of the pion are $m_\pi \simeq 140$ MeV$/c^2$. The pions exist in three varieties ($\pi^-, \pi^0, \pi^+$) but we will not take this fact into account[10]. We will take the interaction Hamiltonians as

$$\hat{H}_{s,\,\text{int}} = -g_s \sum_N \int d^3r : \hat{\bar{\psi}}_N(x)\hat{\psi}_N(x) : \hat{\varphi}_s(x), \tag{10.5.1}$$

$$\hat{H}_{Y,\,\text{int}} = -ig_Y \sum_N \int d^3r : \hat{\bar{\psi}}_N(x)\gamma_5\hat{\psi}_N(x) : \hat{\varphi}_\pi(x). \tag{10.5.2}$$

(cf. (9.8.19), (9.8.20)). The difference betweeen (10.5.1) and (10.5.2), the $i\gamma_5$ in (10.5.2), is due to the fact that pions are pseudoscalar. The sum in $N$ runs over $N = n, p$ (neutron and proton). In this section we will find the potentials equivalent to (10.5.1), (10.5.2) in the leading order of $1/c$, i.e., in the nonrelativistic limit. Corrections of relative order $1/c^2$ can be evaluated with the methods of the preceding sections and are left as an exercise.

Let us start by considering pion–nucleon interactions, as given by (10.5.2). We have first to evaluate the scattering amplitude for a two-nucleon collision; to simplify matters, we take these nucleons to be a neutron and a proton. We have the states, and the matrix element in the Born approximation,

---

[10] The structure of the system of nucleons, with two particles $n, p$ of nearly the same mass, and the pions, where also the masses are very similar, is not difficult to take into account and may be found, for example, in dedicated nuclear physics texts such as Brown and Jackson (1976), Elton (1959) or Blatt and Weisskopf (1952).

$$|i\rangle = \hat{b}_p^+(p, \lambda)\hat{b}_n^+(p_n, \lambda_n)|0\rangle,$$

$$\langle f| = \langle 0|\hat{b}_n(p_n', \lambda_n')\hat{b}_p(p', \lambda'); \tag{10.5.3}$$

$$\langle f|\hat{S}|i\rangle_{\text{Born}} = g_Y^2 \int d^4x_1 d^4x_2 \langle f|T(: \hat{\bar{\psi}}_p(x_1)\gamma_5\hat{\psi}_p(x_1) :$$

$$\times \hat{\varphi}_\pi(x_1)\hat{\varphi}_\pi(x_2) : \hat{\bar{\psi}}_n(x_2)\gamma_5\hat{\psi}_n(x_2) :)|i\rangle. \tag{10.5.4}$$

Since there are no pions in the initial or final state we may replace

$$T\hat{\varphi}_\pi(x_1)\hat{\varphi}_\pi(x_2) \to \Delta(x_1 - x_2),$$

where the pion propagator (cf. (9.7.4)) is

$$\Delta(x) = \int \frac{d^4k}{(2\pi)^4} e^{-ik\cdot x} \frac{i}{k\cdot k - m_\pi^2 + i0}.$$

Integrating the space part, and letting $x_0 = c\tau$ (we now write $c$ explicitly), yields

$$\Delta(\mathbf{r}, c\tau) = \frac{-i}{2(2\pi)^2 r} \int_{-\infty}^{+\infty} dk_0 e^{-ick_0 r} e^{ir\sqrt{k_0^2 - \mu_\pi^2}},$$

$$\mu_\pi \equiv \frac{m_\pi c}{\hbar}.$$

Substituting into (10.5.4) and operating in a manner similar to that of Sect. 10.2, we find that

$$\langle f|\hat{S}|i\rangle_{\text{Born}} = g_Y^2 \frac{1}{(2\pi)^6} \int d^3r_1 d^3r_2 u^+(p_n', \lambda_n')\gamma_0\gamma_5 u(p_n, \lambda_n)$$

$$\times u^+(p', \lambda')\gamma_0\gamma_5 u(p, \lambda) e^{i(\mathbf{P}_n - \mathbf{P}_n')\mathbf{r}_1 + i(\mathbf{p} - \mathbf{p}')\mathbf{r}_2} \tag{10.5.5}$$

$$\times \frac{-i}{2(2\pi)} \int dt e^{it(E_f - E_i)} \frac{\exp\frac{ir}{\hbar c}\sqrt{(E(p) - E(p'))^2 - m_\pi^2 c^4}}{r};$$

$$\exp\frac{ir\sqrt{(E(p) - E(p'))^2 - m_\pi^2 c^4}}{\hbar c}$$

$$\simeq e^{-r/r_\pi}\left\{1 + \frac{r}{r_\pi}\frac{(E(p) - E(p'))^2}{2m_\pi^2 c^2}\right\}, \quad r_\pi = \frac{1}{\mu_\pi} = \frac{\hbar}{m_\pi c}. \tag{10.5.6}$$

The form of (10.5.6) already indicates that the potential equivalent to the interaction, called the *Yukawa potential*, will be a short-range one, because of the exponential $e^{-r/r_\pi}$ in (10.5.6). We could now pursue a path like that followed in Sect. 10.2 and find the leading potential and the corrections. The calculation is somewhat cumbersome: because the leading term is $O(1/c^2)$, we have to keep terms of this order everywhere. A simpler method is desirable, and will be given below.

The result is

$$V_{Y\,\mathrm{eff}} = \frac{-\hbar^2 g_Y^2/4\pi}{4m_N^2 c^2}\,\frac{e^{-r/r_\pi}}{r_\pi^2 r}\left\{\frac{1}{3}\boldsymbol{\sigma}_n\boldsymbol{\sigma}_p + \left(\frac{1}{3} + \frac{r_\pi}{r} + \frac{r_\pi^2}{r^2}\right)S_{12}\right\}$$

$$+0(1/c^4) + \text{terms with } \delta(r);$$
(10.5.7a)

$$S_{12} = \frac{3}{r^2}(\mathbf{r}\boldsymbol{\sigma}_n)(\mathbf{r}\boldsymbol{\sigma}_p) - \boldsymbol{\sigma}_n\boldsymbol{\sigma}_p.$$

We do not specify the terms in $\delta(\mathbf{r})$ because the Yukawa potential is a reasonable approximation only at "long" distances, $r > r_\pi$. A remarkable fact is that the coupling constant in (10.5.7a) can be written as

$$f_Y^2 \equiv \frac{\hbar^2 g_Y^2/4\pi}{4m_N^2 r_\pi^2 c^2} = \frac{m_\pi^2}{4m_N^2}g_Y^2/4\pi \simeq \left(\frac{1}{14}g_Y\right)^2/4\pi.$$
(10.5.7b)

The potential is thus characterized by a small coupling, $f_Y^2 \simeq 0.1$, although the value of $g_Y^2, g_Y^2 \simeq 170$ is enormous.

For the scalar case, a similar evaluation gives

$$V_{s\,\mathrm{eff}} \simeq -\frac{g_s^2}{4\pi}\,\frac{e^{-r/r_s}}{r},\; r_s = \hbar/m_s c.$$
(10.5.8)

Let us return to the nucleon–pion interaction, (10.5.4). Replacing $T\hat{\varphi}_\pi\hat{\varphi}_\pi$ by the propagator, and using the momentum representation of the latter, and after some manipulations like those of the previous sections, we find the fully relativistic Born amplitude for nucleon–nucleon scattering,

$$\langle f|\hat{S}|i\rangle_{\mathrm{Born}} = (2\pi)^4\delta_4(p_f - p_i)\frac{\overline{u}(p_n',\lambda_n')}{(2\pi)^{3/2}}g_Y\gamma_5\frac{u(p_n,\lambda_n)}{(2\pi)^{3/2}}$$

$$\times\frac{i}{(p-p')^2 - m_\pi^2}\frac{\overline{u}(p',\lambda')}{(2\pi)^{3/2}}g_Y\gamma_5\frac{u(p,\lambda)}{(2\pi)^{3/2}}.$$
(10.5.9)

If we denote by a dotted line the propagation of a pion, then we can put (10.5.9) in correspondence with the diagram of Fig. 10.5.1.

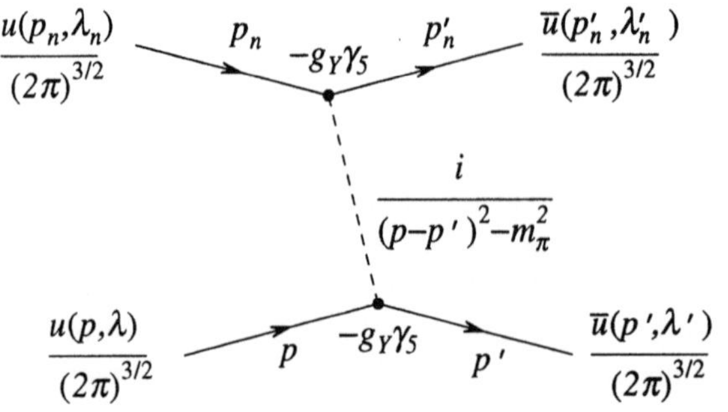


Fig. 10.5.1. Nucleon–nucleon collision with pion exchange.

The potential in (10.5.7) can be most easily obtained from the NR limit of (10.5.9). The NR scattering amplitude corresponding to (10.5.9) is

$$T_{NR}^B = \frac{g_Y^2}{(2\pi)^2} \frac{1}{(2m_N c^2)^{4/2}}$$

$$\times \bar{u}(p_n', \lambda_n') \gamma_5 u(p_n, \lambda_n) \frac{1}{(p-p')^2 - m_\pi^2 c^2} \bar{u}(p', \lambda') \gamma_5 u(p, \lambda).$$

Now,

$$\gamma_0 \gamma_5 = \begin{pmatrix} 0 & 1 \\ -1 & 0 \end{pmatrix},$$

so that

$$\bar{u}(p', \lambda') \gamma_5 u(p, \lambda) = u_b^+(p', \lambda') u_s(p, \lambda) - u_s^+(p', \lambda') u_b(p, \lambda)$$

$$\simeq c\chi'(\lambda')^+ (\mathbf{p} - \mathbf{p}')\boldsymbol{\sigma}\chi(\lambda),$$

and hence

$$T_{NR}^B = \frac{g_Y^2}{(2\pi)^2} \frac{1}{(2m_N)^2} \frac{1}{c^2}$$

$$\times \chi_n^+(\lambda_n')(\mathbf{p}_n - \mathbf{p}_n')\boldsymbol{\sigma}_n \chi_n(\lambda_n) \frac{1}{(p-p')^2 - m_\pi^2 c^2} \chi_p^+(\lambda')(\mathbf{p} - \mathbf{p}')\boldsymbol{\sigma}_p \chi_p(\lambda)$$

$$\simeq \frac{g_Y^2}{4\pi^2} \frac{1}{4m_N^2 c^2} \frac{1}{(\mathbf{p} - \mathbf{p}')^2 + m_\pi^2 c^2}$$

$$\times \chi_n^+(\lambda')(\mathbf{p} - \mathbf{p}')\boldsymbol{\sigma}_n \chi_n(\lambda_n)\chi_p^+(\lambda')(\mathbf{p} - \mathbf{p}')\boldsymbol{\sigma}_p \chi_p(\lambda);$$

here we have approximated

$$\frac{1}{(p-p')^2 - m_\pi^2 c^2} \simeq \frac{-1}{(\mathbf{p} - \mathbf{p}')^2 + m_\pi^2 c^2},$$

and used momentum conservation to replace $\mathbf{p}_n - \mathbf{p}_n' = \mathbf{p}' - \mathbf{p}$. Inverting the Fourier transform (10.1.7c) now,

$$\hat{V}(\mathbf{r}) = -\frac{1}{2\pi} \int d^3q \, e^{-i\mathbf{q}\mathbf{r}/\hbar} T_{NR}^B(\mathbf{p} - \mathbf{p}'),$$

with $\mathbf{q} = \mathbf{p} - \mathbf{p}'$, and using the formulas

$$\int d^3q \, e^{-i\mathbf{q}\mathbf{r}/\hbar} \frac{1}{\mathbf{q}^2 + m_\pi^2 c^2} = \frac{(2\pi\hbar)^3}{2\hbar} \frac{1}{r} e^{-r/r_\pi}, \quad r_\pi = \hbar/m_\pi c,$$

$$\int d^3q \, q_k q_j e^{-i\mathbf{q}\mathbf{r}/\hbar} \frac{1}{\mathbf{q}^2 + m_\pi^2 c^2} = \frac{-(2\pi\hbar)^3}{2\hbar} \left( \nabla_k \nabla_j e^{-r/r_\pi}/r \right),$$

we can obtain (10.5.7) immediately.

## 10.6 Weak Neutral Currents. Parity Violation in Atoms

Unified theories of weak and electromagnetic interactions[11] contain terms given, at low energy, by the effective interaction Hamiltonian,

$$
\hat{H}_{NC} = -G \int d^3r \sum_{\mu\nu} g_{\mu\nu} \sum_j : \hat{\bar{\psi}}_j(x)(v_j\gamma_u + a_j\gamma_\mu\gamma_5)\hat{\psi}_j(x) :
$$
$$
\times : \sum_i \hat{\bar{\psi}}_i(x)(v_i\gamma_\nu + a_i\gamma_\nu\gamma_5)\hat{\psi}_i(x) : .
$$

$(10.6.1)$

The indices $i, j$ run over various elementary fermions (electrons, muons, neutrinos, quarks). The interactions described by (10.6.1) are called *neutral current interactions* for reasons that we need not go into here. Because ordinary and axial vectors appear simultaneously in (10.6.1), it follows that these interactions violate parity conservation.

Although (10.6.1) holds strictly speaking for elementary particles, an interaction like (10.6.1) can also be written as an effective interaction for *low-energy* interactions of nonelementary objects, such as nucleons or even nuclei.

Let us now consider the contribution of (10.6.1) to the scattering amplitude for electrons $(e)$ and other fermions, which we denote by $N$, and assume also to be of spin $1/2$:

$$
|i\rangle = \hat{b}_e^+(p_1, \lambda_1)\hat{b}_N^+(p_2, \lambda_2)|0\rangle,
$$

$$
\langle f| = \langle 0|\hat{b}_N(p_2', \lambda_2')\hat{b}_e(p_1', \lambda_1').
$$

In the Born approximation,

$$
\langle f|\hat{S}|i\rangle_{\text{Born}} = 2iG(2\pi)^4\delta_4(p_f - p_i)
$$

$$
\times \sum_{\mu\nu} g_{\mu\nu}\frac{\bar{u}_e(p_1', \lambda_1')}{(2\pi)^{3/2}}(v_e\gamma_\mu + a_e\gamma_\mu\gamma_5)\frac{u_e(p_1, \lambda_1)}{(2\pi)^{3/2}}
$$

$$
\times\frac{\bar{u}_N(p_2', \lambda_2')}{(2\pi)^{3/2}}(v_N\gamma_\nu + a_N\gamma_\nu\gamma_5)\frac{u_N(p_2, \lambda_2)}{(2\pi)^{3/2}}.
$$

The most interesting application for us will be to evaluate neutral current interaction effects for atoms. In this case we may assume that $m_N \gg m_e$, and we obtain the amplitude $T_B$ in the form

$$
T_B = 2G\frac{1}{(2\pi)^2}
$$

$$
\times \left\{ u_{NR}^+(p_1', \lambda_1')(v_e + a_e\gamma_5)u_{NR}(p_1, \lambda_1)v_N\chi_N^+(\lambda_2')\chi_N(\lambda_2) \right.
$$

$(10.6.2)$

$$
\left. - u_{NR}^+(p_1', \lambda_1')(v_e\boldsymbol{\alpha} + a_e\boldsymbol{\alpha}\gamma_5)u_{NR}(p_1, \lambda_1)a_N\chi_N^+(\lambda_2')\boldsymbol{\sigma}_N\chi_N(\lambda_2) \right\}.
$$

---

[11] Details about these theories may be found, for example, in Kiesling (1988).

Here $u_{NR} \equiv u/\sqrt{2p_0}$, $\chi$ are Pauli spinors and we have used (10.2.14), (10.1.6). Introducing now a factor

$$\int d^3r \; e^{i(\mathbf{p_1}-\mathbf{p_1'})\mathbf{r}}\delta(\mathbf{r}) = 1,$$

expanding the $u_{NR}^+, u_{NR}$ as in (10.2.14), taking into account that, in the Pauli realization,

$$\gamma_5 = \begin{pmatrix} 0 & 1 \\ 1 & 0 \end{pmatrix}, \; \boldsymbol{\alpha}\gamma_5 = \begin{pmatrix} \boldsymbol{\sigma} & 0 \\ 0 & \boldsymbol{\sigma} \end{pmatrix},$$

and writing factors of $\hbar, c$ explicitly, we get, from (10.6.2), the expression

$$T_B^{NR} \simeq \frac{2G}{(2\pi)^2} \int d^3r \; e^{-i\mathbf{p_1'r}/\hbar}$$

$$\times \left\{ \chi_e^+(\lambda_1') \left[ v_e \left( \delta(\mathbf{r}) + \frac{\hbar^2}{8m^2c^2}(\triangle\delta(\mathbf{r}) + \delta(\mathbf{r})\triangle) \right. \right. \right.$$

$$\left. - \frac{\hbar^2}{4m^2c^2}(\boldsymbol{\sigma}_e\boldsymbol{\nabla})\delta(\mathbf{r})(\boldsymbol{\sigma}_e\boldsymbol{\nabla}) \right)$$

$$\left. + a_e \frac{-i\hbar}{2mc} \Big( \delta(\mathbf{r})\boldsymbol{\sigma}_e\boldsymbol{\nabla} + \boldsymbol{\sigma}_e\boldsymbol{\nabla}\delta(\mathbf{r}) \Big) \right] \chi_e(\lambda_1)v_N\chi_N^+(\lambda_2')\chi_N(\lambda_2)$$

$$-\chi_e^+(\lambda_1') \left[ a_e\delta(\mathbf{r})\boldsymbol{\sigma}_e + \frac{\hbar^2}{8m^2c^2}(\triangle\delta(\mathbf{r}) + \delta(\mathbf{r})\triangle)\boldsymbol{\sigma}_e \right. \qquad (10.6.3)$$

$$- \frac{\hbar^2}{4m^2c^2}(\boldsymbol{\sigma}_e\boldsymbol{\nabla})\delta(\mathbf{r})\boldsymbol{\sigma}_e(\boldsymbol{\sigma}_e\boldsymbol{\nabla})$$

$$\left. + v_e \frac{-i\hbar}{2mc} \Big( \delta(\mathbf{r})\boldsymbol{\sigma}_e(\boldsymbol{\sigma}_e\boldsymbol{\nabla}) + (\boldsymbol{\sigma}_e\boldsymbol{\nabla})\boldsymbol{\sigma}_e\delta(\mathbf{r}) \Big) \right]$$

$$\times \chi_e(\lambda_1)a_N\chi_N^+(\lambda_2')\sigma_N\chi_N(\lambda_2) \bigg\} e^{i\mathbf{p_1r}/\hbar}$$

$$+ \left( \text{relative order} \frac{1}{c^2} \right).$$

We have replaced $\mathbf{p}_1$ by $-i\hbar\boldsymbol{\nabla}$ to the right of $\delta(\mathbf{r})$, and $\mathbf{p}'$ by $-i\hbar\boldsymbol{\nabla}$ located now to the left of $\delta(\mathbf{r})$, in accordance with the prescription (10.2.18). The reduction to an equivalent potential is straightforward from (10.6.3); it is left to the reader. We will only consider the piece that violates parity, $\hat{V}_{AV}$, which is that containing $a_ev_N$ or $a_Nv_e$. After a simple calculation we get,

$$\hat{V}_{AV} = \frac{-G\hbar}{mc} \left\{ ia_Nv_e\Big( \delta(\mathbf{r})\boldsymbol{\sigma}_N\boldsymbol{\nabla} + \boldsymbol{\sigma}_N\boldsymbol{\nabla}\delta(\mathbf{r}) \Big) \right.$$

$$\left. + a_Nv_e(\boldsymbol{\nabla}\delta(\mathbf{r}))\boldsymbol{\sigma}_e \times \boldsymbol{\sigma}_N - ia_ev_N\Big( \delta(\mathbf{r})\boldsymbol{\sigma}_e\boldsymbol{\nabla} + \boldsymbol{\sigma}_e\boldsymbol{\nabla}\delta(\mathbf{r}) \Big) \right\}; \qquad (10.6.4)$$

$\boldsymbol{\sigma}_e$ acts on $\chi_e$, and $\boldsymbol{\sigma}_N$ on $\chi_N$. Expression (10.6.4) is a contact potential: particles only interact through it when they are at the same point[12]. For the particular case where the particles $N$ are nuclei, the interaction (10.6.4), which appears in addition to the familiar electromagnetic one, leads to parity violation effects in atoms (Bouchiat, 1974). In the standard theory of electroweak interactions $v_e$ is small as compared to $a_e$. Moreover, $v_N$ is large for a heavy nucleus since its value is the coherent sum of the $v_q$ of the quarks that make it up. Because of this we will study in detail only the term in $a_e v_N$ of (10.6.4), that is to say, we consider the approximate potential

$$\hat{V}_{\mathrm{approx}} = \frac{iG\hbar}{mc} a_e v_N \Big( \delta(\mathbf{r})\boldsymbol{\sigma}_e\boldsymbol{\nabla} + \boldsymbol{\sigma}_e\boldsymbol{\nabla}\delta(\mathbf{r}) \Big). \tag{10.6.5}$$

In the standard model,

$$G = \frac{G_F}{2\sqrt{2}}, \;\; a_e = -1, \;\; G_F = (1.015 \pm 0.03) \times 10^{-5}\frac{\hbar^3}{m_p^2 c},$$

where $m_p$ is the mass of the proton. $v_N$ depends on the nucleus; we have $v_N \sim -\mathcal{N}$, with $\mathcal{N}$ the number of neutrons in the nucleus. (For quarks of type $u$, $v_u = 1 - (8/3)\sin^2\theta_W$; for $d$-type ones, $v_d = -1 + (4/3)\sin^2\theta_W$, where $\sin^2\theta_W \simeq 1/4$.)

The presence of (10.6.5) induces transitions between states with opposite parities. A simple such case is that of $2p - 2s$ transitions in hydrogenlike atoms (nuclear charge $Ze$). The corresponding matrix element is

$$\delta_{\mathrm{sp}}E = \langle 2\mathrm{p}|\hat{V}_{\mathrm{approx}}|2\mathrm{s}\rangle.$$

To negligible $O((Z\alpha)^2)$ corrections we may use nonrelativistic wave functions. The calculation is elementary. If we denote by $M = \pm 1, 0$ the third component of the orbital angular momentum of the $2p$ state, and define

$$\sigma_M = \left\{ \begin{array}{l} \sigma_e, M = 0, \\[4pt] \dfrac{-M}{\sqrt{2}}(\sigma_1 + iM\sigma_2), \;\; M = \pm 1, \end{array} \right.$$

then we find

$$\delta_{\mathrm{sp}}E = -(m_e c^2)(Z\alpha)^4 \frac{v_N a_e}{16\pi} \frac{1.015 \times 10^{-5}}{2\sqrt{2}} \frac{m_e^2}{m_p^2} \tag{10.6.6}$$

$$\times \chi_e^+(\lambda_{2\mathrm{p}}) i\sigma_M \chi_e(\lambda_{2\mathrm{s}});$$

$\lambda_{2\mathrm{p}}, \lambda_{2\mathrm{s}}$ are the third components of the electron spin in the $2p, 2s$ states.

The effect is very small. The fine-structure splitting for these same levels, and for $j(2\mathrm{p}) = 3/2$, is

$$\delta_{\mathrm{fine}}E = (m_e c^2)(Z\alpha)^4 \frac{1}{16},$$

---

[12] Of course, this is an approximation, albeit a very good one: the range of these interactions is of some $10^{-16}$ cm, much smaller than even nuclear radii (of the order of $10^{-13}$ cm).

so the ratio is, approximating $\chi_e^+ i\sigma_M \chi_e \sim 1$, $|v_N a_e| \sim \mathcal{N}$,

$$\left| \frac{\delta_{\rm sp} E}{\delta_{\rm fine} E} \right| \sim \mathcal{N} \times 10^{-12}.$$

The effect is thus minute, in spite of which it is observable (and *has* been observed) under favourable circumstances.


## Problems

**P.10.1.** Evaluate the effects proportional to $m_e/m_p$, $m_e^2/m_p^2$ for the potential in the hydrogen atom.

**P.10.2.** Evaluate the corrections $1/c^4$ for the Yukawa potential, and $1/c^2$ for the scalar one.

**P.10.3.** Consider the electromagnetic interaction between spinless particles and photons as given by (9.8.18). If $e_1, e_2$ are the charges of the scalar particle, show that, in the NR limit, the interaction reduces to a Coulomb potential $V(r) = e_1 e_2/r$. Evaluate the relativistic corrections, order $1/c^2$. Show that, as advanced in Sect. 2.5, they vanish if one particle is very massive. Evaluate the next order ones, $O(1/c^4)$.

# 11. Relativistic Collisions in Field Theory. Feynman Rules. Decays

## 11.1 Electron–Positron Annihilation into Two Photons, and Pair Creation by Two-Photon Collisions

### 11.1.1 $e^+e^-$ Annihilation into $2\gamma$

The initial and final states for this process are,

$$|i\rangle = \hat{b}^+(p_1,\lambda_1)\hat{d}^+(p_2,\lambda_2)|0\rangle, \langle f| = \langle 0|\hat{a}(k_1,\eta_1)\hat{a}(k_2,\eta_2), \qquad (11.1.1)$$

where the $\eta$ are the helicities of the photons and we define photon states without a factor $1/\sqrt{n!}$, as we did for electron states. The interaction Hamiltonian is still given by (10.4.1). In the Born approximation,

$$\langle f|\hat{S}|i\rangle_{\text{Born}} = \frac{i^2 e^2}{2!}\int d^4x_1 d^4x_2 \sum_{\mu\nu} g_{\mu\mu}g_{\nu\nu}\langle 0|\hat{a}(k_1,\eta_1)\hat{a}(k_2,\eta_2)$$

$$\times T(:\hat{\bar{\psi}}(x_1)\gamma_\mu\hat{\psi}(x_1): \hat{A}_\mu(x_1)\hat{A}_\nu(x_2): \hat{\bar{\psi}}(x_2)\gamma_\nu\hat{\psi}(x_2):) \qquad (11.1.2)$$

$$\times \hat{b}^+(p_1,\lambda_1)\hat{d}^+(p_2,\lambda_2)|0\rangle.$$

Let us write

$$\hat{a}(k_2,\eta_2)\hat{A}_\mu(x_1) = \hat{A}_\mu(x_1)\hat{a}(k_2,\eta_2) + [\hat{a}(k_2,\eta_2),\hat{A}_\mu(x_1)],$$

$$\hat{a}(k_1,\eta_1)\hat{A}_\nu(x_2) = \hat{A}_\nu(x_2)\hat{a}(k_1,\eta_1) + [\hat{a}(k_1,\eta_1),\hat{A}_\nu(x_2)];$$

because there are no photons in the initial state, only the commutator will produce a nonzero value. The same is true if we had associated $\hat{a}(k_2,\eta_2)$ to $\hat{A}_\nu(x_2)$ and $\hat{a}(k_1,\eta_1)$ to $\hat{A}_\mu(x_1)$; this last contribution is obtained by applying a permutation to the former and we will denote it simply by (perm).

Using (9.6.7), and working in the Heaviside system of units and with $\hbar = c = 1$, a convention that will be followed throughout this whole chapter, we have for the commutator

$$\left[\hat{a}(k,\eta),\hat{A}_\lambda(x)\right] = \frac{1}{(2\pi)^{3/2}}e^{ik\cdot x}\epsilon_\lambda^*(k,\eta),$$

so that (11.1.2) becomes

$$\langle f|\hat{S}|i\rangle_{\text{Born}} = i^2 e^2$$

$$\times \sum_{\mu\nu} g_{\mu\mu} g_{\nu\nu} \frac{\epsilon_\nu^*(k_1,\eta_1)}{(2\pi)^{3/2}} \frac{\epsilon_\mu^*(k_2,\eta_2)}{(2\pi)^{3/2}} \frac{1}{2!} \int d^4x_1 d^4x_2 e^{ik_1\cdot x_1} e^{ik_2\cdot x_2} \tag{11.1.3}$$

$$\times \langle 0|T(: \hat{\bar{\psi}}(x_1)\gamma_\mu\hat{\psi}(x_1) :: \hat{\bar{\psi}}(x_2)\gamma_\nu\hat{\psi}(x_2) :)\hat{b}^+(p_1,\lambda_1)\hat{d}^+(p_2,\lambda_2)|0\rangle$$

$$+ \text{(perm)},$$

and, as announced, the term (perm) is obtained from the preceeding one just by exchanging the two photons. This, of course, was to be expected: owing to Bose statistics the scattering amplitude must be symmetric under exchange of the photons.

If we now split $\hat{\bar{\psi}}, \hat{\psi}$ as in (10.4.4), (10.4.6), we get

$$\langle 0|T : \hat{\bar{\psi}}(x_1)\gamma_\mu\hat{\psi}(x_1) :: \hat{\bar{\psi}}(x_2)\gamma_\nu\hat{\psi}(x_2) : \hat{b}^+(p_1,\lambda_1)\hat{d}^+(p_2,\lambda_2)|0\rangle$$

$$= \langle 0|T : \hat{\bar{\psi}}\gamma_\mu\hat{\psi} :: \hat{\bar{\psi}}(x_2)\gamma_\nu\hat{\psi}^{(-)}(x_2) : \hat{b}^+(p_1,\lambda_1)\hat{d}^+(p_2,\lambda_2)|0\rangle$$

$$+\langle 0|T : \hat{\bar{\psi}}\gamma_\mu\hat{\psi} :: \hat{\bar{\psi}}(x_2)\gamma_\nu\hat{\psi}^{(+)}(x_2) : \hat{b}^+(p_1,\lambda_1)\hat{d}^+(p_2,\lambda_2)|0\rangle$$

$$\equiv \mathcal{M}_{1\mu\nu} + \mathcal{M}_{2\mu\nu}.$$

We consider only the first term explicitly; the second is quite similar. Then,

$$\mathcal{M}_{1\mu\nu} = \sum_\lambda \int \frac{d^3p}{2p_{10}} \frac{\bar{v}(p,\lambda)}{(2\pi)^{3/2}} e^{-ip\cdot x_1}$$

$$\times \langle 0|\hat{d}(p,\lambda)\gamma_\mu T(\hat{\psi}(x_1)\hat{\bar{\psi}}(x_2))\gamma_\nu \hat{d}^+(p_2,\lambda_2)|0\rangle$$

$$\times \frac{u(p_1,\lambda_1)}{(2\pi)^{3/2}} e^{-ip_1\cdot x_2}.$$

The term $\hat{d}\ldots\hat{d}^+$ can be written as

$$\hat{d}\ldots\hat{d}^+ = -\hat{d}^+\ldots\hat{d} + \ldots\{\hat{d},\hat{d}^+\}.$$

Only the anticommutator will survive, and thus

$$\mathcal{M}_{1\mu\nu} = \frac{\bar{v}(p_2,\lambda_2)}{(2\pi)^{3/2}} \gamma_\mu \langle 0|T\psi(x_1)\bar{\psi}(x_2)|0\rangle \gamma_\nu \frac{u(p_1,\lambda_1)}{(2\pi)^{3/2}} e^{-ip_2\cdot x_1 - ip_1\cdot x_2}$$

$$= \int \frac{d^4k}{(2\pi)^4} e^{-ip_2\cdot x_1 - ip_1\cdot x_2} e^{-ik\cdot(x_1-x_2)}$$

$$\times \frac{\bar{v}(p_2,\lambda_2)}{(2\pi)^{3/2}} \gamma_\mu \frac{i(\gamma\cdot k + m)}{k^2 - m^2 + i0} \gamma_\nu \frac{u(p_1,\lambda_1)}{(2\pi)^{3/2}}.$$

Here we have substituted the fermion propagator,

$$\langle 0|T\hat{\psi}_a(x_1)\hat{\bar{\psi}}_b(x_2)|0\rangle = S_{ab}(x_1 - x_2),$$

and the expression (9.7.6) for $S_{ab}$.

The integrals in $d^4x_1 d^4x_2$ are easily performed since all the dependence on $x_1, x_2$ is contained in the exponentials. The integral in $d^4k$ is also straightforward. The final result is

$$\langle f|\hat{S}|i\rangle_{\text{Born}} = (2\pi)^4 \delta_4(p_f - p_i) \sum_{\mu\nu} g_{\mu\mu} g_{\nu\nu}$$

$$\times \left\{ \frac{\epsilon_\nu^*(k_1,\eta_1)}{(2\pi)^{3/2}} \frac{\epsilon_\mu^*(k_2,\eta_2)}{(2\pi)^{3/2}} \frac{\bar{v}(p_2,\lambda_2)}{(2\pi)^{3/2}} ie\gamma_\mu \frac{i}{\not{k} - m + i0} ie\gamma_\nu \frac{u(p_1,\lambda_1)}{(2\pi)^{3/2}} \right. \tag{11.1.4a}$$

$$\left. + \frac{\epsilon_\nu^*(k_2,\eta_2)}{(2\pi)^{3/2}} \frac{\epsilon_\mu^*(k_1,\eta_1)}{(2\pi)^{3/2}} \frac{\bar{v}(p_2,\lambda_2)}{(2\pi)^{3/2}} ie\gamma_\mu \frac{i}{\not{k}' - m + i0} ie\gamma_\nu \frac{u(p_1,\lambda_1)}{(2\pi)^{3/2}} \right\}$$

$$k \equiv p_1 - k_1, \; k' \equiv p_1 - k_2,$$

$$p_i = p_1 + p_2, \; p_f = k_1 + k_2. \tag{11.1.4b}$$

We can associate the various terms in (11.1.4) with the elements of the diagrams in Fig. 11.1.1, as shown there.

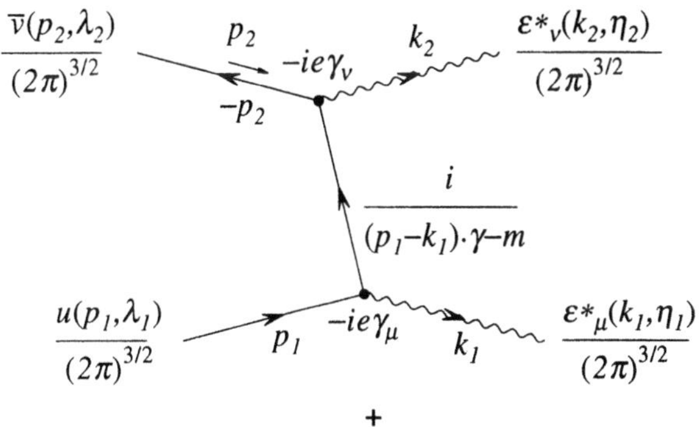


$$+$$

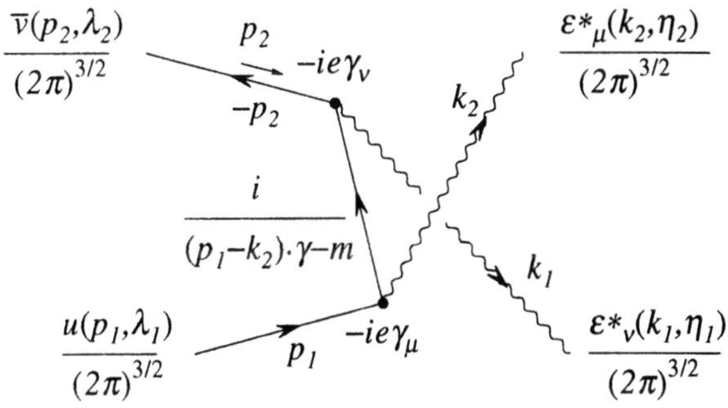


**Fig. 11.1.1.** Electron–positron annihilation into two photons.

### 11.1.2 Creation of an $e^+e^-$ Pair by Two Photons

The initial and final states are now

$$|i\rangle = \hat{a}^+(k_2, \eta_2)\hat{a}^+(k_1, \eta_1)|0\rangle,$$

$$\langle f| = \langle 0|\hat{d}(p_2, \lambda_2)\hat{b}(p_1, \eta_1).$$

The computation is like the previous one and we will not repeat it. The result, which can be associated with the diagrams of Fig. 11.1.2, is

$$\langle f|\hat{S}|i\rangle_{\text{Born}} = (2\pi)^4 \delta_4(p_f - p_i) \sum_{\mu\nu} g_{\mu\mu} g_{\nu\nu}$$

$$\times \left\{ \frac{\overline{u}(p_1, \lambda_1)}{(2\pi)^{3/2}} ie\gamma_\mu \frac{i}{\not{k} - m + i0} ie\gamma_\nu \frac{v(p_2, \lambda_2)}{(2\pi)^{3/2}} \frac{\epsilon_\mu(k_1, \eta_1)}{(2\pi)^{3/2}} \frac{\epsilon_\nu(k_2, \eta_2)}{(2\pi)^{3/2}} \right. \qquad (11.1.5a)$$

$$+ \frac{\overline{u}(p_1, \lambda_1)}{(2\pi)^{3/2}} ie\gamma_\mu \frac{i}{\not{k}' - m + i0} ie\gamma_\nu \frac{v(p_2, \lambda_2)}{(2\pi)^{3/2}} \frac{\epsilon_\mu(k_2, \eta_2)}{(2\pi)^{3/2}} \frac{\epsilon_\nu(k_1, \eta_1)}{(2\pi)^{3/2}} \left. \right\}$$

with

$$k \equiv p_1 - k_1, \ k' \equiv p_1 - k_2; \ p_i = k_1 + k_2, \ p_f = p_1 + p_2. \qquad (11.1.5b)$$

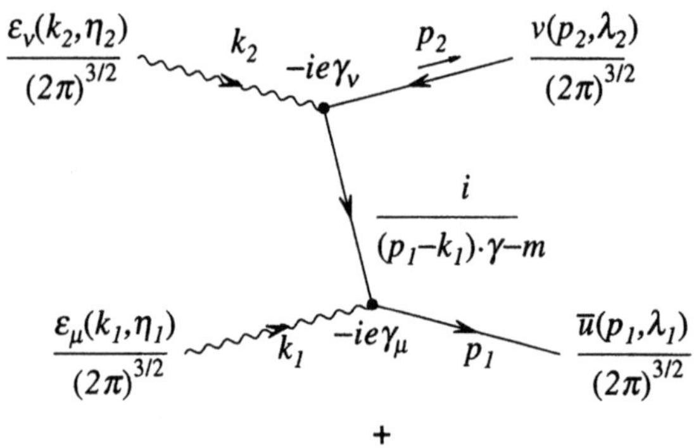

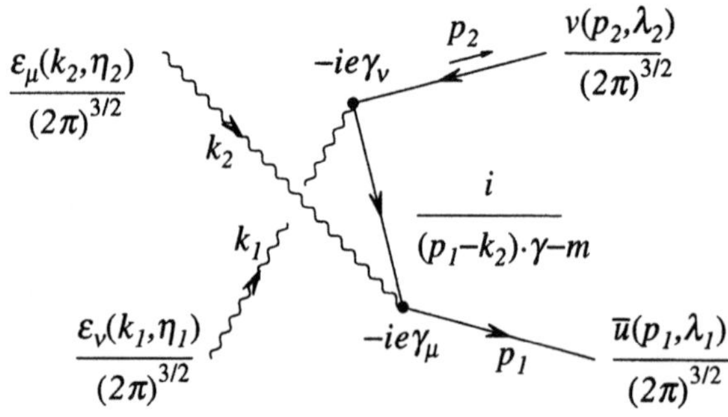

Fig. 11.1.2. Materialization of two photons into an electron–positron pair.

The similitude between (11.1.4), (11.1.5) (and between Figs. 11.1.1 and 11.1.2) is apparent. It can be related to the principle of detailed balance, or the *substitution rule* (also called *crossing symmetry*), which is even more general. This symmetry relates Feynman diagrams corresponding to different processes and can be found in dedicated field-theoretic textbooks (such as Bjorken and Drell, 1964, 1965).

A peculiarity of external antifermions which is depicted in Figs. 11.1.1 and 2 as well as in Fig. 10.4.1 and in other forthcoming figures is that the momentum carried by the external antifermion lines, $-p_2$ in Figs. 11.1.1 and 2, is *opposite* to the physical momentum $p_2$, depicted accompanied by a small arrow there.

## 11.2 Feynman Rules. Gauge Invariance

### 11.2.1 Feynman Rules for the Evaluation of Transition Amplitudes

When evaluating the $S$ matrix elements, in the Born approximation in Sects. 10.2 to 11.1 we saw that such transition amplitudes could be put in one-to-one correspondence with certain graphs (*Feynman graphs or diagrams*). Conversely, it can be proved that, for any process, the matrix element $\langle f|\hat{S}|i\rangle$ can be obtained, in perturbation theory, from a set of diagrams and corresponding rules (*Feynman rules*), which we will describe below. This is true for the Born approximation as well as for higher ones. The proof may be obtained by using Green function properties of propagators, which is how the rules were first derived by Feynman, or through systematic use of Wick's theorem. The first method is explained in detail in the treatise of Bjorken and Drell (1964); for the second, see for example Bogoliubov and Shirkov (1959), or Schweber (1961). We will not present either proof, although, after we have worked out the preceeding examples, the correctness of the method is almost self-evident. So we turn directly to stating in detail the Feynman rules.

1. Every Feynman diagram must be drawn as consisting of connected *external lines, or legs* (either *incoming lines* or *outgoing lines*), *vertices* and *propagators*. Vertices are points where three (or more, for some interactions) lines meet. We will denote fermions by continuous lines; for external fermions these include an arrow in the direction of time for particles (time flows from left to right) or against it for antiparticles. For internal fermions, we associate the arrows with the flow of charge. Spin 0 particles will be distinguished by a dotted line, and vector particles (in particular photons) by a wavy line. These lines are depicted in Fig. 11.2.1.
   The indices in the diagram are *linked*. Thus a vector particle propagator $(-ig_{\mu\nu}/(k^2 - M^2 + i0)$ must join the vertices with indices $\mu, \nu$; and a fermion propagator $(i/(\not{p} - m + i0))_{ab}$ will run from $\gamma_{\mu a'a}$ to $\gamma_{\nu bb'}$.
   The expression for the vector boson propagator in Fig. 11.2.1 corresponds to covariant quantization and Heaviside units.

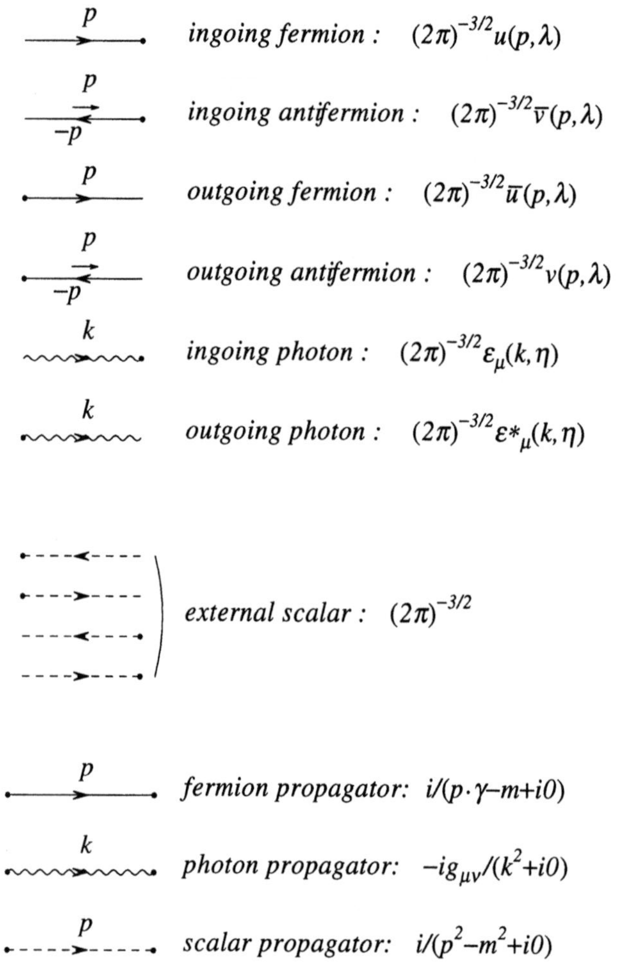

$p$     *ingoing fermion* :    $(2\pi)^{-3/2}u(p,\lambda)$

$p$     *ingoing antifermion* :    $(2\pi)^{-3/2}\bar{v}(p,\lambda)$

$p$     *outgoing fermion* :    $(2\pi)^{-3/2}\bar{u}(p,\lambda)$

$p$     *outgoing antifermion* :    $(2\pi)^{-3/2}v(p,\lambda)$

$k$     *ingoing photon* :    $(2\pi)^{-3/2}\varepsilon_\mu(k,\eta)$

$k$     *outgoing photon* :    $(2\pi)^{-3/2}\varepsilon^*{}_\mu(k,\eta)$

*external scalar* :    $(2\pi)^{-3/2}$

$p$     *fermion propagator:* $i/(p\cdot\gamma-m+i0)$

$k$     *photon propagator:* $-ig_{\mu\nu}/(k^2+i0)$

$p$     *scalar propagator:* $i/(p^2-m^2+i0)$

**Fig. 11.2.1.** Lines external and internal (propagators). For fermions (photons) $\lambda(\eta)$ is the spin (helicity). The propagator for a *massive* spin 1 particle is $i(-g_{\mu\nu} + k_\mu k_\nu/m^2)/(k^2 - m^2 + i\,0)$.

The vertices correspond to interactions; they are shown in Fig. 11.2.2 for a few important examples. The *order* of a diagram, with respect to an interaction (say, electromagnetic) is given by the number of times the coupling constant ($e$, in the electromagnetic case) appears in the diagrams, if we count powers of the coupling (like $e^2$ in the last diagram of Fig. 11.2.2) as many times as the exponent of the power.

2. To find the expression corresponding to a Feynman diagram, we substitute each element by its associated factors, as given in Figs. 11.2.1, 11.2.2. The order of the factors is important. It is obtained by following fermionic lines (for either particles or antiparticles) *against* the arrows.

3. The four-momenta of the internal lines (propagators) are found by requiring energy–momentum conservation at *each* vertex.

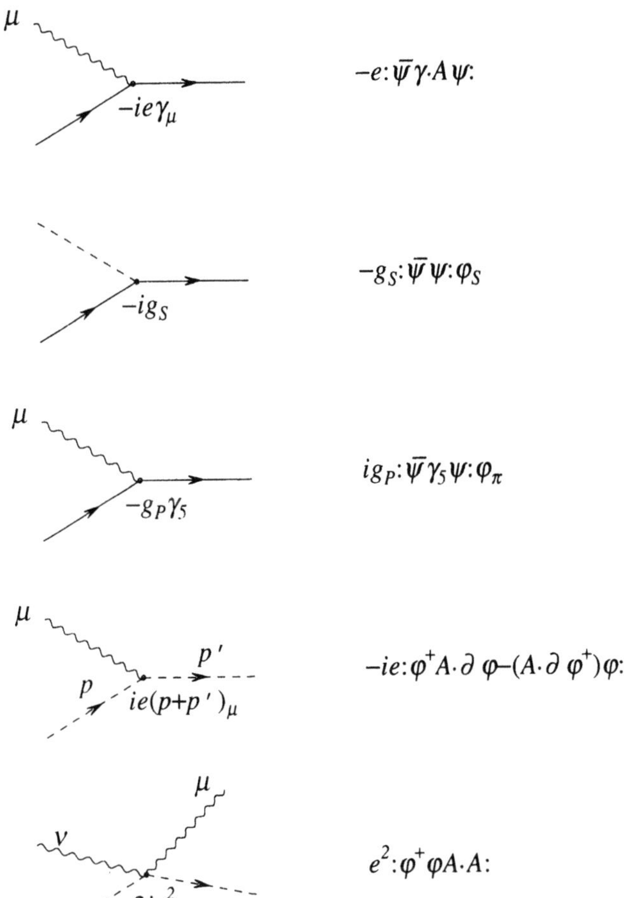



Fig. 11.2.2. Vertices and their corresponding interaction Lagrangians.

4. An overall factor $(2\pi)^4\delta_4(p_f - p_i)$ coming from global four-momentum conservation has to be included.

5. When there are two, or more, fermions (particles or antiparticles) in the final state the sign of the diagram is not fixed by the preceding rules. This is related to the fact that vectors like $\hat{b}_1^+\hat{b}_2^+|0\rangle$ and $\hat{b}_2^+\hat{b}_1^+|0\rangle$ both correspond to the same state, yet they differ by a sign. Once one convention has been chosen, it follows that, given a diagram, another one obtained from it by the exchange of two fermions will get a relative $(-)$ sign. This is obvious when both are in the initial or final state, owing to Fermi statistics; it is also true when one is in the initial state and the other in the final state, although we will not give the proof here. We can use this *sign rule* to find the signs of classes of diagrams, given that of one of them (which will have to be found by detailed calculation). For example, with the definition of initial and final states discussed in Sec. 10.2.1, the sign of the diagram (A) in Fig. 11.2.3 is $(+)$. The sign rule allows us to find the signs of the rest of

the diagrams there (Fig. 11.2.3). The reader may verify the coincidence, sign included, of the expressions obtained with the Feynman rules and the results we have, (10.2.5), (10.3.3), (10.4.12) and (10.4.14), and Figs. 10.2.1, 10.3.1 and 10.4.1, for the lowest-order processes $e^- e^- \to e^- e^-$, and $e^+ e^- \to e^+ e^-$.

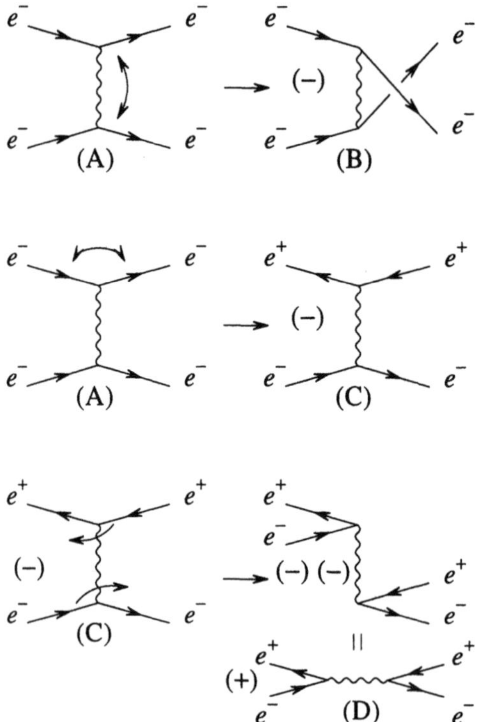

Fig. 11.2.3. Illustration of the sign rule.

Transition amplitudes are obtained, at a given order $n$ in perturbation theory, by adding coherently (i.e., taking signs into account) *all* Feynman diagrams, of order $n$ in the appropriate coupling constant, with the same *external* legs, both incoming and outgoing, whatever the internal structure. Thus both diagrams in Fig. 11.1.2, or in Fig. 11.1.1, are to be added to obtain the full matrix element. *Only diagrams that are topologically different should be written down, and added.*

In the lowest order (Born approximation), all the momenta in the internal lines are determined from the momenta in the external ones by four-momentum conservation at each vertex. In higher orders, however, closed lines (*loops*) appear and some of the momenta are not determined. For such diagrams we have the following extra rules:

6. The momenta that are not fixed by energy–momentum conservation, $k$, should be integrated upon with a factor $(2\pi)^{-4}$; that is to say, we integrate

$$\int \frac{d^4k}{(2\pi)^4},$$

the expression obtained from rules 1–5.

7. Fermion loops with an odd number of attached photons are identically zero (Furry's theorem).

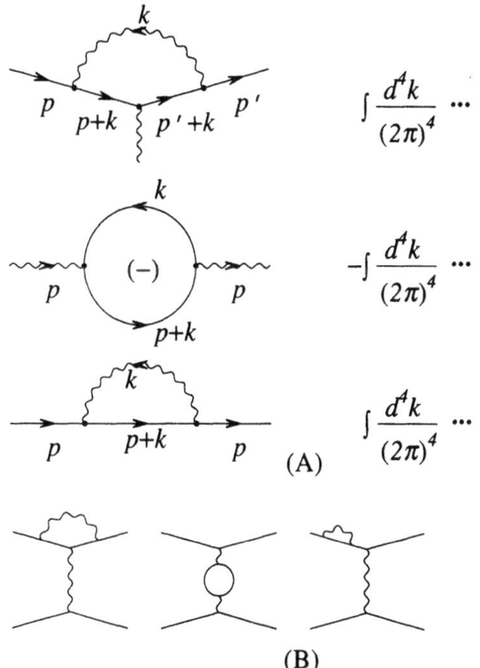

$$\int \frac{d^4k}{(2\pi)^4} \cdots$$

$$-\int \frac{d^4k}{(2\pi)^4} \cdots$$

$$\int \frac{d^4k}{(2\pi)^4} \cdots$$

(A)

(B)

**Fig. 11.2.4.** Higher-order corrections. (**A**) Some elementary loops. (**B**) Loops as part of diagrams.

8. Fermion loops carry an extra $(-)$ sign. Note that in fermion loops one should *not* count twice, once for a particle loop and another for an antiparticle loop, since both are topologically equivalent: a single diagram is enough. Also, the rule that indices must be linked implies a trace over the $\gamma$ matrices that appear in the loop.

Some examples of higher-order corrections can be found in Fig. 11.2.4. They are intended merely as an illustration; the evaluation of loop corrections (also called *radiative corrections*) requires an elaboration that lies outside the scope of our book.

### 11.2.2 Gauge Invariance

For electromagnetic interactions the rules we have just introduced present two types of ambiguity, both connected with gauge invariance. In the first place, the photon propagator is gauge dependent. In general, the momentum space propagator will be of the form

$$i\frac{-g_{\mu\nu} + k_\mu f_\nu(k) + k_\nu f_\mu(k)}{k^2 + i0} \equiv i\frac{N_{\mu\nu}(k)}{k^2 + i0}, \tag{11.2.1}$$

and the function $f$ will depend on the gauge we use. Then, the polarization vectors may be altered. As we saw in Sect. 9.6, if we change

$$\epsilon_\mu(k,\eta) \rightarrow \epsilon_\mu(k,\eta) + k_\mu\varphi(k), \tag{11.2.2}$$

with any $\varphi$, the corresponding physical state remains unaltered. Because the Lagrangian for electromagnetic interactions is gauge invariant, we expect these arbitrarinesses will have no effect on transition amplitudes. We will not give a general proof of this (for which we again refer to books on field theory) but will verify the invariance in detail in some typical situations.

For (11.2.1) we analyse collisions $e^+e^- \rightarrow e^+e^-$. It is, in general, essential to consider all diagrams (at a given order) together; but in our particular case each of the diagrams in Fig. 10.4.1 gives a result that is invariant under gauge changes in (11.2.1). Adding (10.4.12) and (10.4.14), we have

$$\langle f|\hat{S}|i\rangle_{\text{Born}} = ie^2(2\pi)^{-2}\delta_4(p_f - p_i)\sum_{\mu\nu} g_{\mu\mu}g_{\nu\nu}$$

$$\times \left\{ \overline{u}(p_1')\gamma_\mu u(p_1)\frac{N_{\mu\nu}(p_2 - p_2')}{(p_2 - p_2')^2}\overline{v}(p_2)\gamma_\nu v(p_2') \right. \tag{11.2.3}$$

$$\left. - \overline{u}(p_1')\gamma_\mu v(p_2')\frac{N_{\mu\nu}(p_1 + p_2)}{(p_1 + p_2)}\overline{v}(p_2)\gamma_\nu u(p_1) \right\};$$

we have not specified the spins because they play no role in what follows. We have replaced $-g_{\mu\nu} \rightarrow N_{\mu\nu}$, where $N_{\mu\nu}$ is the numerator in (11.2.1), so (11.2.3) is valid in an arbitrary gauge. Start now, for example, with the last term on the right-hand side of (11.2.3). The part of $N_{\mu\nu}(p_1 + p_2)$ proportional to $(p_1 + p_2)_\nu$ yields a contribution

$$\sum_\nu \dots g_{\nu\nu}(p_1 + p_2)_\nu\overline{v}(p_2)\gamma_\nu u(p_1) = \dots \overline{v}(p_2)(\gamma \cdot p_1 + \gamma \cdot p_2)u(p_1).$$

Because $u, v$ satisfy the Dirac equation, we have

$$\gamma \cdot p_1 u(p_1) = mu(p_1),$$

$$\overline{v}(p_2)\gamma \cdot p_2 = -m\overline{v}(p_2), \tag{11.2.4}$$

and the term vanishes. The piece of $N_{\mu\nu}$ containing a factor $(p_1' + p_2')_\mu$ will give a term

$$\dots \overline{u}(p_1')\gamma_\mu v(p_2')(p_1' + p_2')_\mu \dots,$$

which will be seen to vanish in the same manner. The piece of $N_{\mu\nu}$ with a term in $(p_1 + p_2)_\mu$ can be converted into a piece with $(p_1' + p_1')_\mu$ using four-momentum conservation; likewise, $(p_1' + p_2')_\nu = (p_1 + p_2)_\nu$. Now, all terms in $N_{\mu\nu}$ are of one of these forms *except* the piece $-g_{\mu\nu}$. Repeating the analysis for the first term on the right-hand side of (11.2.3), we get that, in both this one and the second we can replace

$$N_{\mu\nu} \to -g_{\mu\nu}, \tag{11.2.5}$$

since all the other pieces of $N_{\mu\nu}$ give zero. This is what was to be shown.

Let us now turn to the ambiguity (11.2.2). Consider a process with any number of external photons; and let us isolate a specific one, whose momentum and helicity will be denoted by $k, \eta$. To a given order, the corresponding transition amplitude $\mathcal{M}$ will be the sum of a certain number of Feynman diagrams. Redefining dumb indices, if necessary, we can always write this amplitude as

$$\mathcal{M} = \sum_{\mu} g_{\mu\mu} M_{\mu} \epsilon_{\mu}(k, \eta) \tag{11.2.6}$$

($\epsilon$ may represent $\epsilon^*$ if the photon is in the final state). Gauge invariance implies that

$$\sum_{\mu} g_{\mu\mu} k_{\mu} M_{\mu} = 0, \tag{11.2.7}$$

so that the indeterminacy (11.2.2) does not affect $\mathcal{M}$ as given by (11.2.6). The general proof may be found in, for example, Bjorken and Drell (1965); here we will check (11.2.7) for $e^+ e^- \to \gamma\gamma$, in the Born approximation.

Summing the contributions of both diagrams in Fig. 11.1.1, we get (11.1.4), which we now rewrite as

$$\langle f|\hat{S}|i\rangle_{\text{Born}} = \delta_4(p_f - p_i) \sum_{\mu} g_{\mu\mu} \epsilon_{\mu}^*(k_1, \eta_1) M_{\mu}, \tag{11.2.8a}$$

$$M_{\mu} = -ie^2 (2\pi)^{-2} \sum_{\nu} \epsilon_{\nu}^*(k_2, \eta_2) g_{\nu\nu}$$

$$\times \left\{ \bar{v}(p_2)\gamma_{\nu} \frac{1}{\not{p}_1 - \not{k}_1 - m} \gamma_{\mu} u(p_1) \right. \tag{11.2.8b}$$

$$\left. + \bar{v}(p_2)\gamma_{\mu} \frac{1}{\not{p}_1 - \not{k}_2 - m} \gamma_{\nu} u(p_1) \right\}.$$

Let us now define $T_{\nu}$ by

$$\sum g_{\mu\mu} k_{\mu} M_{\mu} = -ie^2 (2\pi)^{-2} \sum \epsilon_{\nu}^*(k_2, \eta_2) g_{\nu\nu} T_{\nu},$$

$$T_{\nu} = \bar{v}(p_2)\gamma_{\nu} \frac{1}{\not{p}_1 - \not{k}_1 - m} (k_1 \cdot \gamma) u(p_1) \tag{11.2.9}$$

$$+ \bar{v}(p_2)(k_1 \cdot \gamma) \frac{1}{\not{k}_1 - \not{p}_2 - m} \gamma_{\nu}(p_1),$$

where we have used momentum conservation to replace

$$p_1 - k_2 = k_1 - p_2.$$

Using then the Dirac equation, $(-m + \not{p}_1)u(p_1) = 0$, we have

$$(k_1 \cdot \gamma)u(p_1) = (\not{k}_1 - \not{p}_1 + m - m + \not{p}_1)u(p_1)$$

$$= -(\not{p}_1 - \not{k}_1 - m)u(p_1),$$

so that the first term on the right-hand side of (11.2.9) becomes

$$\overline{v}(p_2)\gamma_\nu \frac{1}{\not{p}_1 - \not{k}_1 - m}(k_1 \cdot \gamma)u(p_1) = -\overline{v}(p_2)\gamma_\nu u(p_1);$$

the second is

$$\overline{v}(p_2)(k_1 \cdot \gamma)\frac{1}{\not{k}_1 - \not{p}_2 - m}\gamma_\nu u(p_2) = \overline{v}(p_2)\gamma_\nu u(p_1).$$

The sum vanishes, and so does $T_\nu$, and hence $\sum g_{\mu\mu}k_{1\mu}M_\mu$, as was to be proved. Note that here the verification of gauge invariance has required the cooperation of two Feynman diagrams. This, in particular, shows that the decomposition of an amplitude into the various Feynman diagrams that contribute to it is *not* gauge invariant.

**Exercise.** Check gauge invariance, in the forms associated with (11.2.1), (11.2.2), for all the processes considered in this and the preceding chapters ∎

It is important to realize that the proof of gauge invariance, in both forms (associated with (11.2.1), (11.2.2)) hinges decisively on the fact that vertices are of the form $\overline{w}_1\gamma_\mu w_2$, with $w = u$ or $v$, and that, because of the Dirac equation, we have

$$\sum g_{\mu\mu}(p_1 - p_2)_\mu \overline{u}(p_2)\gamma_\mu u(p_1) = 0.$$

This is intimately connected with the form of the interaction,

$$\sum g_{\mu\mu}\hat{J}_\mu(x)\hat{A}_\mu(x), \ \hat{J}_\mu(x) = \hat{\overline{\psi}}(x)\gamma_\mu\hat{\psi}(x)$$

with a conserved current $\partial \cdot \hat{J} = 0$, as could have been expected, closely related to gauge invariance.

## 11.3 Polarized and Unpolarized Cross-sections. Sums Over Polarizations

The formulas for cross-sections and decay rates, deduced in Sect. 7.4, such as

$$d\sigma = \frac{2\pi^2}{\sqrt{\lambda(s, m_1^2, m_1^2)}}|F(i \to f)|^2\delta_4(p_f - p_i)\frac{d^3p'_1}{2p'_{10}} \cdots \frac{d^3p'_n}{2p'_{n0}}, \tag{11.3.1}$$

$$\lambda(a, b, c) \equiv a^2 + b^2 + c^2 - 2ab - 2ac - 2bc.$$

$$d\Gamma(i \to f) = \frac{1}{4\pi m}\delta(m - E_f)\delta(p_f)|F(i \to f)|^2\frac{d^3p'_1}{2p'_{10}} \cdots \frac{d^3p'_n}{2p'_{n0}}, \tag{11.3.2}$$

with $F$ given by

$$\langle p_1', \lambda_1'; \ldots; p_n', \lambda_n' | \hat{S} | p_1, \lambda_1; \ldots; p_i, \lambda_i \rangle$$

$$= \langle f|i \rangle + i\delta_4(p_f - p_i)F(i \to f),$$

are, as they stand, defined for the case where the initial $(\lambda_1, \ldots)$ and final $(\lambda_1', \ldots)$ spin variables are measured. Let the initial and final particles have total spin $s_1, \ldots; s_1', \ldots, s_n'$. If the initial spin variables of certain particles, $\lambda_{i_1}, \ldots, \lambda_{i_a}$, are not measured, we have to *average* over such spins. If the particles are massive, there are $(2s_{i_1} + 1) \ldots (2s_{i_a} + 1)$ possible states for such spins, so we have to consider

$$|F|^2 \to \frac{1}{(2s_{i_1} + 1) \ldots (2s_{i_a} + 1)} \sum_{\lambda_{i_1} \ldots \lambda_{i_a}} |F|^2. \tag{11.3.3a}$$

If the particles are massless, then the weight $1/(2s + 1)$ has to be replaced; for instance, for photons by $1/2$; for neutrinos, by unity.

If the spin variables $\lambda_{f_1}', \ldots, \lambda_{f_n}'$ of some final particles are not measured, we have to *sum* over such variables:

$$|F|^2 \to \sum_{\lambda_{f_1}' \ldots \lambda_{f_n}'} |F|^2. \tag{11.3.3b}$$

Equation(11.3.3b) is valid independently of whether the final-state particles are massive or massless. The totally unpolarized cross sections (or decay rates) are obtained by summing over all final spin variables, and averaging over all initial ones. In the rest of these sections we will present formulas to make sums such as (11.3.3) easier to perform.

Consider first a Feynman amplitude containing a spin 1 particle. It may be written as

$$F = \sum_{\mu} g_{\mu\mu} \epsilon_\mu(k, \eta) M_\mu; \tag{11.3.4a}$$

if the particle is in the final state, $\epsilon_\mu$ is to be replaced by $\epsilon_\mu^*$, but the overall result does not change. From (11.3.4a),

$$\sum_\eta |F|^2 = \sum_{\mu\nu} g_{\mu\mu} g_{\nu\nu} M_\mu M_\nu^* \sum_\eta \epsilon_\mu(k, \eta) \epsilon_\nu^*(k, \eta). \tag{11.3.4b}$$

If the particle is massive with mass $m \neq 0$,

$$\sum_\eta \epsilon_\mu(k, \eta) \epsilon_\nu^*(k, \eta) = -g_{\mu\nu} + k_\mu k_\nu / m^2,$$

so that

$$\sum_\eta |F|^2 = \sum_{\mu\nu} g_{\mu\mu} g_{\nu\nu} \left( -g_{\mu\nu} + \frac{k_\mu k_\nu}{m^2} \right) M_\mu M_\nu^*, \quad m \neq 0. \tag{11.3.5}$$

For massless spin 1 particles, like photons, the gauge condition (11.2.7) can be used to eliminate terms proportional to $k_\mu, k_\nu$ which give zero contribution, and then,

$$\sum_\eta |F|^2 = \sum_{\mu\nu} (-g_{\mu\nu}) M_\mu M_\nu^*, \quad m = 0. \tag{11.3.6a}$$

We can thus replace, in an expression like (11.3.4b),

$$\sum_\eta \epsilon_\mu(k,\eta)\epsilon_\nu^*(k,\eta) \to -g_{\mu\nu}. \tag{11.3.6b}$$

Next take spin 1/2 fermions, and let us denote by $w(p,\lambda)$ either Dirac spinors, $u$, or conjugate spinors, $v$. A Feynman amplitude contains factors of the type

$$F = \overline{w}(p_1,\lambda_1) M w(p_2,\lambda_2) = \sum_{ab} \overline{w}_a M_{ab} w_b. \tag{11.3.7}$$

From the Feynman rules it follows that $M$ will be a product, $\Gamma_1 \ldots \Gamma_r$, where the $\Gamma_j$ are any of the matrices $1, \gamma_\mu, i\gamma_5$, or $\sigma_{\mu\nu} = (i/2)[\gamma_\mu, \gamma_\nu]$; or more generally, a sum of such products. For all these $\Gamma_j$ one has (in the Pauli or Weyl realizations)

$$\gamma_0 \Gamma_j^+ \gamma_0 = \Gamma_j,$$

so that

$$(\overline{w}_1(p_1,\lambda_1)\Gamma_1 \ldots \Gamma_r w_2(p_2,\lambda_2))^* = \overline{w}_2(p_2,\lambda_2)\Gamma_r \ldots \Gamma_1 w_1(p_1,\lambda_1). \tag{11.3.8a}$$

We have

$$\sum_\lambda w_a(p,\lambda)\overline{w}_b(p,\lambda) = (\not{p} + \delta_w m)_{ab},$$
$$\delta = +1 \text{ if } w = u; \ \delta = -1, \text{ if } w = v. \tag{11.3.8b}$$

The quantity $\sum_\lambda |F|^2$ will be a sum of terms of the form

$$\sum_{\lambda_1\lambda_2} (\overline{w}_1(p_1,\lambda_1)\Gamma_1' \ldots \Gamma_s' w_2(p_2,\lambda_2))^* \overline{w}_1(p_1,\lambda_1)\Gamma_1 \ldots \Gamma_r w_2(p_2,\lambda_2)$$
$$= \mathrm{Tr}\,\{(\not{p}_1 + \delta_{w_1} m_1)\Gamma_1 \ldots \Gamma_r(\not{p}_2 + \delta_{w_2} m_2)\Gamma_s' \ldots \Gamma_1'\}, \tag{11.3.9}$$

and we have used (11.3.8b) to reduce the sum over spins to a trace.

This reduction can be carried out even if we do *not* sum over spins. We can introduce a fictitious sum over spins, *and* a spin projector

$$\Sigma_+(n,p) = \frac{1}{2}(1 + \gamma_5 \not{n}),$$

which, when applied to an arbitrary spinor $u(p,\lambda)$, cancels all values of $\lambda$ except the projection $+1/2$ along $\mathbf{n}$. We may then, in a transition amplitude, replace

$$u(p,n,+1/2) \to \Sigma_+(n,p)u(p,\lambda),$$

and sum over $\lambda$ in the cross-section, because only the appropriate value will survive.

## 11.4 Compton Scattering (Relativistic)

This is not the place to present an exhaustive list of applications of the tools
devised in the previous sections. The interested reader may find many of
them in texts such as Bjorken and Drell (1964 and 1965); Jauch and Rohrlich
(1959); Akhiezer and Berestetskii (1963). In this and the following section we
will merely present a few typical cases with which to illustrate the methods.

Consider the elastic scattering of a photon and an electron that we have
already evaluated, in the nonrelativistic approximation, in Sect. 8.3. The
initial and final states will be

$$|i\rangle = \hat{a}^+(k,\eta)\hat{b}^+(p,\lambda)|0\rangle,$$

$$\langle f| = \langle 0|\hat{b}(p',\lambda')\hat{a}(k',\eta').$$

To the lowest order the scattering amplitude will be given by the diagrams of Fig. 11.4.1. The Feynman rules allow us to write the amplitude
immediately. In the Heaviside system of units, and with $\hbar = c = 1$.

$$F(i \to f) = \frac{-e^2}{(2\pi)^2} \sum_{\mu\nu} g_{\mu\mu}g_{\nu\nu}\epsilon_\mu^*(k',\eta')\overline{u}(p',\lambda')$$

$$\times \left(\gamma_\mu \frac{1}{\not{p}+\not{k}-m}\gamma_\nu + \gamma_\nu \frac{1}{\not{p}-\not{k'}-m}\gamma_\mu\right) u(p,\lambda)\epsilon_\nu(k,\eta).$$

$$(11.4.1)$$

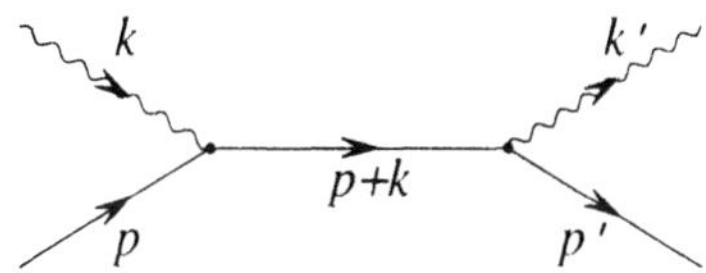


+

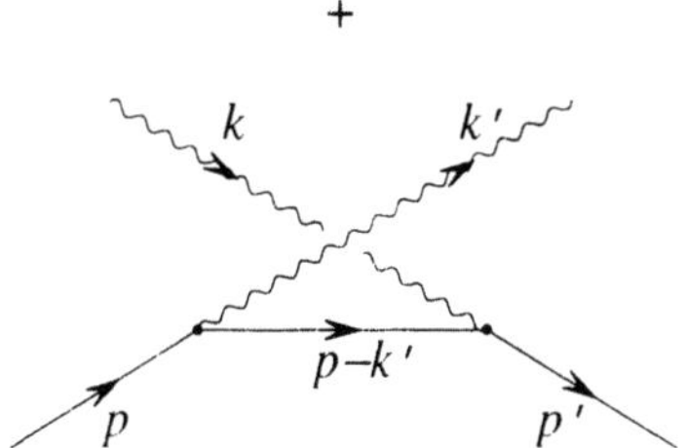


**Fig. 11.4.1.** Diagrams for relativistic Compton scattering.

Assuming that we do not know the initial polarization of the electron,
or measure the final one (but we know those of the photons), we have to
evaluate the centre of mass differential cross-section,

$$\frac{d\sigma}{d\Omega}\bigg|_{c.m.} = \frac{\pi^2}{4s}\frac{1}{2}\sum_{\lambda\lambda'}|F(i \to f)|^2.$$

We may use the result of the previous section to reduce this to traces; but before that, it is convenient to simplify (11.4.1) somewhat. To begin with, we use the fact that $p^2 = m^2$, $k^2 = 0$ to write

$$\frac{1}{\not{p} + \not{k} - m} = \frac{\not{p} + \not{k} + m}{(p+k)^2 - m^2} = \frac{\not{p} + \not{k} + m}{2p \cdot k},$$

$$\frac{1}{\not{p} - \not{k}' - m} = \frac{\not{p} - \not{k}' + m}{(p-k')^2 - m^2} = \frac{\not{p} - \not{k}' + m}{-2p \cdot k}.$$

Then, introducing $e^2/4\pi = \alpha$, we have

$$F(i \to f) = \frac{-\alpha}{\pi}$$

$$\times \overline{u}(p', \lambda') \left\{ \not{\epsilon}^*(k', \eta')(\not{p} + \not{k} + m)\not{\epsilon}(k, \eta)\frac{1}{2k \cdot p} \right. \tag{11.4.2}$$

$$\left. - \not{\epsilon}(k, \eta)(\not{p} - \not{k}' + m)\not{\epsilon}(k', \eta')\frac{1}{2k' \cdot p} \right\} u(p, \lambda).$$

Because of gauge invariance we have the freedom to modify

$$\epsilon_\mu(k', \eta') \to \epsilon_\mu(k', \eta') + k'_\mu f(k'),$$

$$\epsilon_\nu(k, \eta) \to \epsilon_\nu(k, \eta) + k_\nu f(k),$$

with $f$ arbitrary, leaving $F$ invariant. We choose $f$ so that

$$p \cdot \epsilon^* = p \cdot \epsilon = 0. \tag{11.4.3}$$

*This fixes the gauge.* The method is useful when we do *not* sum over $\eta, \eta'$; when summing over $\eta, \eta'$, it is preferable to use gauge arbitrariness to be able to perform the substitution (11.3.6b).

Next, we notice that a term like

$$\ldots (\not{p} + \not{k} + m)\not{\epsilon}(k, \eta)u(p, \lambda)$$

actually vanishes. Indeed, using the anticommutation relations of the $\gamma$, the Dirac equation and (11.4.3), we get

$$(\not{p} + m)\not{\epsilon}(k, \eta)u(p, \lambda) = \not{\epsilon}(k, \eta)(-\not{p} + m)u(p, \lambda)$$

$$+2(p \cdot \epsilon(k, \eta))u(p, \lambda) = 0.$$

In this way we find the rather simple expression

$$F(i \to f) = \frac{-\alpha}{\pi}\overline{u}(p', \lambda')$$

$$\times \left( \frac{\not{\epsilon}^*(k', \eta')\gamma \cdot k\not{\epsilon}(k, \eta)}{2p \cdot k} + \frac{\not{\epsilon}(k, \eta)\gamma \cdot k'\not{\epsilon}^*(k', \eta')}{2p \cdot k'} \right) u(p, \lambda). \tag{11.4.4}$$

Substituting, and then using the trace formulas of the previous section, we obtain

$$\left.\frac{d\sigma}{d\Omega}\right|_{\text{c.m.}} = \frac{\alpha^2}{16s}$$

$$\times \frac{1}{2}\text{Tr}\left\{(\not{p}' + m)\left(\frac{\not{\epsilon}'^*\not{k}\not{\epsilon}'}{k\cdot p} + \frac{\not{\epsilon}\not{k}'\not{\epsilon}'^*}{k'\cdot p}\right)(\not{p}+m)\left(\frac{\not{\epsilon}^*\not{k}\not{\epsilon}'}{k\cdot p} + \frac{\not{\epsilon}'\not{k}'\not{\epsilon}^*}{k'\cdot p}\right)\right\},$$

$$\epsilon \equiv \epsilon(k,\eta),\ \ \epsilon' \equiv \epsilon(k',\eta').$$

The evaluation of the trace is further simplified if we use *real* polarization vectors, $e_\mu$, rather than the complex ones, $\epsilon_\mu$. Of course, both sets form a basis and we can recover the cross-section with given helicity (given by the $\epsilon$) in terms of that with fixed linear polarization (with the $e$). After a simple but boring evaluation, one finds the cross-section in the laboratory reference system (which is where experiments are made),

$$\left.\frac{d\sigma}{d\Omega}\right|_{\text{lab}} = \frac{\alpha^2}{4m^2}\left(\frac{E'_{\gamma l}}{E_{\gamma l}}\right)^2\left\{\frac{E'_{\gamma l}}{E_{\gamma l}} + \frac{E_{\gamma l}}{E'_{\gamma l}} - 2 + 4(e'\cdot e)^2\right\} \tag{11.4.5a}$$

(*Klein–Nishina formula*). Here $E_{\gamma l}$ and $E'_{\gamma l}$ are the laboratory kinetic energies of, respectively, the initial and final photon, related by the expression

$$E'_{\gamma l} = \frac{E_{\gamma l}}{1 + (E_{\gamma l}/m)(1 - \cos\theta_l)}. \tag{11.4.5b}$$

$\theta_l$ is the laboratory angle between initial and final photon momenta. For kinetic energies $E_{\gamma l}, E'_{\gamma l} \ll m$, we have $E'_{\gamma l} \simeq E_{\gamma l}$, and (11.4.5) becomes

$$\left.\frac{d\sigma}{d\Omega}\right|_{\text{lab}} = \frac{\alpha^2}{m^2}(e'\cdot e)^2, \tag{11.4.6}$$

which, of course, agrees with the Thompson formula (Sect. 8.3).

**Exercise.** Consider corrections $O(v^2/c^2)$ to the Thompson formula. Compare with the NR case. Show how the relativistic Feynman diagrams (Fig. 11.4.1) become the NR ones (Sect. 8.3) ∎

## 11.5 Decay of Bound States

### 11.5.1 General Theory

Consider the bound state of two particles, which we start with by taking to be scalar and different. We also assume that their average velocity is sufficiently small to describe the bound state nonrelativistically. The fields associated with the particles will be denoted by $\hat{\phi}_j$, $j = 1, 2$, and the corresponding annihilation operators will be $\hat{a}_j$. We want to study the decay of the bound state into a number of particles with momenta $\mathbf{p}'_1, \ldots, \mathbf{p}'_n$, neglecting the interactions among the latter. The interaction responsible for the decay is given by a Hamiltonian $\hat{H}_I = \int d^3r\,\hat{\mathcal{H}}_I$, and we start by considering the case where $\hat{H}_I$ acts at first order. If $\psi(\mathbf{r})$ is the bound state wave function, normalized to

$$\int d^3r |\psi(\mathbf{r})|^2 = 1, \tag{11.5.1}$$

we will also consider its momentum space counterpart,

$$\tilde{\psi}(\mathbf{k}) = \frac{1}{(2\pi)^{3/2}} \int d^3r e^{-i\mathbf{kr}} \psi(\mathbf{r}); \quad \int d^3k |\tilde{\psi}(\mathbf{k})|^2 = 1. \tag{11.5.2}$$

(Throughout this section we work in natural units, $\hbar = c = 1$ and in the Heaviside system, $\alpha = e^2/4\pi$.)

Let $\mathbf{p}$ be the total three-momentum of the bound state; later we will go to its rest system, i.e., letting $\mathbf{p} = 0$. The bound state will be characterized by the state vector (initial state)

$$|B(\mathbf{p})\rangle = \sqrt{2p_0} \int d^3k \frac{\tilde{\psi}(\mathbf{k})}{2\sqrt{E_1(\mathbf{p}+\mathbf{k}/2)E_2(\mathbf{p}-\mathbf{k}/2)}} \tag{11.5.3}$$

$$\times \hat{a}_1^+ (\mathbf{p}+\mathbf{k}/2)\, \hat{a}_2^+ (\mathbf{p}-\mathbf{k}/2)|0\rangle,$$

where, if $m_B$ is the mass of the bound state and $m_j$ are the masses of its constituents, then

$$p_0 = \sqrt{m_B^2 + \mathbf{p}^2}, \; E_j(\mathbf{k}) = \sqrt{m_j^2 + \mathbf{k}^2}.$$

The factors in (11.5.3) are chosen so that, given (11.5.2) and the commutation relations

$$[\hat{a}_j(\mathbf{k}), \hat{a}_l^+(\mathbf{k}')] = 2\delta_{jl} k_0 \delta(\mathbf{k} - \mathbf{k}'),$$

the bound state vector is normalized relativistically, as is customary in the present text:

$$\langle B(\mathbf{p})|B(\mathbf{p}')\rangle = 2p_0 \delta(\mathbf{p} - \mathbf{p}').$$

The transition amplitude is given, in our approximation, by

$$-\langle f(\mathbf{p}_1', \dots, \mathbf{p}_n')|i \int d^4x\, \hat{\mathcal{H}}_I(x)|B(\mathbf{p})\rangle \tag{11.5.4}$$

$$= i\delta_4(p - \sum p_i') F(B \to f).$$

One may use translational invariance to write

$$\hat{\mathcal{H}}_I(x) = e^{-i\hat{P}\cdot x}\hat{\mathcal{H}}_I(0)e^{i\hat{P}\cdot x} \tag{11.5.5}$$

$$\to e^{i(p-\sum p_i')\cdot x}\mathcal{H}_I(0),$$

the latter expression valid when sandwiched as in (11.5.4). One can then integrate on $d^4x$, getting

$$F(B \to f) = -(2\pi)^4 \langle f(\mathbf{p}_1', \dots, \mathbf{p}_n')|\hat{\mathcal{H}}_I(0)|B(\mathbf{p})\rangle. \tag{11.5.6}$$

The differential decay rate is (in the centre of mass system, i.e., with $\mathbf{p} = 0$)

$$d\Gamma(B \to f(\mathbf{p}'_1, \ldots, \mathbf{p}'_n))$$

$$= \frac{1}{4\pi m_B}\delta(\sum \mathbf{p}'_i)\delta(m_B - \sum p'_{i0})|F(B \to f)|^2 \frac{d^3 p'_1}{2p'_{10}} \cdots \frac{d^3 p'_n}{2p'_{n0}}. \tag{11.5.7}$$

Let us return to the state $|B(\mathbf{p})\rangle$. If the system is weakly bound, we may neglect the average (relative) momentum squared $\langle \mathbf{k}^2 \rangle_B$ compared to the masses $m_j^2, m_B^2$. Thus, (11.5.3) may be written, in this approximation, as

$$|B(\mathbf{0})\rangle = \frac{m_B^{1/2}}{\sqrt{2m_1 m_2}}\left(\int d^3 k\, \tilde{\psi}(\mathbf{k})\right) \hat{a}_1^+(\mathbf{0})\hat{a}_2^+(\mathbf{0})|0\rangle$$

$$= \frac{(2\pi)^{3/2} m_B^{1/2}\psi(\mathbf{0})}{\sqrt{2m_1 m_2}}\hat{a}_1^+(\mathbf{0})\hat{a}_2^+(\mathbf{0})|0\rangle,$$

where, in the last step, we have used the Fourier transformation (11.5.2). Substituting this into (11.5.6), we see that we have

$$F(B \to f) = \frac{(2\pi)^{3/2} m_B^{1/2}\psi(\mathbf{0})}{\sqrt{2m_1 m_2}}F(1(\mathbf{0}) + 2(\mathbf{0}) \to f), \tag{11.5.8}$$

where

$$F(1(\mathbf{0})+2(\mathbf{0}) \to f) = -(2\pi)^4 \langle f(\mathbf{p}'_1, \ldots, \mathbf{p}'_n)|\hat{\mathcal{H}}_I(0)\hat{a}_1^+(\mathbf{0})\hat{a}_2^+(\mathbf{0})|0\rangle \tag{11.5.9}$$

is the *scattering* amplitude for the *constituent* particles 1, 2 (with zero three-momentum) to scatter into the final state $f$. We have reduced the problem of the decay of a weakly bound system to the calculation of scattering of its constituent particles, plus knowledge of the bound state wave function at the origin.

Equation (11.5.8) was deduced using the approximation of treating the interaction responsible for the decay to first order, but it is obvious that the result is quite general. Also, we have used the static approximation, neglecting $\langle \mathbf{k}^2 \rangle_B$ against $m_j^2, m_B^2$; corrections to this may be evaluated by expanding in powers of $\mathbf{k}$.

### 11.5.2 Decays of Positronium

Positronium is a weakly bound system, the binding energy being $m\alpha/4n^2 \ll m$. The previous considerations therefore apply with a few complications that we now discuss.

Consider the decay of states $ns$ ($n$ being the principal quantum number, $s$ the total spin equal to 0 or 1), with $l = 0$. Neglecting relativistic corrections, which as we know are of order $\langle v^2 \rangle \sim \alpha^2$, we find that the wave function is independent of the spin. The decay, however, depends on the spin for the following reasons. The parity of $e^+e^-$ is $(-1)$ (cf. Problem 9.14) and the spatial parity of a state $nl$ is $(-1)^l$. Thus our states have

$$\eta_P^{e^+e^-}(s) = -1. \tag{11.5.10}$$

This does not pose any constraint; but the situation is different in what regards charge conjugation parity. This operation exchanges $e^+$ and $e^-$. Because a state with $s = 0$ is antisymmetric on the spins of $e^+e^-$, and one with $s = 1$ is symmetric, it follows that, for $l = 0$ states,

$$\eta_C^{e^+e^-}(s) = -(-1)^s. \tag{11.5.11}$$

The charge parity of a photon is $(-1)$ (Problem 9.15). For $n$ photons, then,

$$\eta_C^{n\gamma} = (-1)^n.$$

Therefore the state with $s = 0$ (*parapositronium*) can only decay into an *even*, and that state with $s = 1$ (*orthopositronium*) can only decay into an *odd*, number of photons. To lowest order we then have the decays

$$\text{para} \to 2\gamma, \quad \text{ortho} \to 3\gamma.$$

**Exercise.** Verify that the decay ortho $\to \gamma$ is forbidden by energy–momentum conservation ∎

The first nonzero contribution to para decay is of second order in $e$, and that of ortho decay is of order $e^3$. They are associated, respectively, with the diagrams of Fig. 11.5.1 (A) and (B). The analogue of the equation (11.5.8) still holds.

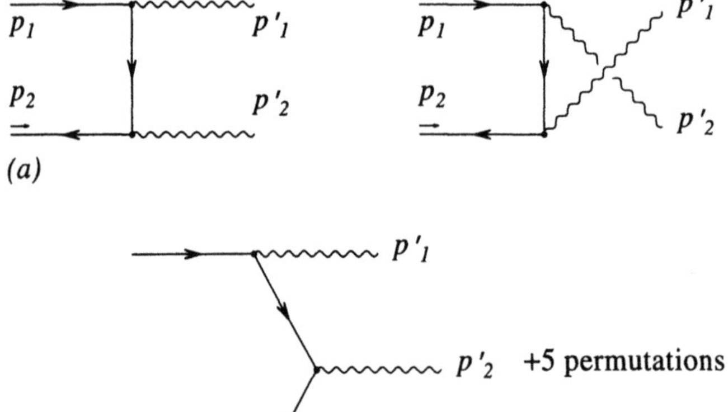


Fig. 11.5.1. Diagrams for the decay of positronium. (a) Parapositronium. (b) Orthopositronium.

**Exercise.** Repeat the deduction of (11.5.8) and verify this ∎

The last complication is a spin complication. This is easily dealt with since the selection rules just discussed pick a definite spin state of positronium.

We will perform the detailed calculation for parapositronium decay; ortho decay involves much more algebra and its details may be found in Berestetskii, Lifshitz and Pitaevskii (1979) or Jauch and Rohrlich (1959), who also give selection rules for decays of higher $e^+e^-$ states. From the evaluation of $e^+e^- \to 2\gamma$ of Sect. 11.1.1, equation (11.1.4),

$$F(e^+(\mathbf{0})e^-(\mathbf{0}) \to \gamma(\mathbf{p}\,'_1, \lambda'_1)\gamma(\mathbf{p}\,'_2, \lambda'_2))$$

$$= -\frac{\alpha}{\pi} \sum g_{\mu\mu} g_{\nu\nu} \epsilon_\mu^*(\mathbf{p}\,'_1, \lambda'_1) \epsilon_\nu^*(\mathbf{p}\,'_2, \lambda'_2) \bar{v}(\mathbf{0}, \lambda_2)$$

$$\times \left\{ \gamma_\nu \frac{1}{\slashed{q} - m} \gamma_\mu + \gamma_\mu \frac{1}{\slashed{q}\,' - m} \gamma_\nu \right\} u(\mathbf{0}, \lambda_1), \qquad (11.5.12)$$

$$q = p_1 - p'_1, \quad q' = p_1 - p'_2.$$

Next we use (11.5.8), (11.5.7). We may sum over the spins of $e^+$, $e^-$: as stated, the correct total spin state (of which there is only one, as we have $s = 0$) is picked automatically. Therefore, and for the total decay rate,

$$\Gamma(\text{para} \to 2\gamma) = \int \frac{d^3 p'_1 d^3 p'_2}{4 p'_{10} p'_{20}} \delta_4\left(p - \sum p'_i\right)$$

$$\times \frac{1}{4\pi m_B} \frac{(2\pi)^3 m_B |\psi_{n0}(\mathbf{0})|^2}{2m^2} \sum_{\lambda_1 \lambda_2 \lambda'_1 \lambda'_2} |F|^2. \qquad (11.5.13)$$

After a simple calculation,

$$\int \frac{d^3 p'_1 d^3 p'_2}{4 p'_{10} p'_{20}} \delta_4\left(p - \sum p'_i\right) = \frac{\pi}{2},$$

and a somewhat more involved exercise in gamma gymnastics gives

$$\sum_{\text{spins}} |F|^2 = \frac{\alpha^2}{\pi^2} \sum g_{\mu\rho} g_{\nu\sigma} \text{Tr}\, (m - \slashed{p}_2)$$

$$\times \left\{ \gamma_\nu \frac{1}{\slashed{p}_1 - \slashed{p}'_1 - m} \gamma_\mu + \gamma_\mu \frac{1}{\slashed{p}_1 - \slashed{p}'_2 - m} \gamma_\nu \right\} (m + \slashed{p}_1) \{\nu \leftrightarrow \rho, \ \mu \leftrightarrow \sigma\}$$

$$= \frac{16\alpha^2}{\pi^2}.$$

Finally, we get

$$\Gamma(\text{para} \to 2\gamma) = \frac{4\pi \alpha^2 |\psi_{n0}(\mathbf{0})|^2}{m^2} = \frac{\alpha^5}{2n^3} m, \qquad (11.5.14)$$

the last expression using the formula

$$|\psi_{n0}(\mathbf{0})| = \frac{(m\alpha)^{3/2}}{\sqrt{8\pi n^3}}. \tag{11.5.15}$$

Experimentally, and for the case $n = 1$, one has $\tau = 1/\Gamma = 1.25 \times 10^{-10}$ s, in agreement with the theoretical result (11.5.14). For orthopositronium,

$$\Gamma(\text{ortho} \to 3\gamma) = \frac{2(\pi^2 - 9)\alpha^6}{9\pi}m, \tag{11.5.16}$$

and the lifetime is now $1.39 \times 10^{-7}$ s.

### 11.5.3 Decay of Muonium into $e^+e^-$. Decays of Quarkonium

Muonium is a bound state of a *muon* ($\mu^-$) and on antimuon ($\mu^+$), particles with properties identical to $e^-, e^+$ except that their mass is some 200 times larger. We consider the decay of orthomuonium; this is different from the decay of orthopositronium because we have now the possibility of decay into a *virtual* photon, which then materializes into an $e^+e^-$ pair, as in the diagram of Fig. 11.5.2. The calculation is similar to the one we have made, but some care has to be exercised now with the spin. The state with spin $s = 1$ for orthomuonium is picked automatically by the selection rules; but we now have three possibilities for the third spin component, $s_z = 0, \pm 1$. If we sum over the spins of the muons, we are also summing over $s_z$: to obtain the decay rate we have to divide by a factor of 3. (Note that because of rotational invariance, $\Gamma$ is independent of $s_z$.) We have

$$F(\mu^+(\mathbf{0})\mu^-(\mathbf{0}) \to e^+(\mathbf{k}_1\lambda_1')e^-(\mathbf{k}_2\lambda_2'))$$

$$= -\frac{\alpha}{\pi}\sum_{\mu\nu}\bar{u}(\mathbf{0},\lambda_1)\gamma_\mu v(\mathbf{0},\lambda_2)\frac{g_{\mu\nu}}{(k_1 + k_2)^2}\bar{v}(\mathbf{k}_2,\lambda_2')\gamma_\nu u(\mathbf{k}_1,\lambda_1').$$

The sum, and average, over spins involves

$$\frac{1}{3}\sum_{\lambda_1\lambda_2\lambda_1'\lambda_2'}|F|^2 = \frac{\alpha^2}{3\pi^2}\frac{1}{16m_\mu^4}\sum[\text{Tr}\,(\not{p}_1 + m_\mu)\gamma_\rho(\not{p}_1 - m_\mu)\gamma_\sigma]$$

$$\times g_{\rho\rho}g_{\sigma\sigma}\text{Tr}\,\not{k}_2\gamma_\rho\not{k}_1\gamma_\sigma = \frac{8\alpha^2}{3\pi^2},$$

where we have neglected $m_e$. Finally,

$$\Gamma(\text{orthomuonium} \to e^+e^-) = \frac{16\pi\alpha^2}{3m_B^2}|\psi_{n0}(\mathbf{0})|^2 = \frac{\alpha^5}{6n^3}m_\mu. \tag{11.5.17}$$

*Quarkonium* is a bound state of quark–antiquark. The charge of a quark $q$ may be $2/3$ or $-1/3$, in units of the proton charge; this will be denoted by $Q_q$. The spin of a quark is $1/2$. For heavy quarks we may assume that their motion is negligible. The states with $l = 0$ may be classified as ortho or paraquarkonium if the total spin is $s = 1$ or $0$. Quarkonium may decay electromagnetically: ortho into $e^+e^-$, para into $2\gamma$. The calculations are identical

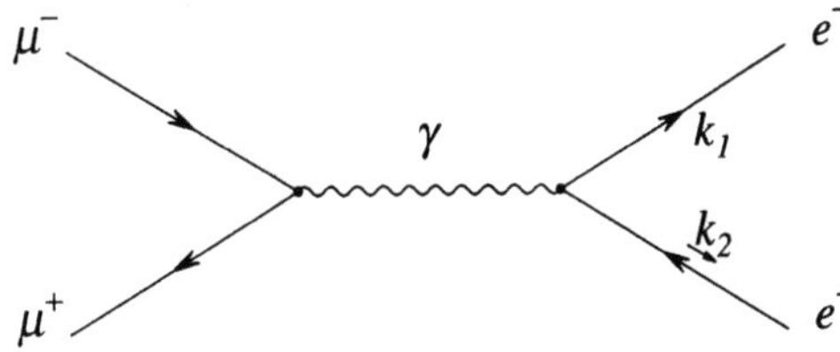

Fig. 11.5.2. Diagram involved in the decay of muonium into $e^+e^-$.

to the previous ones with two provisos. First, we do not have a simple expression like (11.5.15) for the quark–antiquark wave function. Secondly, quarks have an internal degree of freedom (called *colour*), which may take on three values: this yields an extra factor of three. To see this, denote by $q_a$ a quark with colour $a$. A properly normalized state is then

$$\frac{1}{\sqrt{3}} \sum_{a=1}^{3} |q_a \bar{q}_a\rangle. \tag{11.5.18}$$

The electromagnetic interactions of quarks are independent of their colour. The amplitudes involving (11.5.18) thus add coherently giving an amplitude $3 \times 1/\sqrt{3}$ times what one would have if there were no colour: when squaring to get the decay rate this gives the factor of 3 announced. Recalling (11.5.14) and (11.5.17), we then immediately get,

$$\Gamma(\text{orthoquarkonium} \to e^+e^-) = 3Q_q^2 \frac{16\pi\alpha^2}{3m_B^2} |\psi_{\text{ortho}}(\mathbf{0})|^2, \tag{11.5.19}$$

$$\Gamma(\text{paraquarkonium} \to 2\gamma) = 3Q_q^4 \frac{16\pi\alpha^2}{m_B^2} |\psi_{\text{para}}(\mathbf{0})|^2. \tag{11.5.20}$$

## Problems

**P.11.1.** Calculate the cross-sections, unpolarized and in the centre of mass system, for the processes $\gamma\gamma \to e^+e^-$, $e^+e^- \to \gamma\gamma$.

The solution may be found in, for example, Bjorken and Drell (1964).

**P.11.2.** Evaluate the transition amplitude and cross-section for the process $e^+e^- \to \mu^+\mu^-$. The particles $\mu^-, \mu^+$ (*muons*) have a mass some 200 times the electron mass, and otherwise the same properties as $e^-, e^+$.

*Solution.* Only the annihilation channel exists. Letting $p_1, p_2$ be the muon momenta, and $k_1, k_2$ the electron ones, we have

$$F(e^+e^- \to \mu^+\mu^-)$$

$$= \frac{-\alpha}{\pi} \sum \bar{u}(p_1, \lambda_1')\gamma_\mu v(p_2, \lambda_2') \frac{g_{\mu\nu}}{s} \bar{v}(k_2, \lambda_2)\gamma_\mu u(k_1, \lambda_1),$$

$$s = (k_1 + k_2)^2.$$

The cross-section is, in the centre of mass system, and with $s = (E_{e^+} + E_{e^-})^2$,

$$\frac{d\sigma}{d\Omega} \simeq \frac{\alpha^2}{4s}(1 + \cos^2\theta),$$

at high energies; one also has

$$\sigma = \frac{\alpha^2}{8s}(3 - v^2)v, \quad v = (1 - 4m_\mu^2/s)^{1/2},$$

for the total cross-section at any energy.

**P.11.3.** Calculate the corrections to the formula of the previous problem for small velocities, owing to the interactions between the muons.

*Solution.* At small velocities the effects of this interaction may be obtained by solving the Schrödinger equation for the motion of $\mu^+$ in the Coulombic field of the $\mu^-$, and vice versa. Then,

$$\sigma \underset{v \to 0}{\simeq} \frac{\alpha^2}{8s}(3 - v^2)v|\Psi(0)|^2,$$

where $|\Psi(0)|$ is the wave function, normalized to a (distorted) plane wave at infinity, so that

$$|\Psi(0)|^2 = \frac{\pi\alpha/v}{1 - e^{-\pi\alpha/v}}.$$

This follows from the results of Sect. 7.8.3 by considering the $\mu^+\mu^-$ Coulombic interaction as the exactly known one.

# 12. Relativistic Interactions with Classical Sources

## 12.1 Interaction with a Fixed (Classical) Potential

### 12.1.1 Scattering by an External Field

We will now consider the scattering of a particle by an external, fixed classical field. For definiteness we consider an electromagnetic field; the results can be immediately generalized to any other potential (e.g., a Yukawa potential). Then, we treat the corresponding four-potential, $A_\mu^{\text{cl}}(x)$, as a *given*, known $c$ number function. It is convenient (although not necessary) to imagine a very heavy particle, which we denote by a cross, as in Fig. 12.1.1, to be the source of the potential.

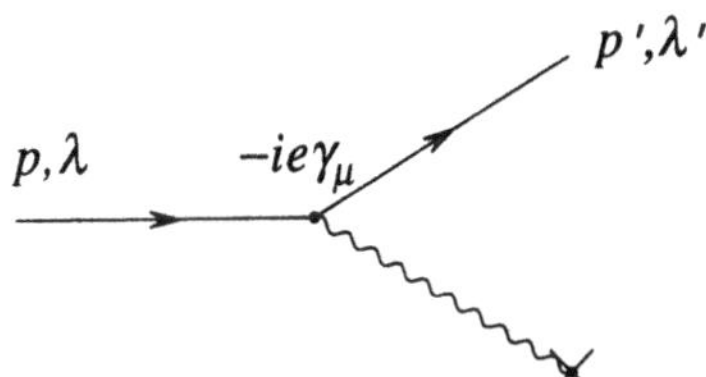


**Fig. 12.1.1.** Scattering by an external electromagnetic field. $\tilde{A}_\mu^{\text{cl}}$ is given by (12.1.3).

We write the interaction Hamiltonian with particles (electrons) of charge $e$ as

$$\hat{H}_{\text{int}} = e \sum_\mu g_{\mu\mu} : \hat{\bar{\psi}}(x)\gamma_\mu\hat{\psi}(x) : A_\mu^{\text{cl}}(x). \tag{12.1.1}$$

The $S$-matrix element for scattering of the particle in state $|p, \lambda\rangle$ into state $|p', \lambda'\rangle$ will be, to lowest order, and if we assume that $p \neq p'$,

$$\langle f|\hat{S}|i\rangle_{\text{Born}} = -ie \sum_\mu \langle 0|\hat{b}(p',\lambda') \int d^4x : \hat{\bar{\psi}}(x)\gamma_\mu\psi(x) : \hat{b}^+(p,\lambda)|0\rangle$$

$$\times g_{\mu\mu} A_\mu^{\text{cl}}(x) \tag{12.1.2}$$

$$= \sum_\mu g_{\mu\mu} \frac{\bar{u}(p',\lambda')}{(2\pi)^{3/2}} (-ie\gamma_\mu) \frac{u(p,\lambda)}{(2\pi)^{3/2}} \tilde{A}_\mu^{\text{cl}}(p' - p),$$

where

$$\tilde{A}_\mu^{\text{cl}}(k) \equiv \int d^4x\, e^{ik\cdot x} A_\mu^{\text{cl}}(x). \tag{12.1.3}$$

This shows how one can modify Feynman's rules to cope with the situation: one should *not* add the factor $(2\pi)^4 \delta_4(p_f - p_i)$; and, on the other hand, the wavy line with the cross at the end (Fig. 12.1.1) contributes a factor

$$\tilde{A}_\mu^{\text{cl}}(k),$$

given by (12.1.3), to the diagram.

The factor $\delta_4(p_f - p_i)$ is absent because we have taken the source of $A^{\text{cl}}$ to be fixed and have therefore failed to take into account that it will absorb energy–momentum as lost by the electron.

If the external field is static, $A^{\text{cl}}(x) = A^{\text{cl}}(\mathbf{r})$ and

$$\tilde{A}_\mu^{\text{cl}}(k) = 2\pi\delta(k_0) \int d^3r\, e^{-i\mathbf{kr}} A_\mu^{\text{cl}}(\mathbf{r}); \tag{12.1.4}$$

if the field is the Coulomb field due to a nucleus with charge $Z|e|$, then we can take (in Heaviside units)

$$\mathbf{A}^{\text{cl}} = 0, \ \ A_0^{\text{cl}}(\mathbf{r}) = \frac{Z|e|}{4\pi r}, \tag{12.1.5a}$$

$$\tilde{A}_\mu^{\text{cl}}(k) = 2\pi\delta(k_0)\frac{Z|e|}{\mathbf{k}^2}\delta_{\mu 0}. \tag{12.1.5b}$$

To evaluate the cross-section for e.g. electron (or positron) scattering by a nuclear field, for example, considered given, one has to be careful because the $S$-matrix element as given in (12.1.2) does not take into account that both initial and final states contain the very heavy source. One can handle this problem by reintroducing the source, with momenta $p_S, p'_S$ before and after the scattering. Because the source is very heavy, we may take $p_{S0} = p'_{S0}$; likewise, we consider that the total energy $s^{1/2}$ can be approximated by $s^{1/2} \simeq p_{S0}$. One can check that this way the factors $p_{S0}$ from the normalization

$$\langle p'_S | p_S \rangle = 2p_{S0}\delta(\mathbf{p}_S - \mathbf{p}'_S)$$

and the $s$ in the expressions for the cross-sections cancel out. The final result is as follows. For *static* potentials, let

$$\tilde{A}_\mu^{\text{cl}}(k) = 2\pi\delta(k_0)\tilde{A}_\mu^{\text{cl}}(\mathbf{k}); \tag{12.1.6a}$$

and let $\overline{F}$ be defined by

$$\langle f|S|i\rangle \equiv i\delta(E_f - E_i)\overline{F},$$

$$|i\rangle = \hat{b}^+(p, \lambda)|0\rangle, \ \ |f\rangle = \hat{b}^+(p', \lambda')|0\rangle. \tag{12.1.6b}$$

Then the cross-section, in the reference system where the source is at rest is

$$\frac{d\sigma(i \to f)}{d\Omega} = \pi^2 |\overline{F}|^2. \tag{12.1.6c}$$

For Coulomb scattering we can use (12.1.5), (12.1.2) to find, in the Born approximation, recalling that in our units $\alpha = e^2/4\pi$,

$$\frac{d\sigma_{\text{Born}}}{d\Omega} = \frac{Z^2\alpha^2}{|\mathbf{p} - \mathbf{p}'|^4} |u^+(p', \lambda')u(p, \lambda)|^2. \tag{12.1.7}$$

In the NR limit, $u^+(p', \lambda')u(p, \lambda) \simeq 2m\delta_{\lambda\lambda'}$. Writing also

$$|\mathbf{p} - \mathbf{p}'|^2 = 4|\mathbf{p}|^2 \sin^2 \theta/2,$$

with $\theta$ the scattering angle, we get

$$\frac{d\sigma_{\text{Born}}}{d\Omega} \underset{\sim}{NR} \frac{Z^2\alpha^2 m^2}{4|\mathbf{p}|^4 \sin^4 \theta/2}\delta_{\lambda\lambda'},$$

in agreement with the well-known *Rutherford formula*.

**Exercise.** Show that (12.1.7) agrees with the leading order in $\alpha_0$ of the equations we found in Sect. 4.3.2 with the wave function formalism ∎

### 12.1.2 Bremsstrahlung

A mixed situation is that in which an electron (say) radiates while being scattered (*bremsstrahlung*), which was evaluated in the nonrelativistic approximation in Sect. 8.4. We will distinguish three cases. (i) The energy $k_0$ of the radiated photon is large. The resulting cross-section, known as the Bethe–Heitler formula, is obtained with a cumbersome but straightforward evaluation like that of the previous subsection. (ii) $k_0$ is small, but still

$$\alpha |\log k_0/m| \ll 1.$$

Here one gets factorization, just as in the nonrelativistic case, Sect. 8.4: if $p_i(p_f)$ is the initial (final) electron momentum, and $k$ the photon momentum,

$$\frac{d\sigma}{d\Omega_f d\Omega_\mathbf{k} dk_0} = \left(\frac{d\sigma^{(0)}}{d\Omega_f}\right) \frac{k_0\alpha}{4\pi^2} \left(\frac{\epsilon \cdot p_f}{k \cdot p_f} - \frac{\epsilon \cdot p_i}{k \cdot p_i}\right)^2; \tag{12.1.8}$$

$d\sigma^{(0)}/d\Omega_f$ is the cross-section without photon emission; $\Omega_f$ are the polar coordinates of $\mathbf{p}_f$.

**Exercise.** Check (12.1.8). Verify that in the nonrelativistic limit it reduces to (8.4.8) ∎

(iii) $k_0$ is so small that

$$\alpha |\log k_0/m| \sim 1.$$

Now the perturbative expansion breaks down, and we get, as in the nonrelativistic case of Sect. 8.4, an infrared catastrophe. Fortunately, when $k_0$ is so small, we may assume that the emission does not affect the electron: this may be treated classically and the problem can be completely solved, which we will do in the next section.

## 12.2 Photon Emission by a Classical Source. The Bloch–Nordsieck Theorem. Classical Limit

### 12.2.1 Classical Radiation

In classical mechanics (see, for example, Landau and Lifshitz, 1951) the electromagnetic energy radiated per unit frequency $\omega$, and solid angle $\Omega$, by a moving particle is

$$\frac{dE^{\mathrm{cl}}}{d\omega d\Omega} = -\frac{c^2 \hbar \alpha}{(2\pi)^2} F,$$

$$F = \mathbf{k}^2 \int_{-\infty}^{+\infty} dt \int_{-\infty}^{+\infty} dt' \left( 1 - \frac{\dot{\mathbf{r}}(t)\dot{\mathbf{r}}(t')}{c^2} \right) e^{i\omega(t-t')-i\mathbf{k}(\mathbf{r}(t)-\mathbf{r}(t'))}.$$

$$(12.2.1)$$

Here $\omega = c|\mathbf{k}|$, and $\mathbf{r}(t)$ describes the trajectory of the particle. We can relate $F$ to the classical current created by the moving particle. This current may be characterized by a four-vector, $j_\mu^{\mathrm{cl}}(x)$, with

$$j_\mu^{\mathrm{cl}}(x) \equiv \left( \begin{array}{c} j_0^{\mathrm{cl}}(x) \\ \mathbf{j}^{\mathrm{cl}}(x) \end{array} \right) = e \left( \begin{array}{c} \delta(\mathbf{r} - \mathbf{r}(t)) \\ c^{-1}\dot{\mathbf{r}}(t)\delta(\mathbf{r} - \mathbf{r}(t)) \end{array} \right).$$

$$(12.2.2)$$

(Throughout this section we use Heaviside units.) It is easy to show that $j_\mu^{\mathrm{cl}}$ satisfies the continuity equation

$$\partial \cdot j^{\mathrm{cl}}(x) = 0.$$

$$(12.2.3)$$

We define the Fourier transform of $j^{\mathrm{cl}}(x)$ by

$$j_\mu^{\mathrm{cl}}(k) = \int d^4 x\, e^{ik \cdot x} j_\mu^{\mathrm{cl}}(x);$$

$$(12.2.4a)$$

note that, because $j_\mu^{\mathrm{cl}}(x)$ is real, we have

$$j_\mu^{\mathrm{cl}}(-k) = j_\mu^{\mathrm{cl}}(k)^*.$$

$$(12.2.4b)$$

By using the explicit expression (12.2.2), we find that

$$j^{\mathrm{cl}}(k) = e \int_{-\infty}^{+\infty} dx_0\, e^{ik_0 x_0 - i\mathbf{k}\mathbf{r}(t)} \left( \begin{array}{c} 1 \\ \dot{\mathbf{r}}(t)/c \end{array} \right).$$

In particular, if we define

$$\tilde{j}_\mu^{\mathrm{cl}}(k) \equiv j_\mu^{\mathrm{cl}}(k)\big|_{k_0 = |\mathbf{k}|},$$

$$(12.2.5a)$$

$$\tilde{j}_\mu^{\mathrm{cl}}(k) = ec \int_{-\infty}^{+\infty} dt\, e^{i\omega t - i\mathbf{k}\mathbf{r}(t)} \left( \begin{array}{c} 1 \\ \dot{\mathbf{r}}(t)/c \end{array} \right),$$

$$(12.2.5b)$$

it then follows that one can write

$$F = \frac{\mathbf{k}^2}{e^2 c^2} \tilde{j}^{\mathrm{cl}}(k) \cdot \tilde{j}^{\mathrm{cl}}(k)^*.$$

$$(12.2.5c)$$

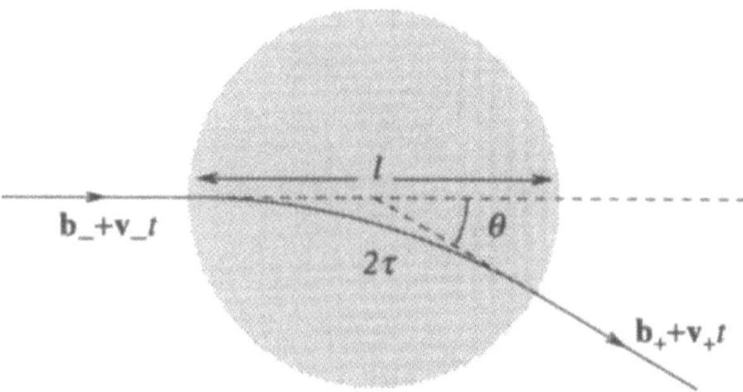


**Fig. 12.2.1.** Trajectory of the radiating particle.

Let us consider the case of very soft photon emission. We assume that the trajectory of the radiating particle satisfies

$$\mathbf{r}(t) \underset{t\to\pm\infty}{\simeq} \mathbf{b}_\pm + \mathbf{v}_\pm t.$$

If the region of interaction is of size $l$ (Fig. 12.2.1) and the time the particle spends there is $2\tau$, we will also assume that

$$|\mathbf{k}|l \ll 1, \ \omega\tau \ll 1. \tag{12.2.6}$$

In this case we may approximate

$$\tilde{j}^{\text{cl}}(k) \simeq ec \left\{ \int_\tau^\infty dt\, e^{i\omega t - i\mathbf{k}\mathbf{v}_+ t - i\mathbf{k}\mathbf{b}_+} \begin{pmatrix} 1 \\ \mathbf{v}_+/c \end{pmatrix} \right.$$

$$+ \int_{-\infty}^{-\tau} dt\, e^{i\omega t - i\mathbf{k}\mathbf{v}_- t - i\mathbf{k}\mathbf{b}_-} \begin{pmatrix} 1 \\ \mathbf{v}_-/c \end{pmatrix}$$

$$\left. + \int_{-\tau}^{+\tau} dt\, e^{i\omega t - i\mathbf{k}\mathbf{r}(t)} \begin{pmatrix} 1 \\ \dot{\mathbf{r}}(t)/c \end{pmatrix} \right\}.$$

As we will see, the resulting $\tilde{j}^{\text{cl}}$ diverges as $|\mathbf{k}| \to 0$; then, the piece $\int_{-\tau}^{+\tau}$ that clearly stays finite in this limit can be neglected. Likewise, the integrations $\int_\tau^\infty$, $\int_{-\infty}^{-\tau}$ can be extended to zero with negligible error. Finally, because $|\mathbf{b}_\pm| < l$, we can replace $\exp(-i\mathbf{k}\mathbf{b}_\pm) \simeq 1$. We then get

$$\tilde{j}^{\text{cl}}(k) \underset{\omega\to 0}{\simeq} ec \int_0^\infty dt \left\{ e^{i(\omega+\mathbf{k}\mathbf{v}_+)t} \begin{pmatrix} 1 \\ \mathbf{v}_+/c \end{pmatrix} + e^{-i(\omega+\mathbf{k}\mathbf{v}_-)t} \begin{pmatrix} 1 \\ \mathbf{v}_-/c \end{pmatrix} \right\}.$$

The integrals can be performed by introducing a convergence factor $e^{-\epsilon t}$ and then letting $\epsilon \to 0$. We find that

$$\tilde{j}^{\text{cl}}(k) \underset{\omega\to 0}{\simeq} \frac{1}{\omega - \mathbf{k}\mathbf{v}_+} \begin{pmatrix} 1 \\ \mathbf{v}_+/c \end{pmatrix} - \frac{1}{\omega - \mathbf{k}\mathbf{v}_-} \begin{pmatrix} 1 \\ \mathbf{v}_-/c \end{pmatrix}, \tag{12.2.7a}$$

and thus

$$F \underset{\omega \to 0}{\simeq} \frac{\mathbf{k}^2}{c^2 \omega^2} \left\{ \frac{|\mathbf{n}(\mathbf{v}_+ - \mathbf{v}_-)|^2}{(1 - \mathbf{n}\mathbf{v}_+/c)^2 (1 - \mathbf{n}\mathbf{v}_-/c)^2} \right.$$

$$\left. - \left( \frac{1}{1 - \mathbf{n}\mathbf{v}_+/c}\mathbf{v}_+ - \frac{1}{1 - \mathbf{n}\mathbf{v}_-/c}\mathbf{v}_- \right)^2 \right\}, \quad \mathbf{n} = \mathbf{k}/|\mathbf{k}|. \tag{12.2.7b}$$

In the NR approximation,

$$F \underset{\substack{\omega \to 0 \\ NR}}{\simeq} \frac{1}{c^4} \left\{ |\mathbf{n}(\mathbf{v}_+ - \mathbf{v}_-)|^2 - |\mathbf{v}_+ - \mathbf{v}_-|^2 \right\}; \tag{12.2.8a}$$

for the energy radiated in all directions we integrate $dE^{\mathrm{cl}}/d\omega d\Omega$ in $d\Omega$ and then get

$$\frac{dE^{\mathrm{cl}}}{d\omega} = \frac{2\hbar\alpha}{3\pi} \left( \frac{\mathbf{v}_+ - \mathbf{v}_-}{c} \right)^2 = \frac{8\hbar\alpha}{3\pi} \frac{v^2}{c^2} \sin^4 \frac{\theta}{2}, \tag{12.2.8b}$$

$\theta$ being the deflection angle (Fig. 12.2.1), and we have set $|\mathbf{v}_\pm| = v$.

### 12.2.2 Photon Emission by a Classical Current

We now consider a classical current, $j_\mu^{\mathrm{cl}}(x)$ (given by (12.2.2), say) in interaction with the quantized photon field, $\hat{A}_\mu(x)$. The interaction Hamiltonian is thus

$$\hat{H}_{\mathrm{int}} = \int d^3r j^{\mathrm{cl}}(x) \cdot \hat{A}(x) \tag{12.2.9}$$

(we work here in natural units, $\hbar = c = 1$). Under these circumstances we have (Appendix A.5)

$$\hat{S} = e^K : \exp(-i \int d^4x j^{\mathrm{cl}}(x) \cdot \hat{A}(x)) :, \tag{12.2.10a}$$

where $K$ is a pure $c$ number,

$$K = \frac{-1}{2} \int d^4x d^4y \sum j_\mu^{\mathrm{cl}}(x) D_{\mu\nu}(x - y) j_\nu^{\mathrm{cl}}(y) g_{\mu\mu} g_{\nu\nu}, \tag{12.2.10b}$$

or, in terms of the Fourier transforms,

$$K = \frac{i}{2(2\pi)^4} \int \frac{d^4k}{k^2 + i0} j^{\mathrm{cl}}(k)^* \cdot j^{\mathrm{cl}}(k).$$

By writing (P.P. denotes the principal value)

$$\frac{1}{k^2 + i0} = \mathrm{P.P.} \frac{1}{k^2} - i\pi\delta(k^2),$$

we can split $K$ into a purely imaginary and a real part,

$$K = i\Phi + R, \tag{12.2.11a}$$

$$\Phi = \frac{1}{2(2\pi)^4}\text{P.P.}\int \frac{d^4k}{k^2} j^{\text{cl}}(k)^* \cdot j^{\text{cl}}(k), \tag{12.2.11b}$$

$$R = \frac{\pi}{2(2\pi)^4}\int d^4k\delta(k^2)j^{\text{cl}}(k)^* \cdot j^{\text{cl}}(k)$$

$$\tag{12.2.11c}$$

$$= \frac{1}{2(2\pi)^3}\int \frac{d^3k}{2k_0}\tilde{j}^{\text{cl}}(k)^* \cdot \tilde{j}^{\text{cl}}(k).$$

To evaluate the probability of emission of $N_{\mathbf{k}\eta}$ photons with wave vector $\mathbf{k}$ and helicity $\eta$ we construct normalizable states. To do so, we proceed as in Sect. 8.5, and enclose the system in a box with side $L$, requiring also periodic boundary conditions:

$$\mathbf{k} = \rho\mathbf{n}, \ \rho = 2\pi/L, \ n_j = \pm 1, \pm 2, \ldots, j = 1, 2, 3.$$

(We exclude the value $n_j = 0$ to avoid divergences; in the limit $L \to \infty$, this will have no effect.) Then, we can go from sums to integrals, and vice versa, with the replacements

$$\int d^3k \leftrightarrow \rho^3 \sum_{\mathbf{n}}, \ \delta(\mathbf{k} - \mathbf{k}') \leftrightarrow \rho^{-3}\delta_{\mathbf{n}\mathbf{n}'}.$$

A state with $N_{\mathbf{k}\eta}$ photons normalized to unity will then be

$$|N_{\mathbf{k}\eta}\rangle = \left(\frac{\rho^3}{2k_0}\right)^{N_{\mathbf{k}\eta}/2}\frac{1}{\sqrt{N_{\mathbf{k}\eta}!}}[\hat{a}^+(\mathbf{k}, \eta)]^{N_{\mathbf{k}\eta}}|0\rangle. \tag{12.2.12}$$

The probability of emission of these $N_{\mathbf{k}\eta}$ photons can be obtained in terms of the $S$-matrix elements:

$$W(N_{\mathbf{k}\eta}) = |\langle N_{\mathbf{k}\eta}|\hat{S}|0\rangle|^2. \tag{12.2.13}$$

Because of the structure of $\hat{S}$, (12.2.10a), the nonvanishing piece there will be the $N_{\mathbf{k}\eta}$th term in the expansion of the exponent, and only the creator part of $\hat{A}(x)$ will contribute. Moreover, the phase $e^{i\Phi}$ will drop when we calculate probabilities. We may then replace $\hat{S}$ by the effective expression

$$\hat{S}_{\text{eff}} = e^R$$

$$\times \exp\left(\frac{-i}{(2\pi)^{3/2}}\int d^4x \sum_{\mu\eta} g_{\mu\mu}j_{\mu}^{\text{cl}}(x)\int \frac{d^3k}{2k_0}e^{ik\cdot x}\epsilon_{\mu}^*(k, \eta)\hat{a}^+(k, \eta)\right) \tag{12.2.14a}$$

$$= e^R\exp\left(\frac{-i}{(2\pi)^{3/2}}\sum_{\eta}\int \frac{d^3k}{2k_0}\epsilon^*(k, \eta) \cdot \tilde{j}^{\text{cl}}(k)\hat{a}^+(k, \eta)\right).$$

In the discrete approximation,

$$\hat{S}_{\text{eff}} = \left\{ \exp\left( \frac{1}{2(2\pi)^3} \rho^3 \sum_{\mathbf{n}} \frac{\tilde{j}^{\text{cl}}(k)^* \cdot \tilde{j}^{\text{cl}}(k)}{2k_0} \right) \right\}$$

$$\times \exp\left( \frac{-i\rho^3}{(2\pi)^{3/2}} \sum_{\eta} \sum_{\mathbf{n}} \frac{\epsilon^*(k,\eta) \cdot \tilde{j}^{\text{cl}}(k)}{2k_0} \hat{a}^+(k,\eta) \right).$$

(12.2.14b)

It is clear that the photon creation will be uncorrelated. For the case of (12.2.13), where only the $N_{\mathbf{k}\eta}$th term in the expansion of the second exponential on the right-hand side of (12.2.14b) will contribute, we get

$$W(N_{\mathbf{k}\eta}) = e^{2R}\overline{W}(N_{\mathbf{k}\eta}),$$

$$\overline{W}(N_{\mathbf{k}\eta}) = \left| (-i)^{N_{\mathbf{k}\eta}} \left( \frac{\rho^3}{(2\pi)^{3/2}} \frac{\epsilon^*(k,\eta) \cdot \tilde{j}^{\text{cl}}(k)}{2k_0} \right)^{N_{\mathbf{k}\eta}} \right|^2$$

(12.2.15a)

$$\times |\langle N_{\mathbf{k}\eta}|[\hat{a}^+(\mathbf{k},\eta)]^{N_{\mathbf{k}\eta}}|0\rangle|^2.$$

Using the definition of $|N_{\mathbf{k}\eta}\rangle$ in (12.2.12), and the discretized commutation relations

$$[\hat{a}(\mathbf{k},\eta), \hat{a}^+(\mathbf{k}',\eta')] = \rho^{-3}2k_0\delta_{\mathbf{nn}'}\delta_{\eta\eta'},$$

we get

$$\langle N_{\mathbf{k}\eta}|[\hat{a}^+(\mathbf{k},\eta)]^{N_{\mathbf{k}\eta}} = \left( \frac{2k_0}{\rho^3} \right)^{N_{\mathbf{k}\eta}/2} \frac{1}{\sqrt{N_{\mathbf{k}\eta}!}} \langle 0|,$$

and therefore

$$\overline{W}(N_{\mathbf{k}\eta})$$

(12.2.15b)

$$= \frac{1}{N_{\mathbf{k}\eta}!} \left( \frac{\rho^3}{2k_0} \frac{\epsilon^*(k,\eta) \cdot \tilde{j}^{\text{cl}}(k)\epsilon(k,\eta) \cdot \tilde{j}^{\text{cl}}(k)^*}{(2\pi)^3} \right)^{N_{\mathbf{k}\eta}}.$$

Although we could perform the calculation for photons of definite helicity, let us consider the case where we are not interested in the polarization of the emitted photons. Writing for ease of notation $N_{\mathbf{k}}^{\eta} \equiv N_{\mathbf{k}\eta}$, we get the probability for production of $N_{\mathbf{k}}$ photons with either helicity,

$$W(N_{\mathbf{k}}) = e^{2R} \sum_{N_{\mathbf{k}}^+ + N_{\mathbf{k}}^- = N_{\mathbf{k}}} \left\{ \overline{W}(N_{\mathbf{k}}^+)\,\overline{W}(N_{\mathbf{k}}^-) \right\}.$$

(12.2.16a)

(Note that the term $e^{2R}$, being a pure $c$-number, is independent of the final state.) Using the formula

$$\sum_{n_1+n_2=n} \left( \frac{1}{n_1!}a_1^{n_1} \frac{1}{n_2!}a_2^{n_2} \right) = \frac{1}{n!}(a_1 + a_2)^n,$$

we have, defining also $\lambda \equiv 2\rho^3/2(2\pi)^3 k_0$,

$$\sum_{N_{\mathbf{k}}^+ + N_{\mathbf{k}}^- = N_{\mathbf{k}}} \left\{ \frac{\lambda^{N_{\mathbf{k}}^+}}{N_{\mathbf{k}}^+!} \left[ (\epsilon^* \cdot \tilde{j}^{\mathrm{cl}})(\epsilon \cdot \tilde{j}^{cl*}) \right]^{N_{\mathbf{k}}^+} \frac{\lambda^{N_{\mathbf{k}}^-}}{N_{\mathbf{k}}^-!} \left[ (\epsilon^* \cdot \tilde{j}^{\mathrm{cl}})(\epsilon \cdot \tilde{j}^{cl*}) \right]^{N_{\mathbf{k}}^-} \right\}$$

$$= \frac{\lambda^{N_{\mathbf{k}}}}{N_{\mathbf{k}}!} \left( \sum_{\eta} \epsilon^*(k,\eta) \cdot \tilde{j}^{\mathrm{cl}}(k) \epsilon(k,\eta) \cdot \tilde{j}^{\mathrm{cl}}(k)^* \right)^{N_{\mathbf{k}}}$$

$$= \frac{\lambda^{N_{\mathbf{k}}}}{N_{\mathbf{k}}!} \left( -\tilde{j}^{\mathrm{cl}}(k) \cdot \tilde{j}^{\mathrm{cl}}(k)^* \right)^{N_{\mathbf{k}}},$$

the last step because $\sum_{\eta} \epsilon_\mu(k,\eta)\epsilon_\nu(k,\eta)^* = -g_{\mu\nu} +$ g.t., where the "g.t." contain terms proportional to $k_\mu$ or $k_\nu$ that yield zero owing to current conservation. Finally,

$$W(N_{\mathbf{k}}) = e^{2R} \frac{1}{N_{\mathbf{k}}!} \left( \frac{-\rho^3}{2k_0} \frac{\tilde{j}^{\mathrm{cl}}(k) \cdot \tilde{j}^{\mathrm{cl}}(k)^*}{(2\pi)^3} \right)^{N_{\mathbf{k}}}. \tag{12.2.16b}$$

Likewise, if we are only interested in the probability of producing $N$ photons with whatever wave vectors, we get the quantity

$$W(N) = e^{2R} \prod_{\mathbf{k}} \frac{1}{N_{\mathbf{k}}!} \left( -\rho^3 \frac{\tilde{j}^{\mathrm{cl}}(k) \cdot \tilde{j}^{\mathrm{cl}}(k)^*}{2(2\pi)^3 k_0} \right)^{N_{\mathbf{k}}}$$

$$= \prod_{\mathbf{k}} \frac{1}{N_{\mathbf{k}}!} e^{-\overline{N}_{\mathbf{k}}} (\overline{N}_{\mathbf{k}})^{N_{\mathbf{k}}}, \tag{12.2.17}$$

$$\overline{N}_{\mathbf{k}} = -\frac{\rho^3}{2(2\pi)^3 k_0} \tilde{j}^{\mathrm{cl}}(k) \cdot \tilde{j}^{\mathrm{cl}}(k)^*, \tag{12.2.18}$$

where we have used the explicit expression (found, for example, in (12.2.11)) for $R$. Not surprisingly, (12.2.17) is a Poisson distribution, so the photons are emitted in a totally uncorrelated manner.

The quantity $\overline{N}_{\mathbf{k}}$ can be interpreted as the average number of photons with wave number $\mathbf{k}$ emitted, since

$$\overline{N}_{\mathbf{k}} = \sum N_{\mathbf{k}} W(N_{\mathbf{k}_1}, \ldots, N_{\mathbf{k}}, \ldots).$$

The energy radiated can then be written as that of a single photon, $\hbar\omega$, times the average number of photons,

$$E = \sum_{\mathbf{n}} \overline{N}_{\mathbf{k}} \hbar\omega, \quad \omega = ck_0,$$

or, in natural units, and going over to the continuum,

$$E = \sum_{\mathbf{n}} \overline{N}_{\mathbf{k}} k_0 = \rho^{-3} \int d^3k\, k_0 \overline{N}_{\mathbf{k}}$$

$$= \frac{1}{2(2\pi)^3} \int d^3k (-\tilde{j}^{\mathrm{cl}}(k) \cdot \tilde{j}^{\mathrm{cl}}(k)^*)$$

$$= \frac{1}{2(2\pi)^3} \int d\omega\, d\Omega_{\mathbf{k}} (-\mathbf{k}^2 \tilde{j}^{\mathrm{cl}}(k) \cdot \tilde{j}^{\mathrm{cl}}(k)^*).$$

Therefore, we have obtained

$$\frac{dE}{d\omega d\Omega} = -\frac{1}{2(2\pi)^3}\mathbf{k}^2\tilde{j}^{\mathrm{cl}}(k)\cdot\tilde{j}^{\mathrm{cl}}(k)^*,\tag{12.2.19}$$

*which agrees with the classical expression,* (12.2.1), *when we recall* (12.2.5c) for $F$ and that $e^2/4\pi = \alpha$.

### 12.2.3 Radiation of Coherent States

In the previous subsection we saw that some of the properties of the radiation from a classical source are like those of *classical* radiation of *classical* electromagnetic fields. We will now complete the demonstration by showing that indeed the products of radiation from a classical source are the coherent states discussed in Sect. 8.5 which, up to corrections of order $\hbar$, correspond to classical radiation, and this in spite of the fact that the photons are emitted in an uncorrelated manner.

We start with a state without photons; more complicated situations may be treated without undue difficulty but present few novel features. At time $t \to -\infty$ we thus begin with the vacuum, $|0\rangle$. At time $t \to +\infty$ this will have evolved into a state that, up to an irrelevant phase, is

$$|0\rangle \to \hat{S}_{\mathrm{eff}}|0\rangle.$$

With the system enclosed in a box of side $L$ with periodic boundary conditions, we may use (12.2.14b) to obtain

$$\hat{S}_{\mathrm{eff}}|0\rangle = \Omega \prod_{\eta\mathbf{n}} e^{\rho^3 \alpha_0 \hat{a}^+(\mathbf{k_n},\eta)}|0\rangle,\tag{12.2.20a}$$

where

$$\Omega = \exp\left(\frac{1}{2(2\pi)^3}\rho^3 \sum_{\mathbf{n}} \frac{\tilde{j}^{\mathrm{cl}}(k)^*\cdot\tilde{j}^{\mathrm{cl}}(k)}{2k_0}\right)\tag{12.2.20b}$$

and the amplitude $\alpha_0$ is

$$\alpha_0 = \alpha_0(k,\eta) = \frac{i\epsilon^*(\mathbf{k},\eta)\cdot\tilde{j}^{\mathrm{cl}}(k)}{2(2\pi)^{3/2}k_0}.\tag{12.2.20c}$$

Equation (12.2.20a), when compared with (8.5.7), clearly shows the state $\hat{S}_{\mathrm{eff}}|0\rangle$ to be a product of coherent states $|\Phi(\mathbf{k},\eta)\rangle$ with amplitudes $\alpha_0(\mathbf{k},\eta)$ given (as in the classical case) by (12.2.20c).

### 12.2.4 The Bloch–Nordsieck Theorem

We showed in previous sections that the perturbative calculation fails for emission of soft photons. This failure occurs when

$$\alpha \log(E_0/E_\gamma) \sim 1,\tag{12.2.21}$$

where $E_0$ is a typical energy. For example, when $E_0$ is of the order of the rest energy of the particle, $mc^2$, condition (12.2.20) will hold for

$$E_\gamma \sim mc^2 \times e \times 10^{-59},$$

for electrons this is of the order of $10^{-54}$ eV.

For such low-frequency photons, the emission of any number of them leaves the motion of the radiating particle unaltered; we may therefore consider this motion as *given*. For the case where there are many radiating particles, we need only be interested in the average motion. Under these circumstances, we can approximate the quantum current of radiating particles by its average, i.e., by the *classical* current generated by a classical particle with a prescribed trajectory. That is to say, we can use as interaction Hamiltonian that of a classical current interacting with a *quantized* radiation field, precisely the situation considered in the previous subsections (see, for example, (12.2.9)). As shown there, one can solve the problem totally, and in particular one finds that the quantum radiation reproduces, for the energy and intensity radiated, the purely classical formulas. This is the result of the Bloch–Nordsieck (1937) theorem[1].

A byproduct of the analysis is that a scattered charge always radiates, and indeed the number of radiated photons tends to infinity as their frequency goes to zero, (12.2.18). Thus, strictly speaking, there are no such things as isolated electrons; to be accurate, one would have to replace the statement "an electron with momentum **p**", by "an electron with momentum **p** accompanied by no photons with frequency above a certain $\omega_{\max}$", $\omega_{\max}$ given, for example, by the threshold for experimental detection.

## 12.3 Propagation of an Electron in a Classical Potential. The Proper-Time Method

In this section we will consider the propagation of a Dirac particle in a classical, static potential. If we denote by $V_\mu^{\rm cl}$ the potential[2], then the propagator will satisfy the differential equation

$$(i\partial\!\!\!/ - m + V\!\!\!\!/^{\rm cl}(x))S^V(x,y) = i\delta(x-y); \tag{12.3.1}$$

note that now $S$ depends both on $x$ and $y$ (and not just on the difference, $x-y$). Equation (12.3.1) may be solved in a number of cases. In the first subsection here we will consider the situation where one can solve the equation for the energies and wave functions, which will be illustrated with the case of an electron in a Coulomb potential. In the next subsections we will develop

---

<sup></sup>[1] The proof we presented here in Section 12.2.2 is not the original one but that due to Glauber (1951).

[2] We assume it to be a four-vector so that the analysis is directly applicable to the case where the particle is an electron, and the potential is the Coulombic one.

a method, the proper-time method, with which the case of a charged particle in a constant field or in a plane electromagnetic wave may be fully solved.

### 12.3.1 Electron in a Coulomb Potential

To solve this problem we recall the solutions for the energy levels, and wave functions, found in Sects. 4.3, 4.4 for the situation where one has an electron in a Coulombic field. We introduce some notation. We denote by $U$, $V$ the corresponding wave functions for particles and antiparticles. To be precise, suppose the potential to be

$$V_\mu^{cl}(x) = (-\alpha_0/r)\delta_{\mu 0}$$

(natural units and the Heaviside system are used throughout). If the charge of the source of the potential is $Z|e|$, then $\alpha_0 = Ze^2$: we take $\alpha_0$ to be positive (potential *attractive* for electrons). We then set

$$U^{(k,j,M\omega)}(x) = \frac{k}{\sqrt{\pi E_k}} e^{-iE_k t} \psi_{k\omega}^{jM}(\mathbf{r}),$$

$$V^{(k,j,M\omega)}(x) = \frac{k}{\sqrt{\pi E_k}} e^{iE_k t} i\gamma_2 \psi^{jM}(\mathbf{r})^*,$$

(12.3.2)

for the continuous spectrum, and one has

$$E_k = \sqrt{m + k^2}, \ k = |\mathbf{k}|.$$

The $V$ will represent positrons. For the discrete spectrum only the $U$ exist:

$$U^{(n',j,M,\omega)}(x) = e^{-iE_{n'}t} \psi_{n'\omega}^{(jM)}(\mathbf{r}),$$

$$E_{n'} = m\left\{ 1 + \left[\alpha_0/(n' + \sqrt{(j+1/2)^2 - \alpha_0^2})\right]^2 \right\}^{-1/2}.$$

(12.3.3)

The $\psi$ may be written as (for, e.g., the continuous spectrum and with $j = l + 1/2$)

$$\psi_{k+}^{(jM)}(\mathbf{r}) = \begin{pmatrix} \mathcal{Y}_M^{l+} f_{l+} \\ \mathcal{Y}_M^{l+1,-} g_{l+} \end{pmatrix},$$

$$\psi_{k-}^{(jM)}(\mathbf{r}) = \begin{pmatrix} \mathcal{Y}_M^{l+1,-} f_{l-} \\ \mathcal{Y}_M^{l+} g_{l-} \end{pmatrix};$$

(12.3.4)

recall (3.6.9). Finally, the $f, g$ are given in (4.2.13), (4.4.8). With the normalization adopted there we have the relation

$$\int d^3 r\, W^{(kjM\omega)}(x)^* W^{(k'j'M'\omega')}(x) = \delta(k - k')\delta_{jj'}\delta_{MM'}\delta_{\omega\omega'},$$

(12.3.5a)

$W = U$ or $V$, and

$$\int d^3 r\, U^{(n'jM\omega)}(x)^* U^{(\overline{n}'j'M'\omega')}(x) = \delta_{n'\overline{n}'}\delta_{jj'}\delta_{MM'}\delta_{\omega\omega'}. \tag{12.3.5b}$$

Because the $U$ form a complete set of solutions of the Dirac equation (including the potential interaction) for electrons, and the $V$ for positrons, we may define a field operator representing electrons and positrons in the Coulombic field by

$$\begin{aligned}
\hat{\psi}^C(x) = \frac{1}{\sqrt{2}} \int_0^\infty dk \sum_{jM\omega} \Big\{ &U^{(kjM\omega)}(x)\hat{b}_C(kjM\omega) \\
+ V^{(kjM\omega)}(x)\hat{d}_C^+(kjM\omega)\Big\} &+ \sum_{n'jM\omega} U^{(n'jM\omega)}(x)\hat{b}_C(n'jM\omega).
\end{aligned} \tag{12.3.6}$$

The operators $\hat{b}_C, \hat{b}_C^+$ destroy and create interacting electrons; and the same for the $\hat{d}_C, \hat{d}_C^+$ and positrons. The canonical commutation relation

$$\left\{ \hat{\psi}_a^C(\mathbf{r},t), \hat{\psi}_{a'}^C(\mathbf{r}',t)^+ \right\} = \delta_{aa'}\delta(\mathbf{r} - \mathbf{r}'), \tag{12.3.7}$$

will be satisfied if we assume the anticommutation relations

$$\left\{ \hat{b}_C(kjM\omega), \hat{b}_C^+(k'j'M'\omega') \right\} = \delta(k - k')\delta_{jj'}\delta_{MM'}\delta_{\omega\omega'}, \tag{12.3.8a}$$

and an identical equation for the $\hat{d}_C, \hat{d}_C^+$, and

$$\left\{ \hat{b}_C(n'jM\omega), \hat{b}_C^+(\overline{n}'j'M'\omega') \right\} = \delta_{n'\overline{n}'}\delta_{jj'}\delta_{MM'}\delta_{\omega\omega'}. \tag{12.3.8b}$$

It is then intuitively evident that the expression for the propagator will be

$$\begin{aligned}
S_{aa'}^C(x,y) = \theta(x_0 - y_0)\underline{\hat{\psi}_a^C(x)\hat{\overline{\psi}}_{a'}^C(y)} \\
+\theta(y_0 - x_0)\underline{\hat{\overline{\psi}}_{a'}^C(y)\hat{\psi}_a^C(x)}.
\end{aligned} \tag{12.3.9}$$

The explicit verification, which involves (12.3.7) and the equation $\delta(x_0-y_0) = \partial\theta(x_0 - y_0)/\partial x_0$, is left to the reader.

Evaluating the contractions in (12.3.9), we can find the explicit expression for the propagator:

$$S^C(x,y) = S_+(x,y) + S_-(x,y), \tag{12.3.10a}$$

$$\begin{aligned}
S_+(x,y) = \int dk \sum_{jM\omega} &U^{(kjM\omega)}(x)\overline{U}^{(kjM\omega)}(y) \\
+ \sum_{n'jM\omega} &U^{(n'jM\omega)}(x)\overline{U}^{(n'jM\omega)}(y),
\end{aligned} \tag{12.3.10b}$$

$$S_-(x,y) = \int dk \sum_{jM\omega} V^{(kjM\omega)}(x)\overline{V}^{(kjM\omega)}(y). \tag{12.3.10c}$$

The solution given by (12.3.10) can only be obtained if the Dirac wave equation can be solved; but a formal expression for $S$, in the general case, may be deduced as follows. We perform a four-dimensional Fourier transform of (12.3.1) both in $x$ *and* $y$. If we define

$$S^V(p,p') = \frac{1}{(2\pi)^4} \int d^4x d^4y \, e^{ip\cdot x} e^{-ip'\cdot y} S^V(x,y), \qquad (12.3.11a)$$

$$V(q) = \frac{1}{(2\pi)^4} \int d^4x \, e^{iq\cdot x} V(x), \qquad (12.3.11b)$$

then (12.3.1) becomes

$$\int d^4p'' \left\{ (\not{p} - m)\delta(p - p'') + V(p - p'') \right\} S^V(p'',p')$$
$$= i\delta(p - p'). \qquad (12.3.12)$$

We may introduce a functional Hilbert space of functions $\varphi(p)$; on it we may interpret (12.3.12) as a relation between kernels of integral operators: $\delta(p-p')$ is the kernel of the unit operator, and $V$ acts as follows:

$$V \; : \; \varphi(p) \to \int d^4p'' V(p - p'')\varphi(p'').$$

Then we write (12.3.12) in operator form as

$$(\not{p} - m + V)S^V = i,$$

whose inverse gives the desired expression for the propagator:

$$S^V = i(\not{p} - m + V)^{-1}. \qquad (12.3.13)$$

This expression is an obvious generalization of the propagator for free particles, (9.7.6).

### 12.3.2 The Proper-Time Method

The expressions we have deduced for the propagator of a fermion in an external field were either formal or in the form of an infinite sum. There are, however, a few situations where closed formulas may be obtained using the Fock–Schwinger *proper-time method*, to which we now turn.

We consider an equation like that satisfied by the fermion propagator, (12.3.1),

$$H(x, i\partial)G(x, y) = \delta(x - y); \qquad (12.3.14)$$

$H$ is a differential operator. The idea is to consider a Hilbert space (similar to the one introduced at the end of the previous subsection) of functions $\varphi(x)$ of the *four* variables $x_\mu$, and to consider $H$ as a Hamiltonian defined in this Hilbert space; then $G$ would be the corresponding Green function. The operators $X_\mu, P_\mu = i\partial_\mu$ are then canonical operators in this Hilbert space; they satisfy the commutation relations

$$[X_\mu, P_\nu] = -ig_{\mu\nu}.$$

A basis in the Hilbert space is that of the states $|x\rangle$, normalized to $\langle x|y\rangle = \delta(x - y)$, and such that

$$X_\mu|x\rangle = x_\mu|x\rangle. \tag{12.3.15}$$

For any state $|\varphi\rangle$, we define the operator $H(X, P)$ so that

$$\langle x|H(X, P)|\varphi\rangle = H(x, i\partial)\varphi(x) = H(x, i\partial)\langle x|\varphi\rangle. \tag{12.3.16}$$

Likewise,

$$\langle x|X_\mu|x_0\rangle = x_\mu\delta(x - x_0), \ \langle x|P_\mu|x_0\rangle = i\partial_\mu\langle x|x_0\rangle = i\partial_\mu\delta(x - x_0). \tag{12.3.17}$$

The Green function, $G$, is defined by[3]

$$G(x, x_0) = \langle x|H(X, P)^{-1}|x_0\rangle. \tag{12.3.18}$$

Indeed: let $H^{-1}|x_0\rangle \equiv |y_0\rangle$. Then, on the one hand,

$$\langle x|H|y_0\rangle = H(x, i\partial)\langle x|y_0\rangle,$$

and on the other,

$$\langle x|H|y_0\rangle = \langle x|y_0\rangle = \delta(x - y_0).$$

Therefore,

$$H(x, i\partial)\langle x|H^{-1}|x_0\rangle = \delta(x - y_0),$$

which agrees with (12.3.14) if we identify $G$ as in (12.2.18).

We now solve the problem as in the standard quantum mechanical case. We introduce an evolution operator $U(\tau) = e^{-i\tau H}$ which evolves in a fictitious time[4] $\tau$, and whose generator is $H$. It satisfies an evolution equation and two boundary conditions. They are best written by defining the functions $U(x, x_0; \tau)$:

$$U(x, x_0; \tau) = \langle x|e^{-i\tau H}|x_0\rangle. \tag{12.3.19}$$

Then, we have

$$i\frac{\partial}{\partial\tau}U(x, x_0; \tau) = H(x, i\partial)U(x, x_0; \tau), \tag{12.3.20a}$$

$$U(x, x_0; 0) = \delta(x - x_0); \tag{12.3.20b}$$

$$\lim_{\tau \to -\infty} U(x, x_0; \tau) = 0. \tag{12.3.20c}$$

---

[3] The inverse $H^{-1}$ is undefined. Later we will use the pertinent boundary conditions to give a prescription to circumvent the singularity.

[4] This is the "proper time", whose classical analogue is indeed the proper time of the particle.

The first two are obvious. The last is related to the regulation necessary to define $H^{-1}$. It implies that we really replace $H$ by $H + i\epsilon$ with $\epsilon > 0$, $\epsilon \to 0$. This replacement should be assumed in both (12.3.18) and (12.3.19). We then have

$$G(x, x_0) = -i \int_{-\infty}^{0} d\tau U(x, x_0; \tau),  \qquad (12.3.21)$$

as may be easily checked.

The difficult part, of course, is the calculation of $U$. For this, we go to a "Heisenberg picture", defining operators and states

$$X_\mu(\tau) = U^{-1}(\tau) X_\mu U(\tau), \;\; P_\mu(\tau) = U^{-1}(\tau) P_\mu U(\tau),  \qquad (12.3.22a)$$

$$|x(\tau)\rangle = U(\tau)|x\rangle.  \qquad (12.3.22b)$$

It is clear that, because

$$P_\mu(\tau) = m \frac{\partial}{\partial \tau} X_\mu(\tau),$$

one has that $P_\mu(\tau)$ is actually a function of $X(0), X_\mu(\tau)$; so we may write

$$H(X(\tau), P(\tau)) \equiv U^{-1}(\tau) H(X, P) U(\tau)$$

$$= F(X(\tau), X(0); \tau).  \qquad (12.3.23)$$

We may then elaborate (12.3.20a) to

$$i\partial_\tau U(x, x_0; \tau) = \langle x | H(X, P) U | x_0 \rangle$$

$$= \langle x | U U^{-1} H U | x_0 \rangle = \langle x(\tau) | U^{-1} H U | x_0 \rangle$$

$$= \langle x(\tau) | F(X(\tau), X(0); \tau) | x_0(0) \rangle$$

$$= F(x(\tau), x(0); \tau) \langle x(\tau) | x_0(0) \rangle.$$

The last step follows if we have rearranged the $X(\tau), X(0)$ of $F$ so that they are ordered, the first to the left of the last. The equation is now an ordinary differential equation, with solution

$$U(x, x_0; \tau) = C(x, x_0) \exp\left( -i \int^{\tau} d\tau' F(x, x_0; \tau') \right).  \qquad (12.3.24)$$

The constant $C$ may be obtained by requiring

$$\langle x(\tau) | P_\mu(\tau) | x_0(0) \rangle = i\partial_\mu \langle x(\tau) | x_0(0) \rangle,$$

$$\langle x(\tau) | P_\mu(0) | x_0(0) \rangle = -i \frac{\partial}{\partial x_{0\mu}} \langle x(\tau) | x_0(0) \rangle.  \qquad (12.3.25)$$

### 12.3.3 Dirac Particle in a Constant Field, or in a Plane Wave

We next consider the case of a particle in a constant electromagnetic field, $F_{\mu\nu} = $ constant. It will prove convenient to consider a matrix, $\underset{\sim}{F}$, with components $F_\mu{}^\nu = \sum_\rho F_{\mu\rho} g_{\rho\nu}$.

A slight complication is that we now have to replace derivatives by covariant derivatives: if $e$ is the charge of the particle,

$$\partial_\mu \to \partial_\mu - eA_\mu = D_\mu;$$

also the momentum should be replaced by the canonical momentum,

$$P_\mu \to \Pi_\mu = P_\mu - eA_\mu.$$

The equation satisfied by the propagator $S$ is

$$(i\slashed{\partial} - m - e\slashed{A})S(x, x_0) = i\delta(x - x_0); \tag{12.3.26}$$

to get rid of spin complications, we define $G$ by

$$G(x, x_0) \equiv \frac{1}{i}(i\slashed{\partial} + m - e\slashed{A})S(x, x_0), \tag{12.3.27}$$

so that $G$ satisfies (for *constant* $F_{\mu\nu}$)

$$HG \equiv (-D^2 - m^2 - \frac{e}{2}\sum g_{\mu\mu}g_{\nu\nu}\sigma_{\mu\nu}F_{\mu\nu})G(x, x_0) = \delta(x - x_0). \tag{12.3.28}$$

The machinery of the previous subsection is now immediately applicable. In our case, the operators $X(\tau), \Pi(\tau)$ satisfy

$$\frac{d}{d\tau}X_\mu(\tau) = i[H, X_\mu(\tau)] = -2\Pi_\mu(\tau),$$

$$\frac{d}{d\tau}\Pi_\mu(\tau) = i[H, \Pi_\mu(\tau)] = -2e\sum g_{\rho\rho}F_{\mu\rho}\Pi_\rho(\tau). \tag{12.3.29}$$

The solution to this is, in matrix notation,

$$\underset{\sim}{\Pi}(\tau) = e^{-2e\tau\underset{\sim}{F}}\underset{\sim}{\Pi}(0),$$

$$\underset{\sim}{X}(\tau) = \underset{\sim}{X}(0) + (e\underset{\sim}{F})^{-1}(e^{-2e\tau\underset{\sim}{F}} - 1)\underset{\sim}{\Pi}(0),$$

so that

$$\underset{\sim}{\Pi}(\tau) = -\frac{e}{2}\underset{\sim}{F}e^{-e\underset{\sim}{F}\tau}(\sinh e\tau\underset{\sim}{F})^{-1}[\underset{\sim}{X}(\tau) - \underset{\sim}{X}(0)].$$

The commutator of $X(\tau), X(0)$ is also easily evaluated to be

$$[X_\mu(\tau), X_\nu(0)] = i\left\{(e\underset{\sim}{F})^{-1}(e^{-2e\tau\underset{\sim}{F}} - 1)\right\}_{\mu\nu}.$$

Using this, $H$ becomes

$$H = \underset{\sim}{X}(t)\underset{\sim}{K}\underset{\sim}{X}(t) - 2\underset{\sim}{X}(t)\underset{\sim}{K}\underset{\sim}{X}(0)$$

$$-\frac{i}{2}\,\mathrm{Tr}[e\underset{\sim}{F}\coth(e\tau\underset{\sim}{F})] - \frac{e}{2}\sum g_{\mu\mu}g_{\nu\nu}\sigma_{\mu\nu}F_{\mu\nu} - m^2,$$

$$\underset{\sim}{K} \equiv \frac{e^2}{4}\underset{\sim}{F}^2[\sinh e\tau\underset{\sim}{F}]^2.$$

After some (not entirely trivial) algebra[5] this leads to

$$U(x, x_0; \tau) = \frac{-i}{(4\pi)^2\tau^2}\exp\left\{-ie\int_{x_0}^{x} dy_\mu A^\mu(y)\right.$$

$$-\frac{1}{2}\mathrm{Tr}\,\log[(e\tau\underset{\sim}{F})^{-1}\sinh \tau e\underset{\sim}{F}]$$

$$\left.+\frac{1}{4}(x - x_0)e\underset{\sim}{F}\coth \tau e\underset{\sim}{F}(x - x_0) + \frac{i\tau}{2}\sigma_{\mu\nu}F^{\mu\nu} + i(m^2 - i0)\tau\right\}.$$

The treatment of a particle in a plane electromagnetic wave is similar; the details may also be found in Schwinger (1951).


## Problems


**P.12.1.** Evaluate the nonrelativistic limit of the propagator for an electron in a classical electromagnetic field, and check that it agrees with the standard nonrelativistic expression.

**P.12.2.** Check the relation

$$j_\mu^{\mathrm{cl}}(x) = \frac{-ie}{(2\pi)^4}\int d^4k\, e^{-ik\cdot x}(p_{i\mu}/p_i\cdot k - p_{f\mu}/p_f\cdot k),$$

for the current corresponding to a particle which suddenly changes its momentum from $p_i$ to $p_f$.

---

[5] Interested readers may write this for themselves, or look for it in the text of Itzykson and Zuber (1980), pp. 100–104.

# Appendices

## A.1 Spherical Harmonics, Clebsch–Gordan Coefficients, Matrix Representations of the Rotation Group

### A.1.1 Spherical Harmonics

The *spherical harmonics* provide a basis for the representations of the rotation group. Let $\mathbf{r}$ be a vector, $\theta$, $\varphi$ its polar angles, and $\Omega$ the corresponding solid angle. We write indiscriminately

$$Y_M^l(\hat{\mathbf{r}}) = Y_M^l(\Omega) = Y_M^l(\theta, \varphi), \quad \hat{\mathbf{r}} = \mathbf{r}/|\mathbf{r}|$$

for the spherical harmonics. The total angular momentum $l$ and its third component $M$ are integers. With the phase conventions of Condon and Shortley (1967), and with the angles defined so that $0 \leq \theta < \pi$, $0 \leq \varphi < 2\pi$,

$$Y_M^l(\theta, \varphi) = \sqrt{\frac{2l+1}{4\pi}} \sqrt{\frac{(l-M)!}{(l+M)!}} e^{iM\varphi} P_l^M(\cos\theta).$$

$P_l^M$ are the *Legendre functions*. The $Y$ satisfy orthogonality and completeness relations:

$$\int d\Omega\, Y_M^l(\Omega)^* Y_{M'}^{l'}(\Omega) = \delta_{ll'}\delta_{MM'},$$

$$\sum_{l=0}^{\infty} \sum_{M=-l}^{l} Y_M^l(\Omega) Y_M^l(\Omega')^* = \delta(\Omega - \Omega'),$$

$$\delta(\Omega - \Omega') \equiv \delta(\cos\theta - \cos\theta')\delta(\varphi - \varphi'),$$

$$\int d\Omega \equiv \int_{-1}^{+1} d\cos\theta \int_0^{2\pi} d\varphi.$$

## A.1.2 Some Specific Values

$$Y_M^l(\theta = \varphi = 0) = \delta_{M0}\sqrt{\frac{2l+1}{4\pi}},$$

$$Y_M^l(\theta, \varphi = 0) = \sqrt{\frac{2l+1}{4\pi}}\sqrt{\frac{(l-M)!}{(l+M)!}}P_l^M(\cos\theta),$$

$$Y_M^l(R^{-1}\hat{\mathbf{r}}) = \sqrt{\frac{2l+1}{4\pi}}D_{M0}^{(l)}(R),$$

with $D^{(l)}$ the matrix element corresponding to angular momentum $l$.

## A.1.3 Multiplication Formulas

$$\sum_{M=-l}^{l} Y_M^l(\hat{\mathbf{r}}_1)Y_M^l(\hat{\mathbf{r}}_2)^* = \frac{2l+1}{4\pi}P_l(\hat{\mathbf{r}}_1\hat{\mathbf{r}}_2),$$

$$Y_{M_1}^{l_1}(\theta, \varphi)Y_{M_2}^{l_2}(\theta, \varphi) = \sum_{l=|l_1-l_2|}^{l_1+l_2} \{(2l_1+1)(2l_2+1)/4\pi(2l+1)\}^{1/2}$$

$$\times (l_1, M_1; l_2, M_2|l)\,(l_1, 0; l_2, 0|l)Y_{M_1+M_2}^l(\theta, \varphi),$$

and the $(\ldots|\ldots)$ are Clebsch–Gordan coefficients.

## A.1.4 Gegenbauer-like Formulas

$$e^{i\mathbf{k}\mathbf{r}} = 4\pi \sum_{l=0}^{\infty} \sum_{M=-l}^{l} i^l Y_M^l(\hat{\mathbf{k}})^* Y_M^l(\hat{\mathbf{r}})j_l(kr);$$

$$z^\nu e^{i\gamma z} = 2^\nu \Gamma(\nu) \sum_{n=0}^{\infty} i^n(\nu+n)C_n^\nu(\gamma)J_{n+\nu}(z).$$

Here $C_n^\nu$ are the *Gegenbauer polynomials*, $C_n^{1/2}(x) = P_n(x)$, and $J_n(j_l)$ are the *Bessel functions (spherical Bessel functions)*.

More formulas may be found in the texts of Galindo and Pascual (1978) and Condon and Shortley (1967).

## A.1.5 Spinor and Vector Spherical Harmonics

The *spinor spherical harmonics* $\mathcal{Y}^{l\pm}$ describe the coupling of angular momentum $l$ and spin $1/2$ to total angular momentum $j = l \pm 1/2$:

$$\mathcal{Y}_M^{l\pm}(\theta, \varphi) = \sum_{s_3+\lambda=M} (l, \lambda; \tfrac{1}{2}, s_3|l \pm 1/2)Y_M^l(\theta, \varphi)\chi(s_3);$$

$\chi$ is a Pauli spinor, and hence the $\mathcal{Y}$ are two-component column matrices; we define

$$\chi(1/2) = \begin{pmatrix} 1 \\ 0 \end{pmatrix}, \quad \chi(-1/2) = \begin{pmatrix} 0 \\ 1 \end{pmatrix}.$$

Likewise, vector spherical harmonics $\mathbf{V}_M^{(l)j}$ appear when we couple angular momentum $l$ and spin 1:

$$\mathbf{V}_M^{(l)j}(\theta,\varphi) = \sum_{\lambda=-1}^{+1} (l, M - \lambda; 1, \lambda|j) Y_{M-\lambda}^l(\theta,\varphi)\chi(\lambda).$$

The vectors $\chi$ are standarized by

$$\chi(0) = \mathbf{n}_3, \quad \chi(\pm 1) = \mp\frac{1}{\sqrt{2}}(\mathbf{n}_1 \pm i\mathbf{n}_2),$$

with $\mathbf{n}_j$ a unit vector along the $Oj$ axis.

The *multipoles* are fixed combinations of $\mathbf{V}$:

$$\mathbf{Y}_M^{(e)l}(\hat{\mathbf{r}}) = \frac{|\mathbf{r}|}{\sqrt{l(l+1)}}\boldsymbol{\nabla}_\mathbf{r}Y_M^l(\mathbf{r})$$

$$= \sqrt{\frac{l}{2l+1}}\mathbf{V}_M^{(l+1)l}(\hat{\mathbf{r}}) + \sqrt{\frac{l+1}{2l+1}}\mathbf{V}_M^{(l-1)}(\hat{\mathbf{r}}),$$

$$\mathbf{Y}_M^{(m)l}(\hat{\mathbf{r}}) = \frac{1}{\sqrt{l(l+1)}}\mathbf{r} \times \boldsymbol{\nabla}_\mathbf{r}Y_M^l(\hat{\mathbf{r}})$$

$$= i\mathbf{V}_M^{(l)l}(\hat{\mathbf{r}}),$$

$$\mathbf{Y}_M^{(L)l}(\hat{\mathbf{r}}) = \frac{1}{|\mathbf{r}|}\mathbf{r}Y_M^l(\hat{\mathbf{r}})$$

$$= -\sqrt{\frac{l+1}{2l+1}}\mathbf{V}_M^{(l+1)l}(\hat{\mathbf{r}}) + \sqrt{\frac{l}{2l+1}}\mathbf{V}_M^{(l-1)l}(\hat{\mathbf{r}}).$$

One has $\mathbf{r}\mathbf{Y}_M^{(e,m)l} = 0$. The orthogonality relations are

$$\int d\Omega\, \mathbf{Y}_M^{(I)l}(\Omega)^*\mathbf{Y}_{M'}^{(I')l'}(\Omega) = \delta_{II'}\delta_{ll'}\delta_{MM'},$$

$$I = e, m, L.$$

## A.1.6 Clebsch–Gordan Coefficients

The *Clebsch–Gordan* or *C–G coefficients* describe the coupling of angular momenta $l', l''$ to get angular momentum $l$:

$$|l, M\rangle = \sum_{\substack{M',M'' \\ M'+M''=M}} (l', M'; l'', M''|l)|l', M'\rangle|l'', M''\rangle.$$

Here $l, l', l''; M, M', M''$ are integer or half integer. It is not necessary to specify the total three-component $M$ in $(\ldots|l)$ because necessarily $M = M' + M''$.

We use the phase conventions of Condon and Shortley (1967); then the C–G matrix is real and orthogonal. Thus one also has

$$|l', M'\rangle |l'', M''\rangle = \sum (l', M'; l'', M''|l)|l, M\rangle.$$

A general expression for the C–G matrix is due to Wigner (1959):

$$(l', M'; l'', M''|l) = \sqrt{2l + 1}$$

$$\times \left\{ \frac{(l + l' - l'')!(l - l' + l'')!(l' + l'' - l)!(l' + M' + M'')!(l - M' - M'')!}{(l + l' + l'' - 1)!(l' - M')!(l' + M')!(l'' - M'')!(l'' + M'')!} \right\}$$

$$\times \sum_{k} \frac{(-1)^{k+l''+M''}(l + l'' + M' - k)!(l' - M' + k)!}{(l - l' + l'' - k)!(l + M' + M'' - k)!k!(k + l' - l'' - M' - M'')!}.$$

Two useful cases are when one of the angular momenta is 1,

$$(l', M'; 1, M''|l' + 1) = \begin{cases} \sqrt{\dfrac{(l' - M' + 1)(l' - M' + 2)}{2(l' + 1)(2l' + 1)}} & , M'' = -1, \\[3ex] \sqrt{\dfrac{(l' - M' + 1)(l' + M' + 1)}{(l' + 1)(2l' + 1)}} & , M'' = 0, \\[3ex] \sqrt{\dfrac{(l' + M' + 1)(l' + M' + 2)}{2(l' + 1)(2l' + 1)}} & , M'' = +1; \end{cases}$$

$$(l', M'; 1, M''|l') = \begin{cases} \sqrt{\dfrac{(l' - M' + 1)(l' + M')}{2l'(l' + 1)}} & , M'' = -1, \\[3ex] \dfrac{M'}{\sqrt{l'(l' + 1)}} & , M'' = 0, \\[3ex] -\sqrt{\dfrac{(l' + M' + 1)(l' - M')}{2l'(l' + 1)}} & , M'' = +1; \end{cases}$$

$$(l', M'; 1, M''|l' - 1) = \begin{cases} \sqrt{\dfrac{(l' + M')(l' + M' - 1)}{2l'(2l' + 1)}} & , M'' = -1, \\[3ex] -\sqrt{\dfrac{(l' - M')(l' + M')}{l'(2l' + 1)}} & , M'' = 0, \\[3ex] \sqrt{\dfrac{(l' - M' - 1)(l' - M')}{2l'(2l' + 1)}} & , M'' = +1. \end{cases}$$

and when it is 1/2:

$$(l, M; 1/2, s_3 | l - 1/2) = \begin{cases} -\sqrt{(l - M)/(2l + 1)}, & s_3 = 1/2, \\ \sqrt{(l + M)/(2l + 1)}, & s_3 = -1/2; \end{cases}$$

$$(l, M; 1/2, s_3 | l + 1/2) = \begin{cases} \sqrt{(l + M + 1)/(2l + 1)}, & s_3 = 1/2, \\ \sqrt{(l - M + 1)/(2l + 1)}, & s_3 = -1/2. \end{cases}$$

### A.1.7 Rotation Matrices

A rotation may be described by the *Euler angles*, $\alpha, \beta, \gamma$ in such a way that

$$R_{\alpha\beta\gamma} = R_z(\alpha)R_y(\beta)R_z(\gamma),$$

where $R_j(\theta)$ is a rotation around axis $Oj$ by the angle $\theta$ (in the corkscrew sense). Denoting by

$$D^{(l)}_{M'M}(R_{\alpha\beta\gamma})$$

the matrix elements corresponding to a particle with angular momentum $l$, integer or half integer, we then have

$$D^{(l)}_{M'M}(R_{\alpha\beta\gamma}) = e^{-i(\alpha M' + \gamma M)} D^{(l)}_{M'M}(R_y(\beta)),$$

$$D^{(l)}_{M'M}(R_y(\beta)) = d^{(l)}_{M'M}(\beta)$$

with

$$d^{(l)}_{M'M}(\beta) = \sum_K (-1)^K \frac{\sqrt{(l + M)!(l - M)!(l + M')!(l - M')!}}{(l - M' - K)!(l + M - K)!K!(K + M' - M)!}$$

$$\times \left(\cos \frac{\beta}{2}\right)^{2l + M - M' - 2K} \left(\sin \frac{\beta}{2}\right)^{2K + M' - M}.$$

The sum runs over all integers, or half integers $K$ (as the case may be), with the convention that $1/(-N)! \equiv 0$ for $N =$ integer $> 0$.

## A.2 Special Functions

### A.2.1 Kummer, or Confluent Hypergeometric Functions

$$_1F_1(a, b, z) \equiv M(a, b, z) \equiv \frac{\Gamma(b)}{\Gamma(a)} \sum_{n=0}^{\infty} \frac{\Gamma(a + n)}{n!\Gamma(b + n)} z^n$$

is the *regular Kummer* function. A *singular* function, $U(a, b, z)$, linearly independent of $M$ and satisfying the same differential equation may be found in specialized textbooks (Abramowicz and Stegun, 1965; Magnus, Oberhettinger and Soni, 1966). As $z \to \infty$,

$$M(a, b, z) \simeq \Gamma(b) \left\{ \frac{1}{\Gamma(b-a)} e^{\pm i\pi a} z^{-a} + \frac{1}{\Gamma(a)} e^z z^{a-b} \right\},$$

the $(+)$ sign if $-\pi/2 \leq \arg z \leq 3\pi/2$; $(-)$ if $-3\pi/2 \leq z \leq -\pi/2$.

A few useful relations are:

$$\frac{\partial^n}{\partial z^n} M(a, b, z) = \frac{\Gamma(b)\Gamma(a+n)}{\Gamma(a)\Gamma(b+n)} M(a+n, b+n, z),$$

$$zM(a, b, z) = (b-1)\left\{ M(a, b-1, z) - M(a-1, b-1, z) \right\},$$

$$(1 + a - b)M(a, b, z) = aM(a+1, b, z) + (1-b)M(a, b-1, z)$$

$$M(a, b, z) = e^z M(b-a, b, -z),$$

(*Kummer transformation*).

The differential equation is

$$zM''(a, b, z) + (b-z)M'(a, b, z) - aM(a, b, z) = 0.$$

Laplace transform:

$$\int_0^\infty dx \, e^{-\lambda x} x^{\gamma-1} M(a, \gamma, kx) = \Gamma(\gamma) \lambda^{a-\gamma} (\lambda - k)^{-a}.$$

Relation with other special functions:

$$M(a, 2a, 2ix) = \Gamma(a + 1/2) e^{ix} \left( \frac{x}{2} \right)^{-a+1/2} J_{a+1/2}(x);$$

$$M(-N, b, z) = \frac{\Gamma(b)\Gamma(N+1)}{\Gamma(N+b)} L_N^{b-1}(z), \quad N = \text{integer} \geq 0.$$

### A.2.2 Bessel Functions

$$J_\nu(z) = \left( \frac{z}{2} \right)^\nu \sum_{n=0}^\infty \frac{(-z^2/4)^n}{n!\Gamma(\nu+n+1)},$$

$$Y_\nu(z) \equiv N_\nu(z) = (\sin \pi\nu)^{-1} \left\{ J_\nu(z) \cos \pi\nu - J_{-\nu}(z) \right\}.$$

Differential equation:

$$z^2 X_\nu''(z) + z X_\nu'(z) + (z^2 - \nu^2) X_\nu(z) = 0,$$

$$X = J \text{ or } Y.$$

For $n = \text{integer} \geq 0$,

$$J_{n+1/2}(x) = \sqrt{\frac{2}{\pi x}} \, x^{n+1} \left( \frac{-1}{x} \frac{d}{dx} \right)^n \frac{\sin x}{x}.$$

Hänkel's formula:

$$\int_0^\infty dx\, x\, J_\nu(kx)J_\nu(k'x) = \frac{1}{k}\delta(k-k').$$

At large $z$,

$$J_\nu(x) \simeq \sqrt{\frac{2}{\pi z}}\left\{\cos\left(z - \frac{\nu\pi}{2} - \frac{\pi}{4}\right) + 0(e^{|Imz|}/z)\right\},$$

$$Y_\nu(z) \simeq \sqrt{\frac{2}{\pi z}}\left\{\sin\left(z - \frac{\nu\pi}{2} - \frac{\pi}{4}\right) + 0(e^{|Imz|}/z)\right\}.$$

### A.2.3 Spherical Bessel Functions

$$j_l(z) = \sqrt{\frac{\pi}{2z}}J_{l+1/2}(z); \quad y_l(z) = \sqrt{\frac{\pi}{2z}}Y_{l+1/2}(z).$$

$$j_n(z) \underset{z\to 0}{\simeq} z^n \frac{\sqrt{\pi}}{2^{n+1}\Gamma(n+3/2)}; \quad j_n(z) \underset{z\to\infty}{\simeq} \frac{\sin(z - n\pi/2)}{z}.$$

### A.2.4 Bessel Functions of the Second Kind

$$k_\lambda(z) = \sqrt{\frac{\pi}{2z}}K_{\lambda+1/2}(z);$$

$$K_\nu(z) = \frac{\pi}{2\sin\pi\nu}\sum_{n=0}^\infty \left(\frac{(z/2)^{-\nu}}{\Gamma(-\nu+n+1)} - \frac{(z/2)^\nu}{\Gamma(\nu+n+1)}\right)\frac{(z^2/4)^n}{n!}.$$

For $n = $ integer $\geq 0$,

$$k_n(z) = \frac{\pi}{2}z^n\left(-\frac{1}{z}\frac{d}{dz}\right)^n \frac{e^{-z}}{z}.$$

A differentiation formula:

$$k_\lambda'(z) = \frac{\lambda}{z}k_\lambda(z) - k_{\lambda+1}(z).$$

### A.2.5 Laguerre Polynomials

$$L_N^\nu(x) = x^{-\nu}\frac{e^x}{N!}\frac{d^N}{dx^N}(x^{N+\nu}e^{-x})$$

$$= \sum_{k=0}^N (-1)^k \frac{\Gamma(N+\nu+1)x^k}{\Gamma(N-k+1)\Gamma(\nu+k+1)k!};$$

$$\int_0^\infty dx\, x^\nu e^{-x} L_N^\nu(x) L_{N'}^\nu(x) = \frac{\Gamma(N+\nu+1)}{N!}\delta_{NN'}.$$

Differential equation:

$$xL_N^\nu(x)'' + (\nu + 1 - x)L_N^\nu(x)' + NL_N^\nu(x) = 0.$$

Two useful integration formulas may be obtained from a formula originally due to Schrödinger (which may be found in Galindo and Pascual, 1978; Bethe and Salpeter, 1974). They are,

$$\int_0^\infty dx\, x^\lambda e^{-x} \left[L_N^k(x)\right]^2$$

$$= \begin{cases} \displaystyle\sum_{r=0}^N \frac{\Gamma(\lambda + 1 + r)\Gamma^2(\lambda + 1 - k)}{\Gamma(r + 1)\Gamma^2(N + 1 - r)\Gamma^2(\lambda + 1 - k - N + r)}, \lambda > k - 1; \\ \displaystyle\sum_{r=0}^N \frac{\Gamma(\lambda + 1 + r)\Gamma^2(\nu + k - \lambda - r)}{\Gamma(r + 1)\Gamma^2(\nu + 1 - r)\Gamma^2(k - \lambda)}, \lambda < k. \end{cases}$$

## A.3 Relation Between the Lorentz Group and the Group $SL(2, C)$

To every Minkowski vector $v$ with components $v_\mu$ we associate the $2 \times 2$ complex matrix

$$\tilde{v} = v_0 + \boldsymbol{\sigma}\mathbf{v} = \sum_{\mu\nu} g_{\mu\nu}\tilde{\sigma}_\mu v_\nu = \begin{pmatrix} v_0 + \sigma_3 & v_1 - iv_2 \\ v_1 + iv_2 & v_0 - v_3 \end{pmatrix}, \tag{A.3.1}$$

$$\tilde{\sigma}_0 = \sigma_0 = 1, \; \tilde{\sigma}_i = -\sigma_i.$$

We have

$$\tilde{\sigma}_\mu = \sum_\nu g_{\mu\nu}\sigma_\nu; \; \text{Tr}\,\tilde{\sigma}_\mu\sigma_\nu = 2g_{\mu\nu};$$
$$\det \tilde{v} = v \cdot v, \; v_\mu = \frac{1}{2}\text{Tr}\,\sigma_\mu\tilde{v}; \tilde{v}^+ = \tilde{v}, \tag{A.3.2}$$

the last relation holding if the $v_\mu$ are real.

For every Lorentz transformation,

$$\Lambda \; : \; v \to \Lambda v \equiv v_\Lambda,$$

we have a corresponding matrix $A$, $A$ in $SL(2, C)$ (i.e., $A$ is a $2 \times 2$ complex matrix of unit determinant). We *define* $A$ by

$$A\tilde{v}A^+ = \tilde{v}_\Lambda = \tilde{\sigma} \cdot \Lambda v. \tag{A.3.3}$$

Actually, both $\pm A$ correspond to the same $\Lambda$. An explicit formula for the correspondence given implicitly by (A.3.3) is obtained as follows.

Choose the vectors $v^{(\alpha)}$ with $v_\mu^{(\alpha)} = \delta_{\alpha\mu}$. Applying (A.3.3) to these, we get immediately

$$\Lambda_{\beta\alpha} = \frac{1}{2}\text{Tr}\,\sigma_\beta A\sigma_\alpha A^+. \tag{A.3.4}$$

The inverse is slightly more difficult to obtain. We will consider separately accelerations $L(v)$ such that

$$L(v)n_t = v; \quad n_{t\mu} = \delta_{\mu 0},$$

and rotations, $R$. For the first, and because $\tilde{n}_t = 1$, (A.3.3) gives

$$A(L(v))A^+(L(v)) = \tilde{v},$$

with solution

$$A(L(v)) = \pm \tilde{v}^{1/2}. \tag{A.3.5a}$$

(Note that $\tilde{v} = L(v)n_t$ is positive definite.) For a pure boost, $A(L(v))^+ = A(L(v))$.

**Exercise.** Prove this ∎

For rotations, $R$, we have $Rn_t = n_t$; hence (A.3.3) gives

$$A(R)A^+(R) = 1,$$

i.e., $A$ is *unitary*. Let $\boldsymbol{\theta}$ be the parameters of $R$. For $\boldsymbol{\theta}$ infinitesimal, and $v_0 = 0$,

$$\tilde{v} \equiv \boldsymbol{\sigma}\mathbf{v} \to \boldsymbol{\sigma}\mathbf{v} + \sum \sigma_j \theta_k v_l \epsilon_{jkl}.$$

If we write

$$A(R) = \exp i\boldsymbol{\theta}\boldsymbol{\lambda} \simeq 1 + i\boldsymbol{\theta}\boldsymbol{\lambda},$$

we then get, from (A.3.3),

$$(1 + i\boldsymbol{\theta}\boldsymbol{\lambda})\boldsymbol{\sigma}\mathbf{v}(1 - i\boldsymbol{\theta}\boldsymbol{\lambda}) \simeq \boldsymbol{\sigma}\mathbf{v} + \sum \epsilon_{jkl}\sigma_j \theta_k v_l,$$

from which

$$[\lambda_j, \sigma_k] = -i\sum \epsilon_{jkl}\sigma_l,$$

and hence $\boldsymbol{\lambda} = -\boldsymbol{\sigma}/2$:

$$A(R(\boldsymbol{\theta})) = \exp \frac{-i}{2}\boldsymbol{\theta}\boldsymbol{\sigma}. \tag{A.3.5b}$$

If the four-vector $v$ is such that $v^2 = 1$, $v_0 > 0$, we define $\boldsymbol{\xi}$ by

$$\cosh \xi = v_0, \quad \sinh \xi = |\mathbf{v}|, \quad \boldsymbol{\xi}/|\boldsymbol{\xi}| = \mathbf{v}/|\mathbf{v}|.$$

Then,

$$\tilde{v}^{1/2} = \cosh \frac{\xi}{2} + \frac{1}{\xi}\boldsymbol{\xi}\boldsymbol{\sigma}\sinh \frac{\xi}{2} = \exp \frac{1}{2}\boldsymbol{\xi}\boldsymbol{\sigma},$$

so that

$$A(L(v)) = \exp \frac{1}{2}\boldsymbol{\xi}\boldsymbol{\sigma}. \tag{A.3.5c}$$

**Exercise.** Prove that $\det A(L(v)) = \det A(R(\theta)) = 1$. Prove that the set

$$A(L(v))A(R(\theta))$$

exhausts the set of all $2 \times 2$ matrices with unit determinant (the group $SL(2, C)$).

*Hint.* Use the polar decomposition: any matrix $A$ may be written as

$$A = HU \tag{A.3.6}$$

with $H$ positive definite and $U$ unitary. If $\det A = 1$, $\det H$, $\det U$ can also be taken to be so. Check that any such $H$ may be written as (A.3.5c), and any such $U$ as in (A.3.5b) ∎

Let us use the notation

$$D_{\alpha\beta}^{(1/2)}(\Lambda) \equiv A_{\alpha\beta}(\Lambda), \tag{A.3.7}$$

$$\tilde{D}_{\dot{\alpha}\dot{\beta}}^{(1/2)}(\Lambda) \equiv (A^{-1+}(\Lambda))_{\dot{\alpha}\dot{\beta}}. \tag{A.3.8}$$

We also define

$$\hat{v} \equiv v_0 - \boldsymbol{\sigma}\mathbf{v} = \sigma \cdot v,$$

$$\hat{v}_\Lambda \equiv \sigma \cdot \Lambda v. \tag{A.3.9}$$

One may check by explicit verification that

$$A^{-1+}\hat{v}A^{-1} = \hat{v}_\Lambda, \tag{A.3.10}$$

a formula which is the counterpart of (A.3.3) and which indeed provides another representation of $\mathcal{L}$ into $SL(2, C)$, inequivalent to that given by (A.3.3).

We link this to the text by noting that, in the Weyl realization,

$$D(\Lambda) = \begin{pmatrix} D^{(1/2)}(\Lambda) & 0 \\ 0 & \tilde{D}^{(1/2)}(\Lambda) \end{pmatrix}$$

$$= \begin{pmatrix} A_{\alpha\beta}(\Lambda) & 0 \\ 0 & (A^{-1+}(\Lambda))_{\dot{\alpha}\dot{\beta}} \end{pmatrix}, \tag{A.3.11a}$$

the $D$ being given in Sect. 3.2. Likewise,

$$\gamma \cdot v = \begin{pmatrix} 0 & \tilde{v} \\ \hat{v} & 0 \end{pmatrix}. \tag{A.3.11b}$$

As an application we prove the transformation properties of the $\gamma$ matrices, equation (3.2.13a) in the main text. In the Weyl realization, and for an arbitrary four-vector $v$,

$$D^{-1}(\Lambda)\gamma \cdot v D(\Lambda) = \begin{pmatrix} A^{-1} & 0 \\ 0 & A^{+} \end{pmatrix} \begin{pmatrix} 0 & \tilde{v} \\ \hat{v} & 0 \end{pmatrix} \begin{pmatrix} A & 0 \\ 0 & A^{-1+} \end{pmatrix}$$

$$= \begin{pmatrix} 0 & A^{-1}\hat{v}A^{-1+} \\ A^{+}\hat{v}A & 0 \end{pmatrix} = \begin{pmatrix} 0 & \tilde{\sigma}\cdot\Lambda^{-1}v \\ \sigma\cdot\Lambda^{-1}v & 0 \end{pmatrix} \qquad \text{(A.3.11c)}$$

$$= \begin{pmatrix} 0 & (\Lambda\tilde{\sigma})\cdot v \\ (\Lambda\sigma)\cdot v & 0 \end{pmatrix} = (\Lambda\gamma)\cdot\sigma,$$

and we have used (A.3.3), (A.3.11a) with $\Lambda^{-1}$ in lieu of $\Lambda$. Because $v$ is arbitrary, (A.3.11c) gives

$$D^{-1}(\Lambda)\gamma_{\mu}D(\Lambda) = \sum \Lambda_{\mu\nu}\gamma_{\nu}.$$

More on the matters treated in this appendix may be found in Bogoliubov, Logunov and Todorov (1975) or Wightman (1960).

## A.4 $\gamma$ Matrices

Independently of the realization, we define the matrices $\gamma$ through their anticommutation relations:

$$\{\gamma_{\mu}, \gamma_{\nu}\} = 2g_{\mu\nu}1, \qquad \text{(A.4.1)}$$

where 1 (which will henceforth be omitted) is the $4 \times 4$ unit matrix.

From the $\gamma_{\mu}$ we may form the following 16 matrices:

$$1, \gamma_5, \gamma_{\mu}, \gamma_{\mu}\gamma_5, \sigma_{\mu\nu}, \qquad \text{(A.4.2)}$$

with

$$\gamma_5 = i\gamma_0\gamma_1\gamma_2\gamma_3, \quad \sigma_{\mu\nu} = \frac{i}{2}[\gamma_{\mu}, \gamma_{\nu}].$$

For all $\mu$, $\{\gamma_{\mu}, \gamma_5\} = 0$; moreover, $\gamma_5^2 = 1$. Denoting by $\Gamma_n$ any of the 16 matrices above, we may easily check that

$$\text{Tr}\,\Gamma_n\Gamma_k = 0 \text{ if } n \neq k;$$

$$\text{Tr}\,1 = 4, \quad \text{Tr}\,\gamma_{\mu}\gamma_{\nu} = 4g_{\mu\nu},$$

$$\text{Tr}\,\gamma_{\mu}\gamma_5\gamma_{\nu}\gamma_5 = -4g_{\mu\nu}, \qquad \text{(A.4.3a)}$$

$$\text{Tr}\,\sigma_{\mu\nu}\sigma_{\alpha\beta} = 4(g_{\mu\alpha}g_{\nu\beta} - g_{\mu\beta}g_{\nu\alpha}).$$

Moreover,

$$\text{Tr}\,\gamma_{\mu_1}\ldots\gamma_{\mu_{2n+1}} = 0; \quad \text{Tr}\,\gamma_5\gamma_{\mu}\gamma_{\nu} = 0;$$

$$\text{Tr}\,\gamma_{\mu}\gamma_{\nu}\gamma_{\alpha}\gamma_{\beta} = 4S_{\mu\nu\alpha\beta}, \quad S_{\mu\nu\alpha\beta} = g_{\mu\nu}g_{\alpha\beta} + g_{\mu\beta}g_{\nu\alpha} - g_{\mu\alpha}g_{\nu\beta}. \qquad \text{(A.4.3b)}$$

Other useful properties are

$$\not{a}\not{a} = a^2,$$

$$\mathrm{Tr}\,\gamma_5\gamma_\mu\gamma_\nu\gamma_\alpha\gamma_\beta = 4i\epsilon_{\mu\nu\alpha\beta},$$

$$\sum_\alpha g_{\alpha\alpha}\gamma_\alpha\gamma_{\mu_1}\cdots\gamma_{\mu_{2n+1}}\gamma_\alpha = -2\gamma_{\mu_{2n+1}}\cdots\gamma_{\mu_1},$$

$$\sum_\alpha g_{\alpha\alpha}\gamma_\alpha\gamma_\mu\gamma_\nu\gamma_\alpha = 4g_{\mu\nu}, \tag{A.4.3c}$$

$$\gamma_\mu\gamma_\alpha\gamma_\nu = \sum_\beta g_{\beta\beta}(S_{\mu\alpha\nu\beta}\gamma_\beta - i\epsilon_{\mu\alpha\nu\beta}\gamma_\beta\gamma_5),$$

$$\gamma_5\gamma_\mu\gamma_\nu = \gamma_5 g_{\mu\nu} - \frac{i}{2}\sum_{\alpha\beta} g_{\alpha\alpha}g_{\beta\beta}\epsilon_{\mu\nu\alpha\beta}\gamma_\alpha\gamma_\beta,$$

$$[\sigma_{\rho\lambda},\gamma_\mu] = 2i(g_{\lambda\mu}\gamma_\rho - g_{\rho\mu}\gamma_\lambda).$$

These relations (A.4.3) may be proved either directly from (A.4.1), or by reducing the trace of $n$ matrices to that of $n-2$ with the following trick: consider

$$t = \mathrm{Tr}\,\gamma_{\mu_1}\cdots\gamma_{\mu_n};$$

anticommuting the last two, we get

$$t = -\mathrm{Tr}\,\gamma_{\mu_1}\cdots\gamma_{\mu_n}\gamma_{\mu_{n-1}} + 2g_{\mu_{n-1}\mu_n}\mathrm{Tr}\,\gamma_{\mu_1}\cdots\gamma_{\mu_{n-2}}.$$

We may go on anticommuting $\gamma_{\mu_n}$ to bring it to the first place in the product. Assuming $n =$ even, we then find that

$$t = -\mathrm{Tr}\,\gamma_{\mu_n}\gamma_{\mu_1}\cdots\gamma_{\mu_{n-1}} + 2g_{\mu_{n-1}\mu_n}\mathrm{Tr}\,\gamma_{\mu_1}\cdots\gamma_{\mu_{n-2}}$$

$$-\ldots + 2g_{\mu_n\mu_1}\mathrm{Tr}\,\gamma_{\mu_2}\cdots\gamma_{n-1}.$$

Because of the cyclic property of the trace,

$$\mathrm{Tr}\,\gamma_{\mu_n}\gamma_{\mu_1}\cdots\gamma_{\mu_{n-1}} = \mathrm{Tr}\,\gamma_{\mu_1}\cdots\gamma_{\mu_n} = t,$$

so that, finally,

$$2t = 2g_{\mu_{n-1}\mu_n}\mathrm{Tr}\,\gamma_{\mu_1}\cdots\gamma_{\mu_{n-2}} - \ldots + 2g_{\mu_1\mu_n}\mathrm{Tr}\,\gamma_{\mu_2}\cdots\gamma_{\mu_{n-1}} \tag{A.4.4a}$$

$$(n = \text{even}),$$

and we have obtained the desired reduction. The same argument shows that

$$t = 0, \text{ if } n = \text{odd}. \tag{A.4.4b}$$

Let us return to the 16 matrices $\Gamma_n$ in (A.4.2). It is easy to check that they are linearly independent, for, if we had

$$\sum \alpha_n \Gamma_n = 0,$$

then multiplying by $\Gamma_k$, taking traces and using the first relation in (A.4.3a) we would see that all $\alpha_k = 0$. Because of this linear independence, it follows that any $4 \times 4$ matrix $M$ may be written as

$$M = a_S + a_P \gamma_5 + \sum a_V^\mu \gamma_\mu + \sum a_A^\mu \gamma_\mu \gamma_5 + \sum a_T^{\mu\nu} \sigma_{\mu\nu}. \tag{A.4.5}$$

The coefficients may obtained using (A.4.3a); for example,

$$a_A^\mu = \frac{-1}{4} \operatorname{Tr} \gamma_\mu \gamma_5 M.$$

All the properties we have seen are *independent* of the specific realization we may use for the matrices $\gamma$. Useful explicit realizations follow.

### A.4.1 The Pauli Realization

$$\gamma_0 = \begin{pmatrix} 1 & 0 \\ 0 & -1 \end{pmatrix}, \; \gamma_j = \begin{pmatrix} 0 & \sigma_j \\ -\sigma_j & 0 \end{pmatrix}, \; \gamma_5 = \begin{pmatrix} 0 & 1 \\ 1 & 0 \end{pmatrix}. \tag{A.4.6a}$$

### A.4.2 The Weyl Realization

$$\gamma_\mu = \begin{pmatrix} 0 & \tilde\sigma_\mu \\ \sigma_\mu & 0 \end{pmatrix}, \; \gamma_5 = \begin{pmatrix} 1 & 0 \\ 0 & -1 \end{pmatrix}, \tag{A.4.6b}$$

the $\sigma_\mu, \tilde\sigma_\mu$ are given in A.3, (A.3.1).

### A.4.3 The Majorana Realization

$$\gamma_0 = i \begin{pmatrix} 0 & \sigma_1 \\ -\sigma_1 & 0 \end{pmatrix}, \; \gamma_1 = i \begin{pmatrix} 1 & 0 \\ 0 & -1 \end{pmatrix}, \; \gamma_2 = i \begin{pmatrix} 0 & -i\sigma_2 \\ i\sigma_2 & 0 \end{pmatrix},$$

$$\gamma_3 = i \begin{pmatrix} 0 & 1 \\ 1 & 0 \end{pmatrix}, \; \gamma_5 = \begin{pmatrix} 0 & \sigma_3 \\ -\sigma_3 & 0 \end{pmatrix}.$$

The interest of this last realization is that in it the Dirac operator $i\gamma \cdot \partial - m$ is real: the matrices $i\gamma_\mu$ are all real.

An important theorem concerning $\gamma$ matrices is the following:

**Theorem (Pauli's Theorem).** *Given any set of matrices $\gamma_\mu$ satisfying $\{\gamma_\mu, \gamma_\nu\} = 2g_{\mu\nu}$, then:*

1. *If the set is irreducible, i.e., if one cannot split all $\gamma_\mu$ as*

$$\gamma_\mu = \begin{pmatrix} A_\mu & 0 \\ 0 & B_\mu \end{pmatrix},$$

   *then necessarily the $\gamma_\mu$ are $4 \times 4$ matrices.*
2. *If we have two sets $\gamma'_\mu, \gamma''_\mu$ satisfying the anticommutation relations and that are irreducible (and hence $4 \times 4$ matrices), then they can be related by a similarity transformation:*

$$\gamma''_\mu = S\gamma'_\mu S^{-1}.$$

*For example, it is easy to check that*

$$S_{PW} \gamma_\mu^{\text{Pauli}} S_{PW}^{-1} = \gamma_\mu^{\text{Weyl}}, \; S_{PW} = (\text{constant})(\gamma_0^{\text{Pauli}} + \gamma_5^{\text{Pauli}}).$$

We will not prove the theorem[6]. As a consequence of it, it follows that matrices $A$, $B$, $C$ exist with the properties

$$A\gamma_\mu A^{-1} = \gamma_\mu^+, \; B\gamma_\mu B^{-1} = \gamma_\mu^T, \; C\gamma_\mu C^{-1} = \gamma_\mu^*. \tag{A.4.7a}$$

In the Pauli or Weyl realizations,

$$A = a\gamma_0, \; B = b\gamma_0\gamma_2\gamma_5, \; C = c\gamma_2\gamma_5, \tag{A.4.7b}$$

where $a$, $b$, $c$ are arbitrary constants.

Using this, one proves that, again in these realizations,

$$\bar{v}(p,\lambda)\gamma_{\mu_1}\ldots\gamma_{\mu_n}v(p',\lambda') = (-1)^{n+1}\bar{u}(p',\lambda')\gamma_{\mu_n}\ldots\gamma_{\mu_1}u(p,\lambda). \tag{A.4.8}$$

## A.5 Three Lemmas on $T$ and Wick Exponentials

**Lemma 1.** *If $\hat{H}(t)$ are operators such that their commutator is a c number, i.e.,*

$$[\hat{H}(t),\hat{H}(t')] = iC(t,t'),$$

*then*

$$Te^{-i\int_{-\infty}^{+\infty} dt\,\hat{H}(t)} = e^{i\varphi}e^{-i\int_{-\infty}^{+\infty} dt\,\hat{H}(t)},$$

*with $e^{i\varphi}$ a phase,*

$$\varphi = \frac{1}{2}\int_{-\infty}^{+\infty} dt \int_{-\infty}^{t} dt'\,C(t,t'). \tag{A.5.1}$$

The proof may be found in Jauch and Rohrlich (1959). Alternatively it may be carried out along the lines of the proof of the next two lemmas.

**Lemma 2.** *One has*

$$\hat{S} \equiv T\exp\left(-i\int d^4x\, j(x)\cdot\hat{A}(x)\right)$$

$$= \left\{\exp\frac{-1}{2}\int d^4x d^4y \sum g_{\mu\mu}g_{\nu\nu}j_\mu(x)D_{\mu\nu}(x-y)j_\nu(y)\right\} \tag{A.5.2}$$

$$\times \; :\exp\left(-i\int d^4x\, j(x)\cdot\hat{A}(x)\right): ,$$

*for any $j_\mu(x)$ which is a c number.*

---

[6] It is actually not difficult; for example, 2. is proved by finding the transformation that connects any realization to the Pauli one. So, from $\gamma_0^2 = 1$, $\text{Tr}\,\gamma_0 = 0$, it follows that the eigenvalues of $\gamma_0$ are $+1, +1, -1, -1$, and hence it is connected through a similarity transformation to $\gamma_0^{\text{Pauli}}$ that has the same eigenvalues, etc.

*Proof.* Expanding the $T$ exponential, we obtain

$$\hat{S} = \sum_{n=0}^{\infty} \frac{(-1)^n}{n!} \int d^4x_1 \dots d^4x_n \, j^{\mu_1}(x_1) \dots j^{\mu_n}(x_n)$$

$$\times T\hat{A}_{\mu_1}(x_1) \dots \hat{A}_{\mu_n}(x_n). \tag{A.5.3}$$

Summing over repeated Minkowski indices will be understood in the remainder of this proof, and we have defined

$$j^\mu(x) \equiv \sum_\mu g_{\mu\mu} j_\mu(x).$$

We may use Wick's theorem to write

$$T\hat{A}_{\mu_1}(x_1) \dots \hat{A}_{\mu_n}(x_n)$$

$$= \sum D_{\mu_i\mu_j}(x_i - x_j) \dots D_{\mu_k\mu_l}(x_k - x_l) : \hat{A}_{\mu_a}(x_a) \dots \hat{A}_{\mu_b}(x_b) :, \tag{A.5.4}$$

where the sum runs over all partitions of the set $\mu_1 \dots \mu_n$ into the sets $\mu_i\mu_j \dots \mu_k\mu_l; \mu_a \dots \mu_b$. Each of these propagators $D_{\mu_r\mu_s}(x_r - x_s)$ is accompanied, in the expression for $\hat{S}$, by $j^{\mu_r}(x_r)j^{\mu_s}(x_s)$ and is integrated in $d^4x_r d^4x_s$, so it contributes a factor

$$K \equiv \int d^4x_r d^4x_s \, j^{\mu_r}(x_r) D_{\mu_r\mu_s}(x_r - x_s) j^{\mu_s}(x_s),$$

which is actually *independent of the indices* $r, s$. Fix now the number of uncontracted $\hat{A}$ in (A.5.4),

$$: \hat{A}_{\mu_a} \dots \hat{A}_{\mu_b} :,$$

to be $N$ (obviously $N \leq n$), and let $n - N = 2\nu$ (necessarily an even number). By reordering the indices, and because $K$ is independent of the indices, we may rewrite a term coming from the sum in (A.5.4) with $N$ uncontracted $\hat{A}$ as

$$\int d^4x_1 \dots d^4x_n \, j^{\mu_1}(x_1) \dots j^{\mu_n}(x_n) D_{\mu_i\mu_j}(x_i - x_j) \dots D_{\mu_k\mu_l}(x_k - x_l)$$

$$\times \; : \hat{A}_{\mu_a}(x_a) \dots \hat{A}_{\mu_b}(x_b) :$$

$$\tag{A.5.5}$$

$$= K^\nu \int d^4x_1 \dots d^4x_N \, j^{\mu_1}(x_1) \dots j^{\mu_N}(x_N) : \hat{A}_{\mu_1}(x_1) \dots \hat{A}_{\mu_N}(x_N) :$$

$$\equiv K^\nu \hat{W}_N.$$

All terms with $n > N$ in the expansion (A.5.3) will contribute a $\hat{W}_N$ factor. For *fixed* $n$, the number of terms like (A.5.5) is, obviously,

$$\frac{n!}{2^\nu \nu! N!} = \frac{(N + 2\nu)!}{2^\nu \nu! N!},$$

so the contribution of the $n$th term in (A.5.3) with $N$ uncontracted $\hat{A}$ is

$$\frac{(-i)^{N+2\nu}}{2^\nu \nu! N!} K^\nu \hat{W}_N.$$

Replacing now the sum in $n$ by sums over $\nu$ and $N$ and the explicit expression for $\hat{W}_N$, we then get

$$\hat{S} = \sum_N \frac{(-1)^N}{N!} \int d^4x_1 \ldots d^4x_N : (j(x_1) \cdot \hat{A}(x_1)) \ldots (j(x_N) \cdot \hat{A}(x_N)) :$$

$$\times \sum_\nu \frac{(-i)^{2\nu}}{2^\nu \nu!} K^\nu$$

$$= e^{-\frac{1}{2}K} : e^{-i \int d^4x j(x) \cdot \hat{A}(x)} :,$$

and using the value of $K$, (A.5.2) follows.

Reasoning along the same lines one may prove the following relation:

**Lemma 3.**

$$e^{-i \int d^4x j(x) \cdot \hat{A}(x)} = e^{-\frac{1}{2}\overline{K}} : e^{-i \int d^4x j(x) \cdot \hat{A}(x)} :,$$

*where*

$$\overline{K} = \int d^4x d^4y j^\mu(x) \overline{D}(x - y) j^\nu(y),$$

$$\overline{D}_{\mu\nu}(x - y) \equiv \hat{A}_\mu(x)\hat{A}_\nu(y) - : \hat{A}_\mu(x)\hat{A}_\nu(y) :,$$

*i.e., the equivalent result holds for ordinary products as (A.5.2) for $T$ products.*

Alternatively, the result may be obtained as a corollary of Lemmas 1 and 2.


## A.6 Physical Quantities

### A.6.1 SI (Gauss) Units

Bohr radius: $a_B = \hbar^2/e^2 m_e = 0.529\,177 \times 10^{-10}$ m.
Electron mass: $9.109\,56 \times 10^{-30}$ kg.
Proton mass: $1\,826.151\,52 \times$ electron mass.
Electron charge: $1.602\,19 \times 10^{-19}$ C.
$\hbar = 1.054\,59 \times 10^{-34}$ J s $= 0.65822\,\text{eV} \times 10^{-15}$ eV s.
Speed of light: $c = 2.997\,924\,58 \times 10^8$ m s$^{-1}$.


### A.6.2 Natural Units: $c = \hbar = 1$

$1\,\text{MeV}^{-1} = 1.973 \times 10^{-11}$ cm $= 6.582 \times 10^{-22}$ s.
$1\,\text{GeV}^{-2} = 3.894 \times 10^{-4}$ barn.
Classical electron radius: $r_e = \alpha/m_e = 2.817\,938 \times 10^{-15}$ m.
Rydberg (energy): Ry $= \frac{1}{2} m_e \alpha^2 = 13.605\,8$ eV.

## A.6.3 Other Relations

$1\,\text{J} = 6.241 \times 10^{18}\,\text{eV}.$
$1\,\text{eV} = 1.60219 \times 10^{-19}\,\text{J} = 2.418 \times 10^{14}\,\text{cycles}\,\text{s}^{-1}.$
$1\,\text{fermi (or femtometer)} = 10^{-15}\,\text{m}.$
$1\,\text{barn} = 10^{-28}\,\text{m}^2.$
Fine-structure constant: $\alpha^{-1} = 137.036\,0.$

# References


Abramowicz, M. and Stegun, I. A. (1965), *Handbook of Mathematical Functions*, Dover.

Akhiezer, A. and Berestetskii, V. B. (1963), *Quantum Electrodynamics*, Wiley.

Álvarez-Estrada, R. F., Fernández, F., Sánchez-Gómez, J. L. and Vento, V. (1986), *Models of Hadron Structure Based on Quantum Chromodynamics*, Springer.

Bargmann, V. and Wigner, E. P. (1948), *Proc. Nat. Acad. Sci. USA* **34**, 211.

Bethe, H. A. and Salpeter, E. E. (1957), *Quantum Mechanics of One and Two Electron Atoms*, Springer.

Berestetskii, V. B., Lifshitz, E. M. and Pitaevskii, L. P. (1979), *Relativistic Quantum Theory*, Pergamon.

Bjorken, J. D. and Drell, S. D. (1964), *Relativistic Quantum Mechanics*, McGraw-Hill.

Bjorken, J. D. and Drell, S. D. (1965), *Relativistic Quantum Fields*, McGraw-Hill.

Blatt, J. M. and Weisskopf, V. F. (1952), *Theoretical Nuclear Physics*, Chapman.

Bleuler, K. (1950), *Helv. Phys. Acta* **23**, 567.

Bloch, F. and Nordsiek, A. (1937), *Phys. Rev.* **52**, 54.

Bogoliubov (Bogolubov), N. N. and Shirkov, D. V. (1959), *Introduction to the Theory of Quantized Fields*, Interscience.

Bogoliubov (Bogolubov), N. N., Logunov, A. A. and Todorov, I. T. (1975), *Axiomatic Quantum Field Theory*, Benjamin.

Bohr, N. and Rosenfeld, L. (1933), *Kgl. Dansk. Vid. Selsk. Mat-Fys. Medd.* **12**, No. 8.

Bohr, N. and Rosenfeld, L. (1950), *Phys. Rev.* **78**, 794.

Bouchiat, M. A. and Bouchiat, C. (1974), *Phys. Lett.* **28B**, 111.

Brown, G. E. and Jackson, A. D. (1976), *The Nucleon–Nucleon Interaction*, North Holland.

Condon, E. U. and Shortley, G. H. (1967), *The Theory of Atomic Spectra*, Cambridge.

Das, T. P. (1973), *Relativistic Quantum Mechanics of Electrons*, Harper and Row.

Elton, L. R. B. (1959), *Introductory Nuclear Theory*, Pitman.

Fermi, E. (1932), *Rev. Mod. Phys.* **4**, 87.

Galindo, A. and Pascual, P. (1978), *Mecánica Cuántica*, Alhambra (English translation: *Quantum Mechanics*, Vols. I, II, Springer).

Goldberger, M. L. and Watson, K. (1965), *Collision Theory*, Wiley.

Goldstein, H. (1965), *Classical Mechanics*, Addison-Wesley.

Gordon, W. (1928), *Z. Phys.* **48**, 11.

Gottfried, K. (1966), *Quantum Mechanics*, Benjamin.

Greiner, W., Mueller, B. and Rafelski, J. (1985), *Quantum Mechanics of Strong Fields*, Springer.

Gupta, S. N. (1950), *Proc. Roy. Soc.(London)* **A63**, 681.

Itzykson, C. and Zuber, J. B. (1980), *Quantum Field Theory*, McGraw-Hill.

Jauch, J. M. and Rohrlich, F. (1959), *The Theory of Photons and Electrons*, Addison-Wesley.

Karplus, R. and Klein, A. (1952), *Phys. Rev.* **87**, 848.

Kiesling, C. (1988), *Tests of the Standard Model of Electroweak Interactions*, Springer.

Landau, L. D. and Lifshitz, E. M. (1975), *Classical Mechanics*, Pergamon.

Landau, L. D. and Lifshitz, E. M. (1951), *The Classical Theory of Fields*, Pergamon.

Landau, L. D. and Lifshitz, E. M. (1958), *Quantum Mechanics (Nonrelativistic Theory)*, Pergamon.

Magnus, W., Oberhettinger, F. and Soni, R. P. (1966), *Formulas and Theorems for the Special Functions of Mathematical Physics*, Springer.

Moussa, P. and Stora, R. (1968), in *Analysis of Scattering and Decay* (Nikolic, ed.), Gordon and Breach.

Newton, T. D. and Wigner, E. P. (1949), *Rev. Mod. Phys.* **21**, 400.

Rarita, W. and Schwinger, J. (1941), *Phys. Rev.* **60**, 61.

Rose, M. E. (1961), *Relativistic Electron Theory*, Wiley.

Sakurai, J. J. (1967), *Advanced Quantum Mechanics*, Addison-Wesley.

Schiff, L. J. (1968), *Quantum Mechanics*, McGraw-Hill.

Schweber, S. S. (1961), *Relativistic Quantum Field Theory*, Row, Peterson.

Schwinger, J. (1951), *Phys. Rev.* **82**, 664

Strocchi, F. (1967), *Phys. Rev.* **162**, 1429, and *Phys Rev.* **D2**, 2334 (1970).

Thirring, W. (1950), *Phil. Mag.* **41**, 113.

Weinberg, S. (1964), in *Brandeis Lectures on Particles and Field Theory*, Vol. 2 (Deser and Ford, eds.), Prentice Hall.

Wightman, A. S. (1960), in *Dispersion Relations*, Les Houches Lectures (de Witt and Omnès, eds.), Wiley.

Wigner, E. P. (1939), *Ann. Math.* **40**, No. 1.

Wigner, E. P. (1959), *Group Theory*, Academic Press.

Wigner, E. P. (1963), in *Theoretical Physics*, IAEA, Vienna.

Ynduráin, F. J. (1971), *Mecánica Cuántica*, Alianza.

Ynduráin, F. J. (1971), *Nuovo Cimento* **2A**, 104.

Zwanziger, D. (1964a), *Phys. Rev.* **113B**, 1036.

Zwanziger, D. (1964b), in *Lectures in Theoretical Physics*, Vol. VIIa, University of Colorado Press.

# Index

MIX
Papier aus verantwortungsvollen Quellen
Paper from responsible sources
**FSC® C105338**

www.fsc.org

If you have any concerns about our products,
you can contact us on
**ProductSafety@springernature.com**

In case Publisher is established outside the EU,
the EU authorized representative is:
**Springer Nature Customer Service Center GmbH
Europaplatz 3, 69115 Heidelberg, Germany**

Printed by Libri Plureos GmbH
in Hamburg, Germany